T0135304

Lecture Notes on Data Engineering and Communications Technologies

214

Series Editor

Fatos Xhafa, *Technical University of Catalonia, Barcelona, Spain*

The aim of the book series is to present cutting edge engineering approaches to data technologies and communications. It will publish latest advances on the engineering task of building and deploying distributed, scalable and reliable data infrastructures and communication systems.

The series will have a prominent applied focus on data technologies and communications with aim to promote the bridging from fundamental research on data science and networking to data engineering and communications that lead to industry products, business knowledge and standardisation.

Indexed by SCOPUS, INSPEC, EI Compendex.

All books published in the series are submitted for consideration in Web of Science.

Leonard Barolli
Editor

Innovative Mobile and Internet Services in Ubiquitous Computing

Proceedings of the 18th International
Conference on Innovative Mobile and Internet
Services in Ubiquitous Computing (IMIS-2024)

 Springer

Editor
Leonard Barolli
Department of Information and Communication
Engineering, Faculty of Information Engineering
Fukuoka Institute of Technology
Fukuoka, Japan

ISSN 2367-4512   ISSN 2367-4520 (electronic)
Lecture Notes on Data Engineering and Communications Technologies
ISBN 978-3-031-64765-9   ISBN 978-3-031-64766-6 (eBook)
https://doi.org/10.1007/978-3-031-64766-6

This Springer imprint is published by the registered company Springer Nature Switzerland AG
The registered company address is: Gewerbestrasse 11, 6330 Cham, Switzerland

If disposing of this product, please recycle the paper.

Welcome Message of IMIS-2024 International Conference Organizers

Welcome to the 18th International Conference on Innovative Mobile and Internet Services in Ubiquitous Computing (IMIS-2024), which will be from July 3–5, 2024, in conjunction with the 18th International Conference on Complex, Intelligent and Software Intensive Systems (CISIS-2024).

This International Conference focuses on the challenges and solutions for Ubiquitous and Pervasive Computing (UPC) with an emphasis on innovative, mobile and internet services. With the proliferation of wireless technologies and electronic devices, there is a fast growing interest in UPC. UPC enables to create a human-oriented computing environment where computer chips are embedded in everyday objects and interact with physical world. Through UPC, people can get online even while moving around, thus having almost permanent access to their preferred services. With a great potential to revolutionize our lives, UPC also poses new research challenges. The conference provides an opportunity for academic and industry professionals to discuss the latest issues and progress in the area of UPC.

For IMIS-2024, we received many paper submissions from all over the world. The papers included in the proceedings cover important aspects of UPC research domain.

We are very proud and honored to have 2 distinguished keynote talks by Prof. Sriram Chellappan, University of South Florida, USA and Prof. Chao-Tung Yang, Tunghai University, Taiwan, who will present their recent work and will give new insights and ideas to the conference participants.

The organization of an International Conference requires the support and help of many people. A lot of people have helped and worked hard to produce a successful IMIS-2024 technical program and conference proceedings. First, we would like to thank all the authors for submitting their papers, the Program Committee Members and the reviewers who carried out the most difficult work by carefully evaluating the submitted papers. We are grateful to Honorary Co-chairs Prof. Makoto Takizawa, Hosei University, Japan, and Prof. Kuo-En Chang, Tunghai University, Taiwan, for their guidance and support.

Finally, we would like to thank Web Administrator Co-chairs for their excellent and timely work.

We hope that all of you enjoy IMIS-2024 and find this a productive opportunity to learn, exchange ideas and make new contacts.

IMIS-2024 Organizing Committee

Honorary Co-chairs

Makoto Takizawa	Hosei University, Japan
Kuo-En Chang	Tunghai University, Taiwan

General Co-chairs

Fang-Yie Leu	Tunghai University, Taiwan
Kangbin Yim	Soonchunhyang University, Korea

Program Committee Co-chairs

Hsing-Chung Chen	Asia University, Taiwan
Kin Fun Li	University of Victoria, Canada

Advisory Committee Members

Vincenzo Loia	University of Salerno, Italy
Arjan Durresi	IUPUI, USA
Kouichi Sakurai	Kyushu University, Japan

Award Co-chairs

Tomoya Enokido	Rissho University, Japan
Lidia Ogiela	AGH University of Krakow, Poland
Hiroaki Kikuchi	Meiji University, Japan

International Liaison Co-chairs

Chao-Tung Yang	Tunghai University, Taiwan
Farookh Hussain	University of Technology Sydney, Australia
Hyunhee Park	Myongji University, Korea

Publicity Co-chairs

Der-Jiunn Deng   National Changhua University of Education,
                             Taiwan
Tomoyuki Ishida   Fukuoka Institute of Technology, Japan
Keita Matsuo   Fukuoka Institute of Technology, Japan

Finance Chair

Makoto Ikeda   Fukuoka Institute of Technology, Japan

Local Arrangement Co-chairs

Jung-Chun Liu   Tunghai University, Taiwan
Yu-Chen Hu   Tunghai University, Taiwan

Web Administrators

Phudit Ampririt   Fukuoka Institute of Technology, Japan
Shunya Higashi   Fukuoka Institute of Technology, Japan
Ermioni Qafzezi   Fukuoka Institute of Technology, Japan

Steering Committee Chair

Leonard Barolli   Fukuoka Institute of Technology, Japan

Track Areas and PC Members

1. Multimedia and Web Computing

Track Co-chairs

Chi-Yi Lin   Tamkang University, Taiwan
Tomoyuki Ishida   Fukuoka Institute of Technology, Japan

PC Members

Tetsuro Ogi	Keio University, Japan
Yasuo Ebara	Osaka Electro-Communication University, Japan
Hideo Miyachi	Tokyo City University, Japan
Kaoru Sugita	Fukuoka Institute of Technology, Japan
Chang-Hong Lin	National Taiwan University of Science and Technology, Taiwan
Chia-Mu Yu	National Chung Hsing University, Taiwan
Ching-Ting Tu	National Chung Hsing University, Taiwan
Shih-Hao Chang	Tamkang University, Taiwan

2. Data Management and Big Data

Track Co-chairs

Been-Chian Chien	National University of Tainan, Taiwan
Akimitsu Kanzaki	Shimane University, Japan
Wen-Yang Lin	National University of Kaohsiung, Taiwan

PC Members

Hideyuki Kawashima	Keio University, Japan
Tomoki Yoshihisa	Shiga University, Japan
Pruet Boonma	Chiang Mai University, Thailand
Masato Shirai	Shimane University, Japan
Bao-Rong Chang	National University of Kaohsiung, Taiwan
Rung-Ching Chen	Chaoyang University of Technology, Taiwan
Mong-Fong Horng	National Kaohsiung University of Applied Sciences, Taiwan
Nik Bessis	Edge Hill University, UK
James Tan	SIM University, Singapore
Kun-Ta Chuang	National Cheng Kung University, Taiwan
Jerry Chun-Wei Lin	Harbin Institute of Technology, China

3. Security, Trust and Privacy

Track Co-chairs

Tianhan Gao	Northeastern University, China
Lidia Ogiela	AGH University of Krakow, Poland
Aida Ben Chehida Douss	SUPCOM, Tunisia

PC Members

Jindan Zhang	Xianyang Vocational Technical College, China
Qingshan Li	Peking University, China
Zhenhua Tan	Northeastern University, China
Zhi Guan	Peking University, China
Nan Guo	Northeastern University, China
Xibin Zhao	Tsinghua University, China
Cristina Alcaraz	Universidad de Málaga, Spain
Massimo Cafaro	University of Salento, Italy
Giuseppe Cattaneo	University of Salerno, Italy
Zhide Chen	Fujian Normal University, China
Clara Maria	Colombini, University of Milan, Italy
Dong Seong Kim	University of Canterbury, New Zealand
Victor Malyshkin	Russian Academy of Sciences, Russia
Barbara Masucci	University of Salerno, Italy
Xiaofei Xing	Guangzhou University, China
Mauro Iacono	University of Campania "Luigi Vanvitelli", Italy
Jordi Casas	Open University of Catalonia, Spain
Jordi Herrera	Universitat Autònoma de Barcelona, Spain
Antoni Martínez	Universitat Rovira i Virgili, Spain
Francesc Sebé	Universitat de Lleida, Spain
Nadia Kammoun	Sup'com University of Carthage Tunis, Tunisia
Ryma Abassi	Sup'com University of Carthage Tunis, Tunisia
Naoures Khairallah	Sup'com University of Carthage Tunis, Tunisia
Amine Hedfi	Sup'com University of Carthage Tunis, Tunisia

Track 4. Modeling, Simulation and Performance Evaluation

Track Co-chairs

Tetsuya Shigeyasu	Prefectural University of Hiroshima, Japan
Bhed Bista	Iwate Prefectural University, Japan
Remy Dupas	University of Bordeaux, France

PC Members

Jiahong Wang	Iwate Prefectural University, Japan
Shigetomo Kimura	University of Tsukuba, Japan
Chotipat Pornavalai	King Mongkut's Institute of Technology Ladkrabang, Thailand
Danda B. Rawat	Howard University, USA
Gongjun Yan	University of Southern Indiana, USA
Sachin Shetty	Old Dominion University, USA
Shinji Sakamoto	Kanazawa Institute of Technology, Japan
Tetsuya Oda	Okayama University of Science, Japan
Makoto Ikeda	Fukuoka Institute of Technology, Japan

5. Wireless and Mobile Networks

Track Co-chairs

Luigi Catuogno	University of Salerno, Italy
Hwamin Lee	Soonchunhyang University, South Korea
Evjola Spaho	Polytechnic University of Tirana, Albania

PC Members

Aniello Del Sorbo	Orange Labs – Orange Innovation, UK
Clemente Galdi	University of Naples "Federico II", Italy
Stefano Turchi	University of Florence, Italy
Ermelindo Mauriello	Deloitte Spa, Italy
Gianluca Roscigno	University of Salerno, Italy
Dae-Won Lee	Seokyoung University, South Korea
Jong-Hyuk Lee	Samsung Electronics, South Korea
Sung-Ho Chin	LG Electronics, South Korea
Ji-Su Park	Korea University, South Korea

Jaehwa Chung	Korea National Open University, South Korea
Massimo Ficco	University of Salerno, Italy
Jeng-Wei Lin	Tunghai University, Taiwan
Admir Barolli	Aleksander Moisiu University of Durres, Albania
Yi Liu	Oita National Colleges of Technology, Japan

6. Intelligent Technologies and Applications

Track Co-chairs

| Yong-Hwan Lee | Wonkwang University, South Korea |
| Jacek Kucharski | Technical University of Lodz, Poland |

PC Members

Gangman Yi	Gangneung-Wonju National University, South Korea
Hoon Ko	J. E. Purkinje University, Czech Republic
Urszula Ogiela	AGH University of Krakow, Poland
Lidia Ogiela	AGH University of Krakow, Poland
Libor Mesicek	J. E. Purkinje University, Czech Republic
Rung-Ching Chen	Chaoyang University of Technology, Taiwan
Mong-Fong Horng	National Kaohsiung University of Applied Sciences, Taiwan
Bao-Rong Chang	National University of Kaohsiung, Taiwan
Shingo Otsuka	Kanagawa Institute of Technology, Japan
Pruet Boonma	Chiang Mai University, Thailand
Izwan Nizal Mohd Shaharanee	University Utara Malaysia, Malaysia

7. Cloud Computing and Service-Oriented Applications

Track Co-chairs

Baojiang Ciu	Beijing University of Posts and Telecommunications, China
Neil Yen	The University of Aizu, Japan
Flora Amato	University of Naples "Frederico II", Italy

PC Members

Ashiq Anjum	University of Derby, UK
Gang Wang	Nankai University, China
Shaozhang Niu	Beijing University of Posts and Telecommunications, China
Jianxin Wang	Beijing Forestry University, China
Jie Cheng	Shandong University, China
Shaoyin Cheng	University of Science and Technology of China, China
Jingling Zhao	Beijing University of Posts and Telecommunications, China
Qing Liao	Beijing University of Posts and Telecommunications, China
Xiaohui Li	Wuhan University of Science and Technology, China
Chunhong Liu	Heinan Normal University, China
Yan Zhang	Yan Hubei University, China
Hassan Althobaiti	Umm Al-Qura University, Saudi Arabia
Bahjat Fakieh	King Abdulaziz University, Saudi Arabia
Jason Hung	National Taichung University of Science and Technology, Taiwan
Frank Lai	University of Aizu, Japan
Julian Supardi	Sriwijaya University, Indonesia
Nguyen Gia Nhu	Duy Tan University, Vietnam
Vinod Kumar Verma	University of Surrey, UK
Chen-Kun Tsung	National Chin-Yi University of Technology, Taiwan

8. Ontology and Semantic Web

Track Co-chairs

Alba Amato	Italian National Research Council, Italy
Fong-Hao Liu	National Defense University, Taiwan
Omar Khadeer Hussain	University of New South Wales, Canberra, Australia

PC Members

Flora Amato	University of Naples "Federico II", Italy
Claudia Di Napoli	Italian National Research Center (CNR), Italy
Salvatore Venticinque	University of Campania "Luigi Vanvitelli", Italy
Marco Scialdone	University of Campania "Luigi Vanvitelli", Italy
Wei-Tsong Lee	Tamkang University, Taiwan
Tin-Yu Wu	National Ilan University, Taiwan
Liang-Chu Chen	National Defense University, Taiwan
Salem Alkhalaf	Qassim University, Saudi Arabia
Osama Alfarraj	King Saud University, Saudi Arabia
Thamer AlHussain	Saudi Electronic University, Saudi Arabia
Mukesh Prasad	University of Technology Sydney, Australia

9. IoT and Social Networking

Track Co-chairs

Sajal Mukhopadhyay	National Institute of Technology, Durgapur, India
Keita Matsuo	Fukuoka Institute of Technology, Japan

PC Members

Animesh Dutta	NIT Durgapur, India
Sujoy Saha	NIT Durgapur, India
Jaydeep Howlader	NIT Durgapur, India
Nanda Dulal Jana	NIT Durgapur India
Banhi Sanyal	NIT Kurukshetra, India
Makoto Ikeda	Fukuoka Institute of Technology, Japan
Evjola Spaho	Polytechnic University of Tirana, Albania
Masaki Kohana	Chuo University, Japan
Jana Nowakova	VSB-Technical University of Ostrava, Czech Republic

10. Embedded Systems and Wearable Computers

Track Co-chairs

Jiankang Ren	Dalian University of Technology, China
Kangbin Yim	SCH University, South Korea
Darshika Perera	University of Colorado at Colorado Spring, USA

PC Members

Yong Xie	Xiamen University of Technology, Xiamen, China
Xiulong Liu	The Hong Kong Polytechnic University, Hong Kong
Shaobo Zhang	Hunan University of Science and Technology, China
Kun Wang	Liaoning Police Academy, China
Fangmin Sun	Shenzhen Institutes of Advanced Technology, Chinese Academy of Sciences, China
Kyungroul Lee	Mokpo National University, South Korea
Keita Matsuo	Fukuoka Institute of Technology, Japan
Tetsuya Oda	Okayama University of Science, Japan

IMIS-2024 Reviewers

Leonard Barolli	Bhed Bista
Fatos Xhafa	Hsing-Chung Chen
Alba Amato	Kin Fun Li
Santi Caballé	Hiroaki Kikuchi
Pruet Boonma	Lidia Ogiela
Isaac Woungang	Nan Guo
Hyunhee Park	Hwamin Lee
Fang-Yie Leu	Tetsuya Shigeyasu
Kangbin Yim	Kosuke Takano
Marek Ogiela	Flora Amato
Makoto Ikeda	Tomoya Enokido
Keita Matsuo	Minoru Uehara
Francesco Palmieri	Tomoyuki Ishida
Massimo Ficco	Hwa Min Lee
Salvatore Venticinque	Jiyoung Lim
Admir Barolli	Tianhan Gao
Arjan Durresi	Farookh Hussain

Omar Hussain
Nadeem Javaid
Chi-Yi Lin
Luigi Catuogno
Akimitsu Kanzaki
Wen-Yang Lin
Tetsuya Oda
Tomoki Yoshihisa
Masaki Kohana
Hiroki Sakaji
Baojiang Cui
Shinji Sakamoto
Massimo Cafaro
Mauro Iacono
Barbara Masucci
Gianni D'Angelo

Aneta Poniszewska-Maranda
Sajal Mukhopadhyay
Tomoyuki Ishida
Yong-Hwan Lee
Lidia Ogiela
Hiroshi Maeda
Evjola Spaho
Jacek Kucharski
Vamsi Paruchuri
Yong-Hwan Lee
Seyed Buhari
Olivia Fachrunnisa
Yoshihiro Okada
Shinji Sakamoto
Sriram Chellappan
Xu An Wang

IMIS-2024 Keynote Talks

Integrating AI, Citizen-Science, Social-Media and Innovative Hardware Tech for Public Health

Sriram Chellappan

University of South Florida, Tampa, FL, USA

Abstract. Among many public health concerns, mosquito-borne diseases are most challenging. The problem is global now. Rising temperatures, floods and mobility are all exacerbating challenges today. Diseases like Zika fever, malaria, dengue and chikungunya have no vaccines or cures, as a result of which around a million people die each year from mosquito-borne diseases with a vast majority of them being children. In this talk, we will present our R&D on a spectrum of solutions geared to combat mosquito-borne diseases. Our technologies combine innovative/explainable AI algorithms, novel methods of citizen-science engagement and systems, social-media data mining and innovative hardware to address a range of problems in mosquito surveillance, control and disease management under outbreaks. Some results of successful deployments will also be highlighted.

Application of Artificial Intelligence and Internet of Things for Building Smart Services

Chao-Tung Yang

Tunghai University, Taichung, Taiwan

Abstract. The integration of artificial intelligence (AI) and Internet of Things (IoT) technologies has revolutionized the concept of smart services. Intelligent systems influence many aspects of daily life. Also, with the emergence of IoT, AI and machine learning (ML) opportunities have been created for smart computing infrastructure. We have proposed Intelligent Sensors, Edge Computing, and Cloud Computing (iSEC) framework. The project deploys a smart cloud edge-computing architecture to provide ML and deep learning in the cloud edge environment. By leveraging the iSEC architecture and real-time streaming services, AI and IoT can be effectively combined to enhance smart services. One prominent application is the utilization of You Only Look Once (YOLO) image recognition and object detection for intelligent service delivery. This approach enables the identification and analysis of objects in real-time, allowing for efficient and accurate decision-making in various smart service scenarios.

Contents

Pull-Type Relief Supplies Request System for Long-Term Evacuation Support

Yuta Seri and Tomoyuki Ishida[✉]

Fukuoka Institute of Technology, Fukuoka 811-0295, Fukuoka, Japan
mgm24103@bene.fit.ac.jp, t-ishida@fit.ac.jp

Abstract. Herein, we proposed a pull-type relief supplies request system for long-term evacuation life. This framework consists of individual goods request and relief supplies request systems for evacuees, individual relief supplies request system for informationally disadvantaged evacuees, and individual relief supplies request management system for system administrators. The individual relief supplies request system enables the evacuees to request for relief supplies from local governments based on their circumstances. In addition, the individual relief supplies request system for informationally disadvantaged evacuees is targeted at informationally disadvantaged evacuees, such as the elderly. Furthermore, the individual relief supplies request management system for system administrators collectively manages information on relief supplies requested for evacuees and for informationally disadvantaged evacuees.

1 Introduction

Natural disasters—such as torrential rains and typhoons—occur almost every year in Japan—a country with many natural disasters because of its geographical location. During a natural disaster, local governments issue evacuation orders or advisories to residents, and their stay in evacuation centers may be prolonged depending on the situation. In such cases, in addition to the relief supplies distributed evenly by the local and national government, evacuees may require relief supplies tailored to their circumstances. For example, some information may be difficult to obtain through push-type support from local governments, such as assistive devices and nursing care supplies for the elderly during long-term evacuation life. Therefore, a system that allows evacuees to request information and relief supplies from local governments according to their circumstances and that allows local governments to immediately receive this information is necessary.

1.1 Related Works

Akasaka et al. [1] developed a refuge management system for persons requiring special care (RMS-PRBC) with the aim of smoothing the management and operation of evacuation centers during large-scale disasters. Persons requiring special care include the elderly, infants, and other individuals who require special attention. Evacuees enter their name, address, gender, age, nationality, blood type, emergency contact information, injuries, and communicate the damage to their home into this system using their communication terminal or that installed at the evacuation center. This method enables effective management of the health status of evacuees.

Hirohara et al. [2] developed a disaster information registration and sharing system to support information sharing and decision-making. This system digitizes disaster information from the local government disaster response headquarters and reflects this information on large displays installed at the headquarters. Through the disaster information registration system, the disaster response headquarters can register disaster information on Web-GIS. Moreover, they can reflect the information in the registration system on the large display for each content through the sharing system.

1.2 Pull-Type Relief Supplies Request System Architecture

Figure 1 depicts the system architecture of the pull-type relief supplies request system. The system architecture consists of an evacuee agent, an informationally disadvantaged evacuee agent, a system administrator agent, an application server, and a database server.

1.3 Evacuee Agent

The evacuee agent encompasses the following:

- Evacuee Agent User Interface

It is a component of the interface for the individual relief supplies request system for evacuees, which involves registering the name, age, gender, evacuation center, email address, password, managing relief supplies, and login/logout functions.

- User Information Registration Manager

It provides a function to register the evacuees' name, age, gender, evacuation center, email address, and password.

- Evacuee Relief Supplies Registration Manager

It provides a function for evacuees to request relief supplies and others from the disaster response headquarters.

- Evacuee Relief Supplies Management Manager

It offers a function to view the history of relief supplies requested by evacuees, estimated arrival dates, and notifications from the system administrator.

- User Information Update Manager

It presents a function to update the user information of evacuees.

1.4 Informationally Disadvantaged Evacuee Agent

The informationally disadvantaged evacuee agent consists of the following:

- Informationally Disadvantaged Evacuee User Interface

It is a component of the interface for the individual relief supplies request system for informationally disadvantaged evacuees. The interface comprises relief supplies registration/management functions.

- Informationally Disadvantaged Evacuee Relief Supplies Registration Manager

It contains a function for informationally disadvantaged evacuees to request relief supplies and others from the disaster response headquarters.

- Informationally Disadvantaged Evacuee Relief Supplies Management Manager

It provides a function to view the history of relief supplies requested by informationally disadvantaged evacuees and estimated arrival dates.

1.5 System Administrator Agent

The system administrator agent consists of the following:

- System Administrator User Interface

It is an element of the individual relief supplies request management system for the system administrator. The interface entails registering administrator information and notifications to evacuation centers, and login/logout functions.

- Announcement Information Registration Manager

It provides a function for system administrators to record announcements to be sent to the evacuee agent.

- Evacuee Request Reception Manager

It offers a function to receive relief supplies and other requests registered by evacuee agents and informationally disadvantaged evacuee agents.

- Relief Supplies Management Manager

It provides a function to register/update the estimated arrival dates of relief supplies recorded by evacuee agents and informationally disadvantaged evacuee agents.

1.6   Application Server

The application server consists of the following:

- Database Edit Manager

   It operates on the database according to the information registered from the evacuee agents, informationally disadvantaged evacuee agents, and system administrator agents; moreover, it returns the execution results.

- Database Output Manager

   It provides information stored in the database to the evacuee, informationally disadvantaged evacuee, and system administrator agents.

1.7   Database Server

The database server stores account information, information on requested relief supplies, and other requested information registered by evacuee agents. Additionally, it saves relief supplies and other request information registered by informationally disadvantaged evacuee agents. Furthermore, it records the evacuation center name, account information, evacuation shelter notification information, and estimated arrival date of relief supplies registered by system administrator agents. The database server operates on the stored information in response to requests from the application server and returns the execution results.

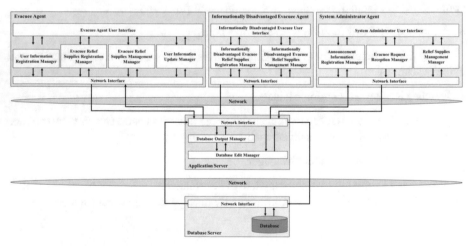

Fig. 1. Pull-type relief supplies request system architecture.

2 Prototype System

2.1 Individual Relief Supplies Request System for Evacuees

The login screen for the individual relief supplies request system for evacuees is shown to the left of Fig. 2. When the evacuee reads the QR code installed at the evacuation center, the login screen is displayed. The evacuee uses the registered email address and password to log in. The new user registration screen for the individual relief supplies request system for evacuees is shown to the right of Fig. 2. The evacuee enters their name, sex, age, email address, evacuation center, and password on the new user registration screen.

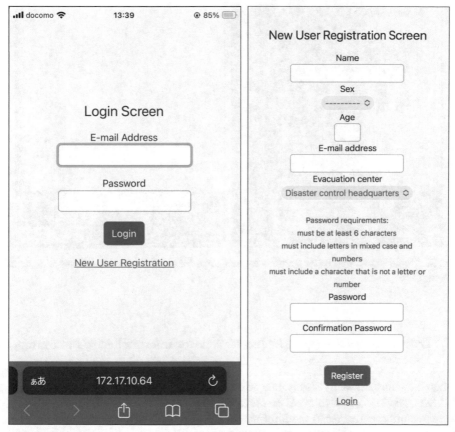

Fig. 2. Login screen and new user registration screen for the individual relief supplies request system for evacuees.

The home screen from when an evacuee selects the hamburger menu of the individual relief supplies request system for evacuees is shown on the left of Fig. 3. The hamburger menu consists of *Home*, *Change Profile*, and *Sign Out*. When an evacuee selects the

Relief Supplies item on the home screen, the screen transitions to the relief supplies request form shown to the right of Fig. 3.

Fig. 3. Home screen and relief supplies request form for the individual relief supplies request system for evacuees.

2.2 Individual Relief Supplies Request System for Informationally Disadvantaged Evacuees

When the informationally disadvantaged evacuee selects Request for Relief Supplies, the following options are displayed: *Medical/Nursing Supplies*, *Food*, *Infant Supplies*, and *Others*. Furthermore, when the informationally disadvantaged evacuee selects *Others*, the following options are displayed: *Clothes*, *Shoes*, *Toilet paper*, *Blanket*, *Portable toilet*, and *Text input* (Fig. 4).

2.3 Individual Relief Supplies Request Management System for System Administrator

On the home screen of the request management system for system administrators, when the item *List of relief supplies request for informationally disadvantaged evacuees* is selected, the system transitions to the confirmation screen for the list of relief supplies request (Fig. 5). On the confirmation screen for the list of relief supplies request, the system administrator can confirm the name, date of birth, age, list of relief supplies, quantity, registration date and time, estimated arrival date, and name of the evacuation center. Moreover, the system administrator can use the filter function on the right side of the screen to refine by name, registration date and time, evacuation center name, and estimated arrival date.

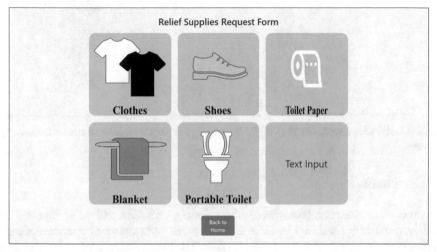

Fig. 4. Relief supplies request form for the individual relief supplies request system for informationally disadvantaged evacuees.

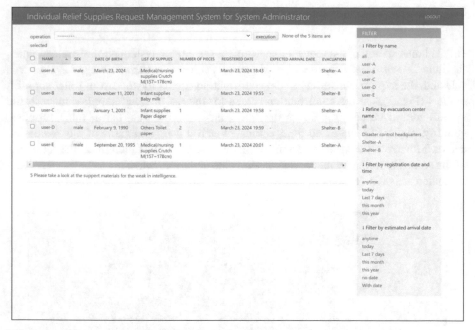

Fig. 5. Confirmation screen for the list of relief supplies request for informationally disadvantaged evacuees.

3 Conclusion

Our proposed system enables evacuees to solicit information and relief supplies from local governments—tailored to their circumstances—via personal communication or tablet terminals installed at evacuation centers. The local government disaster response headquarters can immediately acquire information that is difficult to obtain using push-type support, such as assistive devices and nursing care supplies for the elderly required for long-term evacuation and for evacuees with food allergies. This method is useful for managing relief supplies tailored to evacuees' circumstances during disasters and support for evacuees tailored to their actual situation.

Acknowledgments. This work was supported by JSPS KAKENHI Grant Number JP22K04643.

References

1. Akasaka, K., et al.: Proposal on refuge management system for large scale disaster for persons requiring special care by using triage. In: Proceedings of the 2019 IEEE International Conference on Big Data and Smart Computing (BigComp), pp. 1–7 (2019)
2. Hirohara, Y., Ishida, T., Uchida, N., Shibata, Y.: Proposal of a disaster information cloud system for disaster prevention and reduction. In: Proceedings of the 31st International Conference on Advanced Information Networking and Applications Workshops (WAINA), pp. 664–667 (2017)

Cognicise Virtual Reality System

Kaisei Komoto and Tomoyuki Ishida[(✉)]

Fukuoka Institute of Technology, Fukuoka 811-0295, Fukuoka, Japan
mgm24102@bene.fit.ac.jp, t-ishida@fit.ac.jp

Abstract. In this paper, we propose a virtual reality (VR) system that allows a single person to enjoy "cognicise" activities by combining VR technology and user gesture recognition. Cognicise is a coined term that combines cognition and exercise. In the proposed system, by capturing the user's walking motions, the virtual avatar in the VR space walks in conjunction with the user's walking motion. In addition, by capturing the user's hand motions (rock, paper, and scissors gestures), the user can play the rock-paper-scissors game with the system's virtual avatar in the VR space. The proposed system includes three types of rock-paper-scissors games and a function that allows the user to check their rock-paper-scissors results.

1 Introduction

In Japan, which is a super-aged society, the proportion of elderly people has been increasing consistently since 1950. The proportion of elderly people exceeded 10% in 1985, 20% in 2005, and 29.1% in 2022 [1], and the number of dementia patients has been increasing proportionally. It is predicted that the number of people with dementia will be approximately 6.75 million (prevalence: 18.5%) by 2025, and one in 5.4 people will have dementia [2]. There are various types of dementia prevention methods, and the "cognicise" method [3] is attracting increasing attention. However, most cognicise activities are performed by two or more people, and there is a limited number of cognicise activities that can be performed by only one person. Thus, in this study, we developed a cognicise virtual reality (VR) system that can capture the user's motions using an Intel RealSense [4] depth camera and an Ultraleap 3Di [5] stereo hand tracking camera. In addition, the user experiences the proposed system using a head-mounted display (HMD) device.

2 Related Work

Osawa et al. [6] developed an Android application to improve daily rhythms and prevent dementia in the elderly. This application has three modes, i.e., alarm mode, friendly mode, and game mode. Alarm mode notifies the user of the time to wake up and go to bed using voice and text depending on the time setting. Friendly mode allows the user to feed and pet a three-dimensional (3D) cat model in the application. In game mode, the user can play two types of brain training games.

L. Barolli (Ed.): IMIS 2024, LNDECT 214, pp. 9–18, 2024.
https://doi.org/10.1007/978-3-031-64766-6_2

Matsuoka et al. [7] developed the User eXperience-Trail Making Test (UX-TMT), an Android application that includes cognitive function evaluation and cognitive function improvement training games. The UX-TMT application evaluates cognitive function using the TMT, the N-back task (which is one of the continuous performance tests), and the Stroop task (an executive function test). This system was designed to improve cognitive function through a rock-paper-scissors game.

3   Cognicise VR System Architecture

Figure 1 shows the architecture of the proposed cognicise VR system. As can be seen, the proposed system comprises a VR cognicise control function, a VR cognicise space control function, an HMD function, and landscape storage.

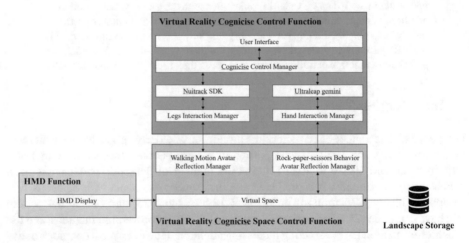

Fig. 1. Architecture of proposed cognicise VR system.

3.1   VR Cognicise Control Function

The VR cognicise control function comprises a user interface, a cognicise control manager, the Nuitrack SDK, an Ultraleap Gemini, a legs interaction manager, and a hands interaction manager.

- User Interface

    The user interface is the interface between the user and the proposed cognicise VR system, and it provides the VR space and virtual avatar to the user.

- Cognicise Control Manager

    The cognicise control manager reflects the user's position information within the cognicise VR system, the 3D landscape selection menu provided to the user, and information about changes in the VR space.

- Nuitrack SDK

    The Nuitrack SDK reflects the user's whole body gesture motions on the virtual avatar in the cognicise VR system.

- Ultraleap Gemini

    Ultraleap Gemini recognizes hand gesture operations using the Ultraleap 3Di.

- Legs Interaction Manager

    The legs interaction manager manages the user's walking motions, which are recognized using the RealSense depth camera.

- Hands Interaction Manager

    The hands interaction manager manages the user's hand gestures, which are recognized using the Ultraleap 3Di.

3.2 VR Cognicise Space Control Function

The VR cognicise space control function includes a rock-paper-scissors behavior avatar reflection manager, a walking motion avatar reflection manager, and a virtual space.

- Rock-paper-scissors Behavior Avatar Reflection Manager

    The rock-paper-scissors behavior avatar reflection manager recognizes the hand gesture (i.e., rock, paper, or scissors) motions using the Ultraleap 3Di and reflects these gestures on the virtual avatar.

- Walking Motion Avatar Reflection Manager

    The walking motion avatar reflection manager recognizes foot gestures (i.e., walking motions) using the RealSense depth camera and reflects them on the virtual avatar.

- Virtual Space

    The virtual space controls the cognicise VR space.

3.3 HMD Function

- HMD Display

    The HMD display presents the cognicise VR space output from the VR space control manager to the user.

3.4 Landscape Storage

The landscape storage stores landscape objects that can be selected by the proposed system.

4 Prototype System

Meta Quest 3 [8] has click operations on the controller and user touch operations. This system allows the user to operate buttons in the VR space using touch operations. When operating a button in VR space, the effect differs depending on the distance between the button and the user's virtual hand and the length of time the button is in contact with the button. In addition, the button's color changes depending on the effect. Table 1 shows the differences in the effects of each button color. Figures 2, 3 and 4 show the differences in effects when performing touch operations using the Meta Quest 3.

Table 1. Meta Quest 3 touch operation details.

Button color	Explanation
Red	• The distance from the button is wide • Nothing happens
Blue	• Contact with button and no time has passed since contact • Scene moves
Yellow	• Contact with button and time has passed since contact • Becomes on standby

Fig. 2. Distance from the button is wide (red).

Fig. 3. Contact with button and no time has passed since contact (blue).

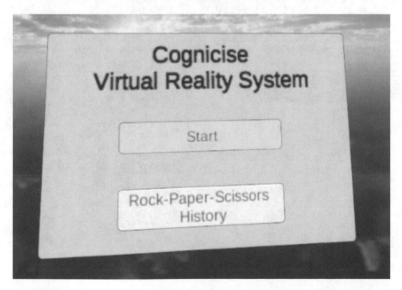

Fig. 4. Contact with button and time has passed since contact (yellow).

As shown in Fig. 5, the landscape selection screen comprises a suburb button, a city center button, and a decision button. Figure 6 shows the landscape when the suburb button is selected.

Fig. 5. Landscape selection screen.

Fig. 6. Landscape displayed when the suburb button is selected.

The walkthrough screen has a walkthrough function that allows the virtual avatar in the VR space to perform a walkthrough in conjunction with the user's walking motions in the real space. Figures 7 and 8 show the landscape before and after movement using the walkthrough function, respectively.

Fig. 7. Landscape before movement using the walkthrough function (suburb).

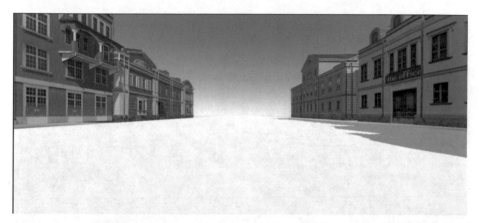

Fig. 8. Landscape after movement using the walkthrough function (suburb).

On the rock-paper-scissors screen, the user can play the rock-paper-scissors game with the system's virtual avatar, as shown in Fig. 9. Figures 10, 11 and 12 show the motions of the virtual hand in the VR space when the user performs the rock, paper, and scissors actions in the real space.

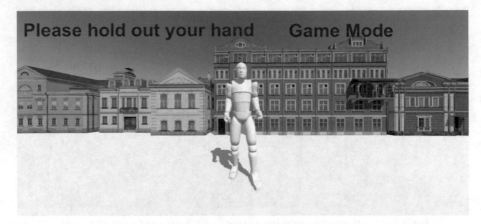

Fig. 9. Rock-paper-scissors scene.

Fig. 10. Virtual hand when the user performs the rock action in the real space

Fig. 11. Virtual hand when the user performs the paper action in the real space.

Fig. 12. Virtual hand when the user performs the scissors action in the real space.

5   Conclusion

This paper has proposed the cognicise VR system, which employs gesture recognition technology. The proposed cognicise VR system includes a walkthrough function and a rock-paper-scissors function in VR space by capturing the user's walking and hand motions (i.e., rock, paper, and scissors gestures) using the RealSense and Ultraleap 3Di devices.

References

1. The Nippon Care-Fit Education Institute: Japan's elderly population is 36.23 million! – Super-aging society and trends in dementia (2023 edition). https://www.carefit.org/liber_carefit/dementia/dementia01.php (2023). Last viewed July 2024
2. Japan Cabinet Office: Annual Report on the Aging Society: 2017 (Summary). https://www8.cao.go.jp/kourei/english/annualreport/2017/2017pdf_e.html (2017). Last viewed July 2024
3. National Center for Geriantrics and Gerontology: Dementia prevention exercise program "Cognicise". https://www.ncgg.go.jp/hospital/kenshu/kenshu/27-4.html (2024). Last viewed July 2024
4. Intel: Depth Camera D435. https://www.intelrealsense.com/depth-camera-d435/ (2024). Last viewed July 2024
5. Ultraleap: Ultraleap 3Di. https://www.ultraleap.com/product/ultraleap-3di/ (2024). Last viewed July 2024
6. Osawa, N., Koizumi, S., Ishii, C.: Development of application program for smartphone and tablet terminal for the sake of life rhythm improvement and dementia prevention for elderly people. In: Proceedings of the 2016 JSME Conference on Robotics and Mechatronics, pp. 1–4 (2016). (in Japanese)
7. Matsuoka, R., et al.: Development of android table based application for neurocognitive assessment. JSiSE Research Report. **31**(5), 89–92 (2017). (in Japanese)
8. Meta: Meta Quest 3: New Mixed Reality VR Headset – Shop Now | Meta Store. https://www.meta.com/jp/en/quest/quest-3/ (2024). Last viewed July 2024

Proposal of a Clinical Training Support System for Nursing Students Using Mixed Reality Technology

Naho Kuriya and Tomoyuki Ishida[✉]

Fukuoka Institute of Technology, Fukuoka 811-0295, Fukuoka, Japan
mgm23105@bene.fit.ac.jp, t-ishida@fit.ac.jp

Abstract. In the medical field, virtual reality technology is increasingly used to acquire skills in clinical practice and nursing schools. Conversely, few systems of nursing education utilize mixed reality (MR) technology based on real space. Therefore, this study proposes a clinical training support system for nursing students using MR technology that superimposes virtual information related to clinical training in real space. This allowed nursing students to deepen their understanding of techniques in clinical practice.

1 Introduction

The development of systems using cross-reality (XR) technology in the medical industry has recently advanced. XR technology creates new experiences by combining real and virtual spaces. It encompasses technologies such as mixed reality (MR), virtual reality (VR), and augmented reality (AR). Currently, these technologies are used in many systems to solve problems in medical practice, with the aim to enriching training and increasing the efficiency of information provision. However, most systems are targeted at medical professionals, such as doctors and nurses, with very few systems targeting nursing students. In addition, issues related to nursing education include a shortage of supervisors and insufficient skill acquisition. Therefore, this study developed a clinical training support system for nursing students using MR technology and created a system that enables them to review materials smoothly during exercise. This system superimposes procedures and precautions related to nursing practice in the real space. Nursing students use MR goggles to view this information and learn techniques for practical exercises.

2 Related Works

Majima et al. [1] developed a wearable learning system for nursing technology education using a head-mounted display (HMD). This system allows users to learn techniques by imitating experts and displaying model videos superimposed in real space.

Shibuya et al. [2] developed a VR teaching material that simulates the nursing techniques of experienced nurses. In this system, learners learn techniques by watching a

L. Barolli (Ed.): IMIS 2024, LNDECT 214, pp. 19–26, 2024.
https://doi.org/10.1007/978-3-031-64766-6_3

first-person model video using an HMD and imitating it. In the evaluations, the system garnered praise for its straightforward procedures and ability to concentrate.

Tokunaga et al. [3] developed a learning system for nursing techniques that learn from mistakes. This system is aimed at nursing college students and is a learning program that incorporates video viewing via e-learning and feedback from other students and teachers. Through this system, nursing students recognize discrepancies between their self-evaluation and others' evaluations, resulting in increased practice hours for self-study.

3 Research Objective

Several systems have been developed to address the shortage of teachers and students' practice time in nursing education. These include systems that enable students to acquire skills by imitating the actions of teachers and instructors as well as systems focusing on skill acquisition through self-reflection during exercises. However, many of these systems require time and effort, and few nursing students can use them easily. To solve the challenges in nursing education, it is necessary for students to engage independently in learning. Therefore, this study developed a system that allows nursing students to review materials smoothly using MR technology within the clinical training support system. Using this system, nursing students can view precautions related to clinical training superimposed on real space via an HMD while working on clinical training. Our proposed system provides nursing students with a deep understanding of techniques used in clinical practice.

4 Clinical Training Support System Configuration

Figure 1 shows the configuration of the proposed clinical training support system. By recognizing the cover of a nursing textbook as an AR marker using Microsoft HoloLens 2 [4], users can review learning materials by superimposing them in real space. This system comprises a mobile agent, marker-management server, and marker-storage database.

4.1 Mobile Agent

The mobile agent targeted nursing students engaged in clinical training tasks such as blood sampling. Nursing students utilized Microsoft HoloLens 2 to recognize the markers displayed in the nursing textbook. Upon recognition of the marker by Microsoft HoloLens 2, learning materials were superimposed in real space, allowing nursing students to work on exercises while viewing them. After Microsoft HoloLens 2 recognizes the marker, data are requested from the marker-management server.

4.2 Marker-Management Server

The marker-management server receives data requests from Microsoft HoloLens 2 and queries the marker-storage database. It also received responses from the marker-storage database and sent the results to Microsoft HoloLens 2.

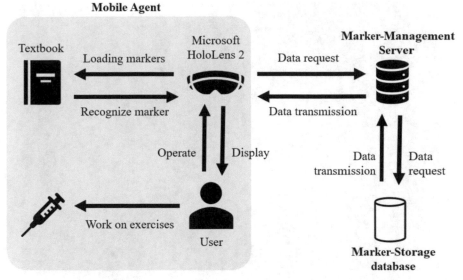

Fig. 1. Clinical training support system configuration.

4.3 Marker-Storage Database

The marker-storage database receives data requests from the marker-management server and provides query result information to the marker-management server. In this study, we created a database on a device using the Vuforia engine [5]. The cover of the nursing textbook used in this system was registered in this database as AR markers.

5 Clinical Training Support System

Our proposed system superimposes virtual information in real space using the cover of a nursing textbook as a marker. In the prototype system, the cover of the nursing textbook, shown in Fig. 2, served as a marker.

Figure 3 shows how the cover of the nursing textbook is registered as a marker by the Target Manager in the Vuforia engine. Here, the image of the "nursing textbook" is registered in the "kyokasyo" database. Upon recognition of the cover of the nursing textbook shown in Fig. 2 by the Microsoft HoloLens 2 camera, the information shown in Fig. 4 is superimposed on the real space.

When this system recognizes the marker, information related to the blood vessels, as shown in Fig. 4, is superimposed in real space. Additionally, a "Cautionary Note" button is displayed on the left of the information for viewing precautions, and a "Video Scene" button was displayed on the right to view the model video. Figure 5 displays the screen scene upon pressing the "Cautionary Note" button. In this scene, precautions regarding blood sampling were superimposed. Furthermore, Fig. 6 displays the screen scene after pressing the "Video Scene" button. In this scene, a model video by an instructor or a similar figure is superimposed.

Fig. 2. The cover of the nursing textbook set as a marker.

When the user presses the "Back" button in the scenes shown in Figs. 5 and 6, the screen returns to Fig. 4. In addition, we added components called "Object Manipulator" and "NearInteractionGrabbable" to the virtual information, allowing users to grab and move superimposed virtual information with their hands. Figure 7 shows how a user grabs and moves virtual information supcrimposed on real space. By realizing the function of grasping and moving virtual information manually, nursing students can review various types of information more smoothly.

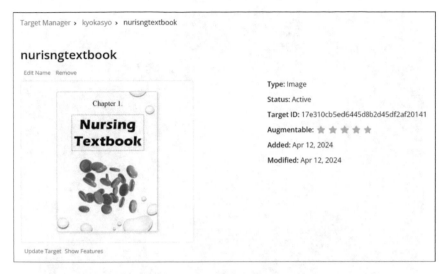

Fig. 3. Marker registration screen.

Fig. 4. Superimposed display screen when loading marker.

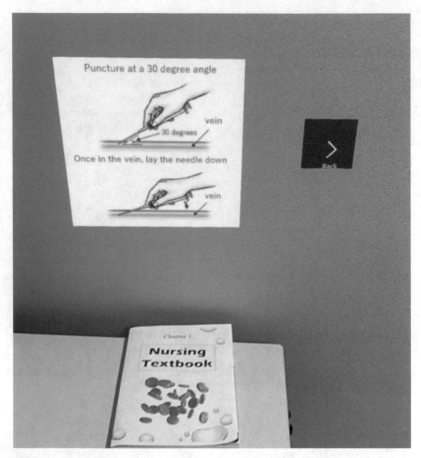

Fig. 5. Superimposed display screen of precaution.

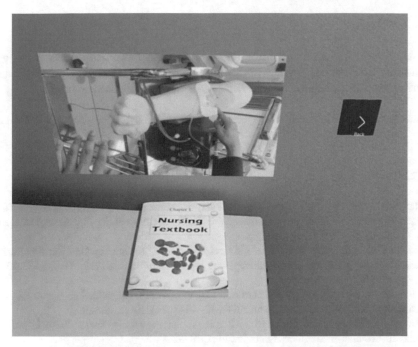

Fig. 6. Superimposed display screen of the model video.

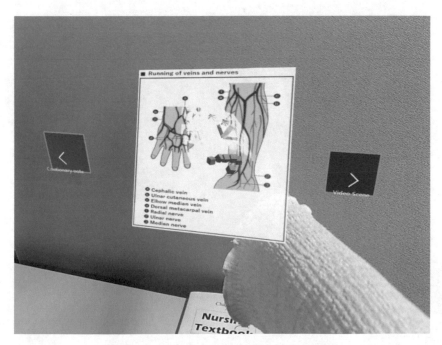

Fig. 7. Grasping and moving virtual information.

6 Conclusion

This study proposed a clinical training support system for nursing students using MR technology. The proposed system allows nursing students to review various information superimposed on real space via the HMD while engaging in clinical practice.

Acknowledgments. This work was supported by JSPS KAKENHI Grant Number JP23K11666.

References

1. Majima, Y., Matsuda, T., Izumi, M., Masuda, S., Maekawa, Y.: Development of wearable learning system for education of nursing skills. In: Proceedings of the 43rd Annual Conference of JSiSE, pp. 69–70 (2018). (in Japanese)
2. Shibuya, H., et al.: Development of VR teaching material for nursing skill learning: usability evaluation using free description analysis. J. Japan Asso. Simulat. Edu. Healthc. Profess. **8**, 21–27 (2020). (in Japanese)
3. Tokunaga, K., Hirano, K.: The effect of watching videos of making mistakes -on nursing skills-. Trans. Japanese Soc. Info. Sys. Edu. **33**(1), 43–46 (2016). (in Japanese)
4. Microsoft: Microsoft HoloLens | Mixed Reality Technology for Business (2024). https://www.microsoft.com/en-us/hololens. Last viewed April 2024
5. PTC: Vuforia Engine Developer Portal (2024). https://developer.vuforia.com/. Last viewed April 2024

A Camera Placement System for Motion Analysis and Object Recognition: System Assessment by Simulations and an Experiment

Kyohei Wakabayashi[1], Chihiro Yukawa[1], Tetsuya Oda[2(✉)], and Leonard Barolli[3]

[1] Graduate School of Science and Engineering, Okayama University of Science (OUS), 1-1 Ridaicho, Kita-ku, Okayama 700-0005, Japan
{r24sdg61d,r24sdf3gn}@ous.jp

[2] Department of Information and Computer Engineering, Okayama University of Science (OUS), 1-1 Ridaicho, Kita-ku, Okayama 700-0005, Japan
oda@ous.ac.jp

[3] Department of Informention and Communication Engineering, Fukuoka Insitute of Technology, 3-30-1 Wajiro-Higashi-ku, Fukuoka 811-0295, Japan
barolli@fit.ac.jp

Abstract. In manufacturing plants, highly technical processes are frequently performed. Hand operated tasks for soldering electronic components to circuit boards are important processes that directly affect the performance and reliability of the products. In many operations performed by people using tools, worker safety is required to prevent burns, cuts and other injuries. Therefore, the workers movement data should be collected and analyzed in order to provide safe environment. Non-contact sensing systems include methods that use cameras to capture images of workers or products and use skeletal estimation and object recognition to detect dangerous movements of workers and product conditions. For carrying out the data analysis, the worker and object should be within the camera viewangle. Therefore, the camera placement is crucial for capturing workers movements and object state accurately. In this paper, we propose a camera placement system in order to carry out motion analysis and object recognition. From the simulation and experimental results, we found that by deciding the camera placement and angle of view, the proposed system can be used for motion analysis and object recognition. Also, the proposed system is an effective system for improving the recognition accuracy and detection rate of humans and objects.

1 Introduction

In electronics manufacturing plants, highly technical processes are frequently performed. In particular, hand operated tasks for soldering electronic components to circuit boards are important processes that directly affect the performance and reliability of the products. In addition to product quality, these industrial technologies are also essential

L. Barolli (Ed.): IMIS 2024, LNDECT 214, pp. 27–38, 2024.
https://doi.org/10.1007/978-3-031-64766-6_4

for ensuring consumer safety. To prevent injuries and human error in the manufacturing plants, workers must be highly experienced and highly skilled, as well as meticulously planned and carefully executed. Operator training and education are the foundation for the proper execution of these complex tasks. In addition, as part of safety measures, instructions and call-outs are utilized as a way to anticipate and mitigate on-the-job accidents and human error. They provide guidelines for workers to operate accurately and quickly and serve as an important role in improving the safety and efficiency of the entire production process.

In many operations performed by people using tools, worker safety is required to prevent burns, cuts and other injuries. In addition, in high quality production, it is important to control the occurrence of defective products, reduce overall costs and improve product quality. Therefore, it is important to collect and analyse the worker movement data by sensing systems [1–3]. However, contact type sensing systems may lead to worker skill degradation. On the other hand, the non-contact sensing systems [4,5] can capture the worker images and the object. Thus, they can detect the hazardous motions or dangerous conditions based on posture estimation and object recognition [6]. But the worker body and the object should be within the camera view-angle. In the case when the operator is working with a tool, it is important that the camera is positioned and angled to capture the tool and the operator movement. When the operator and the tool are out of the view-angle or only one of them is captured, the data for motion analysis are lost [7–11].

In the context of implementing a camera-based system for operational monitoring, the strategic positioning of cameras emerges as a pivotal determinant in capturing both the intricate movements of workers and the nuanced states of objects with utmost accuracy. While the deployment of a multi cameras network offers the potential for comprehensive posture and object characteristic analysis, practical constraints often necessitate the judicious utilization of a singular camera. An optimally configured camera arrangement, thus, becomes instrumental in meticulously capturing and recording the dynamic interplay between worker activities and object states. This granular data acquisition not only facilitates a profound insight into worker performance metrics and safety protocols but also offers invaluable perspectives on product conditions. The detailed data analytics is the foundation for refining and optimizing operational workflows. Furthermore, it enables a comprehensive analysis of the entire work process and the identification of effective improvement measures.

In this paper, we propose a camera placement system in order to carry out motion analysis and object recognition. The proposed system considers camera's performance and the limitation of the available space and equipment for image capture. It is expected to prevent the risk of non analyzable situations and provide high accuracy and reliability in motion analysis and object recognition.

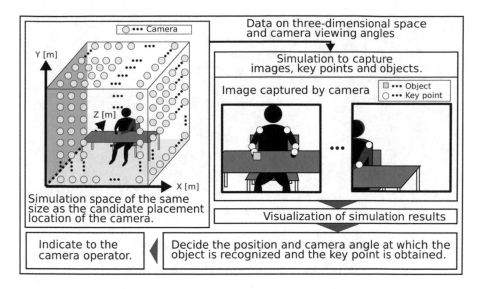

Fig. 1. Proposed system.

The structure of the paper is as follows. In Sect. 2, we describe the proposed system. In Sect. 3, we present the simulation and experimental results. Finally, conclusions and future work are given in Sect. 4.

2 Proposed System

In this section, we present the proposed system. The proposed system is designed to operate a camera based object recognition and motion analysis system for workers using tools at a production site [12, 13]. Therefore, the proposed system can decide the camera angles and positions for setting up the system at construction and production sites to perform object recognition and motion analysis.

Figure 1 shows an overview of the proposed system. The two dimensional key point coordinates obtained from MediaPipe [14, 15] are used to analyze the worker's motion, allowing for posture estimation based on the skeleton. Figure 2 shows the images acquired from the simulation or camera and Table 1 shows the key point numbers corresponding to body part. In addition, the tool held by the worker can be recognized using the object recognition method YOLOv5 [16] and the human posture considering the state of the tool can be analyzed from the bounding box obtained from the images.

The proposed system uses Unity [17] for the virtual space. As inputs are used the width, height and depth of the camera operating environment, the viewing angle of the camera RGB sensor and the depth at which object recognition is capable. The range of three dimensional coordinate systems consisting of XYZ directions represents the decided space for imaging. Also, simulation units are 1.0 [m].

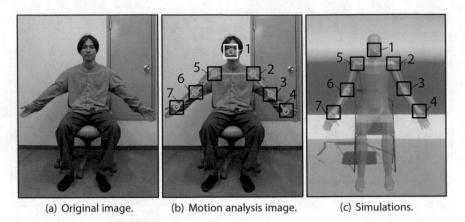

(a) Original image.   (b) Motion analysis image.   (c) Simulations.

Fig. 2. Body parts and key points for the proposed system.

Table 1. Key points corresponding to body joints.

Body parts	Key point numbers (in simulation and motion analysis images)
Nose	1
Left Shoulder	2
Left Elbow	3
Left Hand	4
Right Shoulder	5
Right Elbow	6
Right Hand	7

In the proposed system, the upper body of a worker and a soldering iron are the targets of imaging in order to analyze the motion of the worker during soldering. Therefore, the proposed system uses the coordinate space centered on the working table, which does not move during the operation. The camera rotation angle ranges from 0 [deg.] to 180 [deg.] in the horizontal direction or from 0 [deg.] to 180 [deg.] in the vertical direction and the rotation interval is 45 [deg.]. In the simulation environment, the skeleton is positioned in the same location as the key points based on MediaPipe. The soldering iron for object recognition is placed on the right hand side, assuming right handedness. In addition, a cube is used as an object for object recognition when using YOLOv5. The camera moves 0.5 [m] in the XYZ directions, rotating at each point and capturing images at each location. After deciding on the types, locations and total number of key points, as well as the presence of objects obtained within the camera's field of view, the camera operator gets the camera placement and angle for the object recognition and the number of key points is maximized.

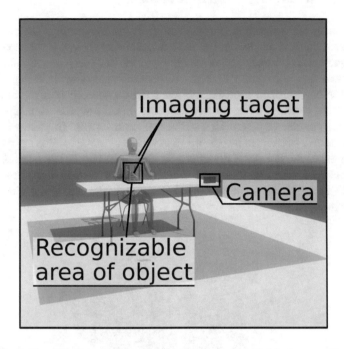

Fig. 3. Simulation environment for motion analysis and object recognition.

Table 2. Parameters and values for simulations.

Functions	Values
Simulation Area Width	2.5 [m]
Simulation Area Hight	2.5 [m]
Simulation Area Depth	2.5 [m]
Number of Divisions for Each Direction	51 [*unit*]
Used Camera	1 [*unit*]
Viewing Angle (Horizontal)	86 [deg.]
Viewing Angle (Vertical)	57 [deg.]
Camera Angle Range (Horizontal)	180 [deg.]
Camera Angle Range (Vertical)	180 [deg.]
Rotation Angle of Camera	45 [deg.]
Number of Upper Body Key Points	7 [*unit*]
Used Object	1 [*unit*]

3   Simulation and Experimental Results

The performance of the proposed system is assessed by carrying out an experiment
a set up a camera in order to capture the images. We investigate whether the object
recognition and body part key points are obtained the same as in the simulation.

We show in Table 2 the simulation parameters and values, while in Fig. 3 we present
the simulation environment used for motion analysis and object recognition. We put
the camera on a working table and we consider that the worker is using a tool. For
simulation, we conider a space with a width and depth of 1.25 [m] and a height of 2.5
[m]. The camera is positioned at 0.5 [m] intervals at a total of 132651 points, then the
camera is moving in each XYZ directions as shown in Fig. 1. The camera angles are
rotated at each position by 45 [$deg.$] in the horizontal and vertical directions. Therefore,
a total of 25 types of images are taken: 5 patterns in the horizontal direction and 5
patterns in the vertical direction.

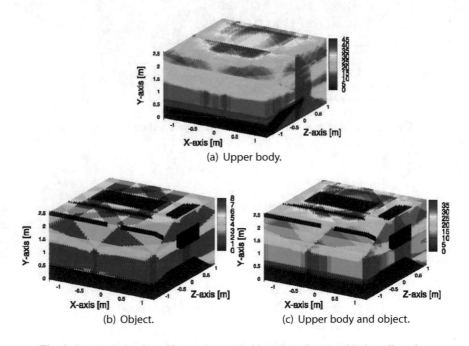

(a) Upper body.

(b) Object.

(c) Upper body and object.

Fig. 4. Summation value of key points and object detection considering all angles.

Figure 4 shows the sum of detected values for key points and objects considering
all angles. Figure 4(a) shows the upper body, Fig. 4(b) shows the object and Fig. 4(c)
shows the upper body results considering the object. For object recognition and key
point obtaining, simulation results show that capturing from a higher viewpoint than
the worker or the working table is most effective. Also, in the case of a camera below
the worktable, it is possible to get key points of the upper body, while it is not possible
to recognize the object.

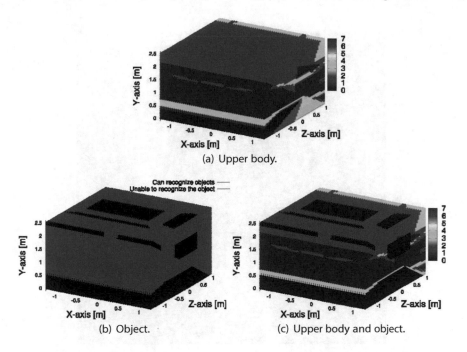

(a) Upper body.

(b) Object.

(c) Upper body and object.

Fig. 5. Highest key points considering all angles and recognizable objects.

Table 3. Highest value of camera placement and angle of view for motion analysis and object recognition.

X-axis [m]	Y-axis [m]	Z-axis [m]	Horizontal Rotation [deg.]	Vertical Rotation [deg.]
−0.6	1.2	−0.35	0	45

Figure 5 shows the highest key points considering all angles and the recognizable objects. Figure 5(a) shows the upper body, Fig. 5(b) shows the object and Fig. 5(c) shows the upper body result considering the object. From Fig. 5(a), it can be seen that the key points of the upper body can be obtained by capturing the image from a position similar to the height of the working table. In addition, Fig. 5(b) and Fig. 5(c) show that object recognition is possible and key points of the upper body are obtained by capturing images from a position similar to the height of the working table and in front of the worker. On the other hand, when the camera is close to the working table and lower than the working table, the object cannot be recognized and the key points of the upper body for a motion analysis considering the object cannot be obtained.

Figure 6 shows the experimental environment in which the object is recognized and the highest key point capture value is achieved. Table 3 shows the camera placement and angle at which the highest value is achieved. Figure 7 shows a comparison of simulated and experimental highest values. We show in Fig. 7(a) and Fig. 7(b) the camera view of the simulated and experimental environment, respectively. Both the upper body and the

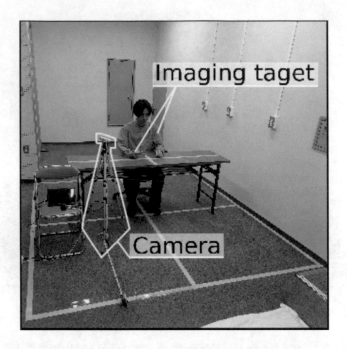

Fig. 6. Experimental environment for highest values.

(a) Camera view of the simulation environ-(b) Camera view of the experimental envi-
ment.   ronment.

Fig. 7. Comparison results of highest values considering object recognition.

Table 4. Median value of camera placement and angle of view for motion analysis and object recognition.

X-axis [m]	Y-axis [m]	Z-axis [m]	Horizontal Rotation [$deg.$]	Vertical Rotation [$deg.$]
−0.8	1.15	0.65	−45	90

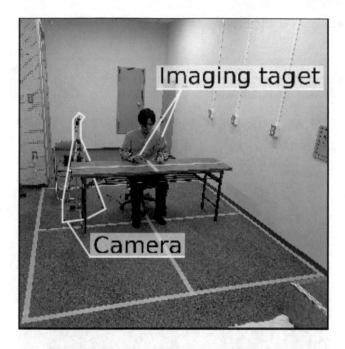

Fig. 8. Experimental environment for median values.

(a) Camera view of the simulation environ-ment.   (b) Camera view of the experimental environment.

Fig. 9. Comparison results of median values considering object recognition.

object are within the camera's angle of view, indicating that all key points and objects have been recognized.

Figure 8 shows the experimental environment when the median key points are taken into account for object recognition. Table 4 shows the median camera placement and angle. Figure 9 shows a comparison of simulated and experimental median values. In Fig. 9(a) and Fig. 9(b), we show the camera view of simulated and experimental environment, respectively. It can be seen that the left elbow and the left wrist connected

Table 5. Lowest value of camera placement and angle of view for motion analysis and object recognition.

X-axis [m]	Y-axis [m]	Z-axis [m]	Horizontal Rotation [$deg.$]	Vertical Rotation [$deg.$]
−0.9	0.5	−0.5	45	90

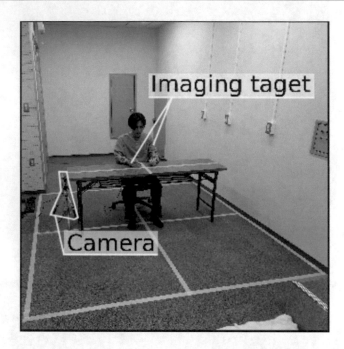

Fig. 10. Experimental environment for lowest values.

from the left elbow, which are shaded by the body in the simulation, are not recognized in the experiment.

Figure 10 shows the experimental environment with minimal object recognition and key point acquisition. Table 5 shows the camera placement and angle at which the lowest is obtained. Figure 11 shows a comparison of simulated and experimental lowest values. Figure 11(a) shows the camera view of the simulated environment and Fig. 11(b) shows the camera view of the experimental environment. From Fig. 11(b) can be seen that objects to be recognized are not within the cameras field of view and the soldering iron is not recognized. Therefore, the key points effective for motion analysis are not correctly obtained in both simulation and experiment. Table 6 shows the comparison results of the number of key points and the presence of recognized objects obtained from the simulation and experiment. We can see that the same results are obtained for the highest, median and lowest values.

From the experimental results, it can be concluded that the proposed system is an effective system for deciding the camera placement and angle for object recognition and motion analysis.

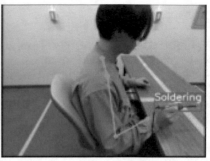

(a) Camera view of the simulation environ- (b) Camera view of the experimental envi-
ment. ronment.

Fig. 11. Comparison results of lowest values considering object recognition.

Table 6. Comparison of number of key points and object recognition.

Classification	Simulation	Recognizable of object	Experiment	Obtained key point number
Highest value	7 [*unit*]	True	7 [*unit*]	1, 2, 3, 4, 5, 6, 7
Median value	5 [*unit*]	True	5 [*unit*]	1, 2, 5, 6, 7
Lowest value	0 [*unit*]	False	0 [*unit*]	Not obtained

4   Conclusions

In this paper, we proposed a camera placement system for motion analysis and object recognition. The proposed system evaluates camera placement and angle of view in order to obtain object recognition and key points. In addition, a comparison was performed between simulation and experiment. From the simulation and experimental results, we conclude as follows.

- The proposed system uses camera imaging simulations to decide on the optimal placement and viewing angle of the camera system for analyzing motion and recognizing objects.
- The proposed system is an effective system for improving the recognition accuracy and detection rate of humans and objects.
- The proposed system can support the operation of camera systems based on object recognition and posture estimation in limited spaces such as construction sites and electronics manufacturing plants.

In the future, we plan to assess the performance of the proposed system through various scenarios. Additionally, we will design a system that considers the environmental and motion limitations.

Acknowledgement. This work was supported by JSPS KAKENHI Grant Number JP24K07993.

References

1. Arpaia, P., et al.: A wearable EEG instrument for real-time frontal asymmetry monitoring in worker stress analysis. IEEE Trans. Instrum. Meas. **69**(10), 8335–8343 (2020)
2. Singh, N., et al.: IoT enabled HELMET to safeguard the health of mine workers. Comput. Commun. **193**, 1–9 (2022)
3. Bortolini, M., et al.: Motion Analysis System (MAS) for production and ergonomics assessment in the manufacturing processes. Comput. Ind. Eng. **139**, 105485 (2020)
4. Kocabas, M., et al.: Vibe: video inference for human body pose and shape estimation. In: Proceedings of the IEEE/CVF Conference on Computer Vision and Pattern Recognition, pp. 5253–5263 (2020)
5. Lu, L., et al.: Flexible noncontact sensing for human-machine interaction. Adv. Mater. **33**(16), 2100218 (2021)
6. Andriyanov, N., et al.: Intelligent system for estimation of the spatial position of apples based on YOLOv3 and real sense depth camera D415. Symmetry **14**(1) (2022)
7. Toshev, A., Szegedy, C.: DeepPose: human pose estimation via deep neural networks. In: Proceedings of the 27-th IEEE/CVF Conference on Computer Vision and Pattern Recognition (IEEE/CVF CVPR-2014), pp. 1653–1660 (2014)
8. Haralick, R., et al.: Pose estimation from corresponding point data. IEEE Trans. Syst. **19**(6), 1426–1446 (1989)
9. Fang, H., et al.: RMPE: regional multi-person pose estimation. In: Proceedings of the IEEE International Conference on Computer Vision, pp. 2334 –2343 (2017)
10. Xiao, B., et al.: Simple baselines for human pose estimation and tracking. In: Proceedings of the European Conference on Computer Vision (ECCV), pp. 466–481 (2018)
11. Martinez, J., et al.: A simple yet effective baseline for 3D human pose estimation. In: IEEE International Conference on Computer Vision, pp. 2640–2649 (2017)
12. Toyoshima, K., et al.: Analysis of a soldering motion for dozing state and attention posture detection. In: Proceedings of the 3PGCIC-2022, pp. 146–153 (2022)
13. Toyoshima, K., et al.: A soldering motion analysis system for monitoring whole body of people with developmental disabilities. In: Proceedings of the AINA-2023, pp. 38–46 (2023)
14. Lugaresi, C., et al.: MediaPipe: a framework for building perception pipelines arXiv preprint arXiv:1906.08172 (2019)
15. Micilotta, A.S., Ong, E.-J., Bowden, R.: Real-time upper body detection and 3D pose estimation in monoscopic images. In: Leonardis, A., Bischof, H., Pinz, A. (eds.) ECCV 2006. LNCS, vol. 3953, pp. 139–150. Springer, Heidelberg (2006). https://doi.org/10.1007/11744078_11
16. Yang, D., et al.: Research of target detection and distance measurement technology based on YOLOv5 and depth camera. In: 2022 4th International Conference on Communications, Information System and Computer Engineering (CISCE), pp. 346–349 (2022)
17. Julian, K., et al.: Digital game-based examination for sensor placement in context of an industry 4.0 lecture using the unity 3d engine-a case study. Procedia Manuf. **55**, 563–570 (2022)

A Real-Time Eye Gaze Tracking Based Digital Mouse

SeHyun Kwak[1], Daeho Lee[1], Siwon Kim[2], and Junghoon Park[1(✉)]

[1] Department of Applied Artificial Intelligence, Ajou University, Suwon, South Korea
{boring1230,daeho5000,stevejobs}@ajou.ac.kr
[2] Department of Computer Science and Engineering, Ajou University, Suwon, South Korea
kimsiw42@ajou.ac.kr

Abstract. Accessibility is a fundamental right for people with disabilities. It is the duty of society to ensure equal opportunities and strong participation in all aspects of society, including education, employment, and cultural activities. Placing people with disabilities side by side with those who are considered 'normal' individuals and perceiving them as 'different' or 'wrong' is a viewpoint that should be clearly rejected. So, this paper is concerned with a solution that enables people with disabilities who have difficulty using their hands, or who are unable to use their hands at all, to use a PC. For this purpose, it proposes a low-cost laptop-based webcam system designed to enable real-time interaction with a computer through eye movements. The system aims to enable people with physical limitations to communicate freely and can be particularly beneficial for patients with amyotrophic lateral sclerosis (ALS), and mobility impairments who prefer to use their eyes rather than their hands.

1 Introduction

Computer vision technology plays a key role in a variety of fields, including robotics, telecommunications, and healthcare [1]. The field of artificial intelligence, especially as it relates to healthcare, is also evolving, with much research being done in the form of alternative technologies to help people with disabilities and vulnerable [2]. However, these results typically served as a high price, which is often considered unaffordable for patients with limited means [3]. To solve this problem, we propose a low-cost iris tracking techniques using inexpensive one web camera. This approach will increase accuracy but reducing cost by utilizing existing image processing techniques. By integrating iris detection and face tracking, it is insensitive to head movement, provides high accuracy, and is expected to benefit such as patients suffering from Lou Gehrig's disease [4, 5]. Assistive technology is technology that helps people who are vulnerable to participate more comfortably and freely in daily, educational, and vocational activities. Therefore, it is often investigated for older or physically disabled people. In a four-year study in Germany, 46% of people with Lou Gehrig's disease needed a communication device, which is a disease that causes muscle weakness and slowed movement, making it difficult for them to communicate with others, even with family members and doctors [4]. The only muscles they can use until the end are the eye muscles, which is why there is a lot

L. Barolli (Ed.): IMIS 2024, LNDECT 214, pp. 39–46, 2024.
https://doi.org/10.1007/978-3-031-64766-6_5

of interest in devising ways to allow people with disabilities to control computers using their eyes, which are used for typing, so they can communicate with their surroundings [5].

There are a few approaches with different form factors, many of which use a camera attached to a wearable device, such as glasses, to capture images of the eye and receive them in high resolution to improve accuracy while reducing the ROI (Region of Interest) near the eye, thus reducing the constraints on head movement [6, 7]. Other forms of research use Corneal Reflection, where IR LEDs are placed, and their light is reflected off the eye [8, 9]. However, these methods require a separate additional hardware-based-device to generate the LED light, as well as other sensing and calculations, which can be expensive. Therefore, it is important to emphasize that gaze estimation techniques that use only conventional web camera, not wearable or mounted devices, using commercially available you can see a low-resolution-web camera or laptop camera. Recently, deep learning techniques have also been used to study gaze tracking, but these methods are slow and dependent on datasets [10].

In this study, we propose a gaze estimation technique using a single camera. It overcomes the limitations of existing methods and uses proposed traditional image processing algorithms rather than like CNN-based deep learning models to overcome the limitations of accurate but expensive deep learning-based methods. The proposed image processing algorithm firstly detects the iris, secondly calculates the gaze vector by recognizing the shape of the detected iris, and then thirdly quickly proceeds to correct the gaze vector through face tracking even if the orientation and position of the head change slightly, and finally proposes a high-accuracy gaze estimation technology that is robust even in a free head situation.

2   Previous System

Electrooculography (EOG):   Electrooculography is a classical method for eye-tracking research, which operates mechanically. It utilizes the difference in electrical charge between the cornea and the retina, known as the positive and negative charges, respectively, to measure the potential difference generated when the eye moves. Accordingly, research has been conducted to record horizontal and vertical eye movements. However, achieving high accuracy has been elusive [11, 12].

Dual Purkinje Image Method:

Various methods have been studied to solve the accuracy problem of the EOG method, one of which is the eye tracking method using Purkinje image. Purkinje, a Czech physiologist, conducted an experiment to measure the change in light reflection of the cornea and lens for near and far focus on subjects. The relationship between the first Purkinje image, which reflects from the anterior surface of the cornea, and the fourth Purkinje image, which reflects from the posterior surface of the lens, was found to be such that the first and fourth Purkinje images move the same distance when the eye moves, such as when the head moves sideways, but they move different distances when the eye rotates. However, these methods are highly dependent on the lighting and environment of the laboratory, making them difficult to generalize.

Fig. 1. An experimental Software-based System using YOLO.

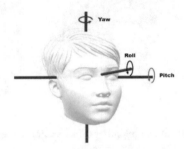

Fig. 2. Yaw, Pitch, Roll for the face.

Requiring Additional Equipment based Gaze Estimation: Classical methods have shown low accuracy, leading to the exploration of gaze tracking research based on Head Mount Systems, where participants wear equipment like glasses. Additionally, studies have been conducted using external IR LEDs and infrared cameras to better capture eye movements [13, 14]. However, these methodologies need more high-cost equipment for purchasing, like the equipment and LEDs.

Deep Learning based Gaze Estimation: To address these issues, research has been conducted to improve accuracy using regression models and CNN models [15, 16]. However, such studies require more expensive GPU-based resources to execute various deep learning models and face limitations in implementing real-time gaze estimation at a low cost.

3 Proposed Algorithm

Previous Systems have employed the Head mount System and Corneal Reflection methods utilizing supplementary apparatus or deep learning. As shown in Fig. 1, this algorithm seeks to overcome these limitations by using only a single camera and image processing algorithms instead of deep learning to provide an iris ellipse-based eye tracking service.

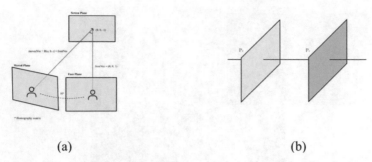

(a)   (b)

Fig. 3. (a) Correcting the location of Face Plane and Screen Plane defined by the Normal Vector of the Face (b) Final placement of Face Plane and Screen Plane.

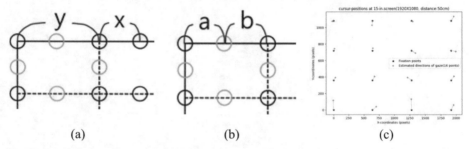

(a)   (b)   (c)

Fig. 4. (a), (b) is the process of calculating process to 16 points using the proportional expression; (c) accuracy calculated after calibration for 16 points.

To achieve this goal, the study explored eye tracking based on iris ellipse configuration, which relies on the complex nature of distinguishing between the pupil and iris, as individuals with unclear and black irises have difficulty distinguishing them via webcam. Initially, the YOLO model is trained to detect eye regions more accurately in the input image based on the environment, and then applied grayscale transformation and histogram equalization to enhance the contrast between dark and light parts of the eye to better extract the iris boundaries, and then applied Gaussian filtering to remove noise and prevent misinterpretation due to edge noise [17]. After masking out areas of misinterpretation that occurred in the eyelid region, the iris boundary was finally extracted using a canny edge detector. Based on the extracted iris, an ellipse was estimated, and the ratio of the major and minor axes of the iris ellipse was used to determine the direction and magnitude of the gaze vector.

To ensure robust gaze estimation despite head movements, as shown as Fig. 2, we employed Head Pose Estimation. In this process, we considered the yaw, pitch, and roll angles of the face to determine its current orientation and direction for calibration. When extracting the gaze vector, it's crucial to correct for variations in the relative position of the pupil due to facial movements. This correction ensures that the initially acquired gaze tracking values retain their significance consistently.

Addressing the computational complexity involved, we computed the yaw, pitch, and roll values based on the current viewpoint and Mean Squared Error (MSE) value.

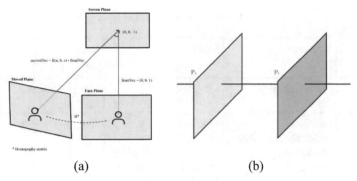

(a)   (b)

Fig. 5. (a) Position correction of the face plane and the screen plane defined by the normal vector of the face; (b) Final placement of the face plane and the screen plane

Additionally, we introduced a parameter α to adjust the update cycle of the timestamp calculation. Larger MSE values result in more frequent updates of head position, while smaller MSE values gradually extend the timestamp to reduce computational overhead for head position updates. Utilizing a facial recognition model, we determined the camera's rotation and translation matrices through face detection and subsequent image cropping. This process relies on the camera matrix and the 2D and 3D coordinates of facial landmarks for estimating the rotation vector and translation matrix.

With the computed facial direction vector, a plane representing the face's gaze direction can be defined. To rectify distortions in gaze vectors due to facial orientation, the face is aligned with the screen of the device, ensuring parallelism, as depicted in Fig. 3. This adjustment helps the computation of gaze vectors for calibration points. Nonetheless, the spherical shape of the eyeball leads to non-linear iris movement and varying degrees of mobility across individuals. In response, this study introduces an interpolation algorithm integrating parameters for individualized eye movement calculation via weighting as shown in Fig. 4. And we calculate the position correction of the face plane and the screen plane by the normal vector of the face as shown in Fig. 5 and expressed in Eq. (1).

$$p = \frac{z'}{z} K \cdot H \cdot K'^{-1} \cdot p' \qquad (1)$$

K: camera Intrinsic matrix
H: Homography Matrix
p: Face Plane Gaze vector
p': Moved Plane Gaze vector
z: z coordinates of p

4 Comparison of Inference Time and Error

The proposed study underwent various comparative processes to accurately detect eyes for enhancing gaze tracking precision and achieving real-time processing on low-spec systems. However, employing large deep learning models for eye detection would hinder the implementation of real-time processing systems. Therefore, we compared various methods to implement our proposed system as shown in Table 1.

Table 1. Comparison of detection algorithms.

Model	Structure	Scale	mAP		FPS	Latency (ms)
			MS	Dev		
YOLOv5x	Convolution + FC Layer	–Single-Scale	42	51.5	62.5	14.2
FPN	ResNet + Faster-R-CNN, RPN (Resion Proposal Network)	–Multi-Scale –Build Feature Pyramid from ResNet –Bottom-up-path + Top-Down-path	50.4	–	26	–
SSD300	VGG16 + Auxiliary Network	–Multi-scale –MultiBox Detector	–	28.8	–	–
EfficientNet	EfficientNet + BiFPN Layer	–Multi-Scale –Compound Scaling –BiFPN	49.6	–	–	42

In this study, instead of extracting key points, a pre-trained Object Detection model, YOLOv5x, was directly used for detecting the entire eye region due to its high accuracy [18]. YOLOv5x demonstrated an optimal balance between accuracy and speed in real-time object detection tasks [19, 20]. Consequently, the proposed system improved the performance of low-cost systems by providing real-time detection speed and reasonable accuracy. To verify the accuracy of the proposed algorithm, gaze point data was collected from users gazing at a 15-inch screen from 50 cm.

When compared to using only one camera without additional equipment, as in previous studies [21], the proposed method showed better accuracy, albeit lower than deep learning-based approaches in previous studies [22]. However, the advantage of this study lies in its ability to perform real-time gaze tracking without employing deep learning.

5   Conclusions

In this study, we propose a conventional webcam-based gaze tracking method for patients with physical disabilities. By utilizing an only-software-based low-cost system compared to conventional eye gaze systems, we reduced costs and increased user convenience. Furthermore, through head direction correction and vector recalibration, we confirmed the scalability of this method not only for patients with immobilized heads but also for general users. Additionally, instead of using deep learning models for real-time gaze tracking, we employed image processing algorithms to provide high performance even in low-cost systems.

References

1. Wiley, V., Lucas, T.: Computer vision and image processing: a paper review. Int. J. Artif. Intell. Res. **2**(1) . STMIK Dharma Wacana, p. 22, Jun. 2018, https://doi.org/10.29099/ijair. v2i1.42

2. Song, K.D., Kim, M., Do, S.: The latest trends in the use of deep learning in radiology illustrated through the stages of deep learning algorithm development. J. Korean Soc. Radiol. **80**(2). The Korean Society of Radiology, p. 202, 2019, https://doi.org/10.3348/jksr.2019.80.2.202

3. Martinez-Martin, E., Costa, A.: Assistive technology for elderly care: an overview. IEEE Access **9**. Institute of Electrical and Electronics Engineers (IEEE), pp. 92420–92430, 2021. https://doi.org/10.1109/ACCESS.2021.3092407

4. Funke, A., et al.: Provision of assistive technology devices among people with ALS in Germany: a platform-case management approach. In: Amyotrophic Lateral Sclerosis and Frontotemporal Degeneration, vol. 19, no. 5–6. Informa UK Limited, pp. 342–350, Jan 2018, https://doi.org/10.1080/21678421.2018.1431786

5. Spataro, R., Ciriacono, M., Manno, C., La Bella, V.: The eye-tracking computer device for communication in amyotrophic lateral sclerosis. Acta Neurol. Scand. **130**(1), 40–45 (2013). https://doi.org/10.1111/ane.12214

6. Ford, A., White, C.T., Lichtenstein, M.: Analysis of eye movements during free search. J. Opt. Soc. Am. **49**, 287 (1959). https://doi.org/10.1364/JOSA.49.000287

7. Li, Z., Guo, P., Song, C.: A review of main eye movement tracking methods. J. Phys: Conf. Ser. **1802**, 042066 (2021). https://doi.org/10.1088/1742-6596/1802/4/042066

8. Jang, J., Jung, J., Park, J.: A study on eye tracking techniques using wearable devices. Korean Inst. Smart Med. **12**, 19–29 (2023). https://doi.org/10.30693/SMJ.2023.12.3.19

9. Park, J., Yim, K.: Technical survey on the real time eye-tracking pointing device as a smart medical equipment. Korean Inst. Smart Media **10**, 9–15 (2021). https://doi.org/10.30693/SMJ.2021.10.1.9

10. Wang, K., Zhao, R., Su, H., Ji, Q.: Generalizing eye tracking with Bayesian adversarial learning. In: IEEE/CVF Conference on Computer Vision and Pattern Recognition (CVPR) (2019). https://doi.org/10.1109/CVPR.2019.01218

11. Elmadjian, C., Shukla, P., Tula, A.D., Morimoto, C.H.: 3D gaze estimation in the scene volume with a head-mounted eye tracker. In: Proceedings of the Workshop on Communication by Gaze Interaction, 2018. https://doi.org/10.1145/3206343.3206351

12. Canny, J.: A Computational Approach to Edge Detection. In: IEEE Transactions on Pattern Analysis and Machine Intelligence, vol. PAMI-8, no. 6. Institute of Electrical and Electronics Engineers (IEEE), pp. 679–698 (1986). https://doi.org/10.1109/TPAMI.1986.4767851

13. Plopski, A., Arno, L., Kashima, W., Taketomi, T., Sandor, C., Kato, H.: Eye-gaze tracking in near-eye head-mounted displays. In 28th Australian Conference on Human-Computer Interaction (OzCHI-2016), November-December (2016)

14. Lee, K.-F., et al.: Eye-wearable head-mounted tracking and gaze estimation interactive machine system for human–machine interface. J. Low Freq. Noise, Vibr. Active Control **40**(1), 18–38 (2019). https://doi.org/10.1177/1461348419875047

15. Nakazawa, A., Nitschke, C.: Point of gaze estimation through corneal surface reflection in an active illumination environment. In: Computer Vision – ECCV, pp. 159–172. Springer, Berlin, Heidelberg (2012). https://doi.org/10.1007/978-3-642-33709-3_12

16. Park, J., Jung, T., Yim, K.: Implementation of an eye gaze tracking system for the disabled people. In: 2015 IEEE 29th International Conference on Advanced Information Networking and Applications. IEEE, Mar (2015). https://doi.org/10.1109/AINA.2015.286

17. Cheng, Y., Wang, H., Bao, Y., Lu, F.: Appearance-based gaze estimation with deep learning: a review and benchmark. arXiv (2021). https://doi.org/10.48550/arXiv.2104.12668

18. Anilkumar, G., Fouzia, M.S., Anisha, G.S.: Imperative methodology to detect the palm gestures (American Sign Language) using Y010v5 and MediaPipe. In: 2nd International Conference on Intelligent Technologies (CONIT). IEEE, Jun (2022). https://doi.org/10.1109/CONIT55038.2022.9847703

46   S. Kwak et al.

19. Zhang, D., Li, J., Shan, Z.: Implementation of Dlib Deep Learning Face Recognition Technology, ICRIS, 7 Nov 2020
20. Ge, Z., Liu, S., Wang, F., Li, Z., Sun, J.: YOLOX: Exceeding YOLO Series in 2021. arXiv: 2107.08430v2 Aug (2021)
21. Cheng, Y., Zhang, X., Lu, F., Sato, Y.: Gaze estimation by exploring two-eye asymmetry. In: IEEE Transactions on Image Processing, vol. 29. Institute of Electrical and Electronics Engineers (IEEE), pp. 5259–5272 (2020). https://doi.org/10.1109/TIP.2020.2982828
22. Su, M.-C., U T.-M., Hsieh, Y.-Z., Yeh, Z.-F., Lee, S.-F., Lin, S.-S.: An eye-tracking system based on inner corner-pupil center vector and deep neural network. Sensors **20**(1), 25 (2019). https://doi.org/10.3390/s20010025

An Evaluation Model of the Effectiveness of College Club Activities Based on Grey Relational Analysis

Qiong Li[✉], Lili Su, Yanyan Zhao, Lixing Li, and Xuan Wang

Foundation Department, Engineering University of PAP, Xi'an 710086, China
573474824@qq.com

Abstract. In order to better improve the level of club activities in college, this study selects evaluation indicators by means of literature reference and questionnaire survey, builds a comprehensive evaluation model that conforms to the characteristics of club activities in college based on grey correlation degree, and carries out empirical analysis to provide a scientific evaluation tool for the future study on the effectiveness of club activities.

Keywords: Rating index · Analytic hierarchy process · Grey correlation degree

1 Introduction

University education should adhere to the correct political direction, adhere to the cultivation of morality, establish advanced educational concepts, promote modern educational methods, comprehensively improve the level of education, and form a comparative advantage in talent competition. As an extension of classroom teaching, student club activities are an important measure for colleges and universities to realize the whole process of education, and play an important role in comprehensively improving the connotative development of students.

The research on club activities mainly focuses on the status quo of club activities, development orientation, management mechanism, education mode and so on [1–5]. The scholars have accumulated some experience in the relevant research on club activities. With the talents training needs of colleges and universities and the issuance of relevant regulations, a college has gradually established more than 50 clubs, involving sports competition, culture and entertainment, scientific and technological innovation and other types. Since the launch of the club activities, the effect has been remarkable, and the comprehensive quality and personality of the students have been comprehensively developed. But there are also many problems about the management of the club. Most clubs operate independently with loose and chaotic management, and the operation mode and management system are not mature enough. In view of the actual situation of this college club, this paper establishes a set of scientific and reasonable effectiveness evaluation model that promotes the further improvement of the club management model and provides theoretical reference for the club management of this college.

L. Barolli (Ed.): IMIS 2024, LNDECT 214, pp. 47–57, 2024.
https://doi.org/10.1007/978-3-031-64766-6_6

Focusing on the requirements of university education and the problems existing in club activities, this paper preliminarily selects indicators based on the relevant evaluation indicators of previous studies of local university associations and the current research results of the institution. In the research process of this paper, we combine various research results and practical experience into one and take their commonalities and characteristics. Through semi-structured interviews with supervision experts, club backbone, responsible person and related organizers, the club effectiveness evaluation indicators were finally determined. Using analytic hierarchy process, combining qualitative and quantitative methods, the weights of each index are given. In the evaluation verification stage, according to the grey comprehensive evaluation method, the weighted average synthesis method is used to calculate the correlation degree, so that the evaluation clubs are ranked.

2 Construction of Club Evaluation Model

2.1 Selection and Establishment of Evaluation Index

With the keyword of "university community evaluation", we searched and sorted out the representative literature in recent years through databases such as CNKI and Wanfang [4–9].The comparative analysis finds that the evaluation indicators of associations mainly include four indicators: association management, association activities, association influence and association evaluation. Table 1 lists the proportion of common indicators in the specific classification of each index.

According to the proportion of evaluation indicators in local colleges and universities, consulting university authoritative experts, club leaders, club backbone and organization managers, we finally establish 4 first-level indicators, 12 s-level indicators and 27 third-level indicators (Table 2).

Table 1. Common evaluation index of local community

Serial number	Specific classification	1	2	3	4	5	6	7	Percent
1	Rules and regulations	√	√	√		√	√	√	86%
2	Organization construction	√	√	√		√	√	√	86%
3	Operation mechanism	√		√		√			43%
4	Routine activity	√	√	√	√	√	√	√	100%
5	Characteristic activity			√	√	√			43%
6	External communication	√		√	√			√	57%
7	Community size	√		√	√	√			57%
8	Propaganda effect	√	√		√	√	√	√	86%
9	Awards and honors	√	√	√		√	√		71%

(continued)

Table 1. (*continued*)

Serial number	Specific classification	1	2	3	4	5	6	7	Percent
10	Scientific research activity		√			√			29%
11	Internal evaluation	√	√	√	√	√			71%
12	External evaluation	√	√	√	√	√		√	86%

Table 2. Evaluation index of club activities in colleges and universities

Primary index	Secondary index	Three-level index
Club management a_1	Rules and regulations b_1	Club constitution f_1
		Club system f_2
	Organization construction b_2	Organizational structure f_3
		Member recruitment f_4
		Change of cadres f_5
		Personnel structure f_6
	Operation mechanism b_3	Plan summary f_7
		Club meeting f_8
		Advisor f_9
		Member management f_{10}
Club activity a_2	Routine activity b_4	Activity form f_{11}
		Number of activities f_{12}
		Activity quality f_{13}
	Characteristic activity b_5	Characteristic activity f_{14}
	External communication b_6	Frequency of organization f_{15}
		Organizational effectiveness f_{16}
Activity effect a_3	Population size b_7	Number of club members f_{17}
	Propaganda effect b_8	Propaganda effect f_{18}
	Awards and honors b_9	Number of winners f_{19}
		Award level f_{20}
	Scientific research activity b_{10}	Scientific research activity f_{21}
Club evaluation a_4	Internal evaluation b_{11}	Responsible person evaluation f_{22}
		Advisor evaluation f_{23}
		Club member evaluation f_{24}

(*continued*)

Table 2. (*continued*)

Primary index	Secondary index	Three-level index
	Student evaluation b_{12}	Management evaluation f_{25}
		Supervisory expert evaluation f_{26}
		Evaluation of the student team f_{27}

2.2 Establishment of Evaluation Index Weight

There are many methods to calculate the weight of evaluation indicator. This paper combines qualitative and quantitative analysis methods, adopts analytic hierarchy process and grey comprehensive evaluation theory to establish the effectiveness evaluation model of university clubs. Analytic hierarchy process (AHP) [6, 10] is an American operations research scientist T.L. Schaty put forward in the early 1970s, it is a complex problem according to the logical relationship of attributes for layer by layer decomposition, forming a hierarchical structure to be analyzed, so as to reduce the difficulty of research problems.

In the research, an expert evaluation team was set up. The expert group compared the relative importance of multiple indicators in the same level according to the scale method from 1 to 9 (Table 3), and constructed a judgment matrix, which was calculated by using the root method and obtained the weight of each indicator after normalization.

Table 3. Nintieth scale of relative importance

Index comparison	Vital	Of great importance	Importance	Slightly important	Equally important	Slightly less important	Insignificance	Very Unimportant	Of no importance
Evaluate scale values	9	7	5	3	1	1/3	1/5	1/7	1/9

Note: 8, 6 ,4, 2, 1/2, 1/4, 1/6, 1/8 is the middle value of the above review

In the actual situation, due to the complexity of the research object and the diversity of people's cognition, the judgment matrix will often have contradictory results, so it is necessary to carry out a consistency test on the judgment matrix. Only when the judgment matrix meets the consistency, can we get an objective and correct conclusion. Instead, the judgment matrix must be modified, and the expert panel members will re-compare the pairs until the consistency test is passed.

2.3 Grey Comprehensive Evaluation Based on Grey Relational Degree Analysis

Grey system theory is a "poor information" uncertain system with "part of information known and part of information unknown" proposed by Professor Deng Julong, a famous scholar, in 1982 [11]. It weakens the randomness and uncertainty of the original data, thus establishing a dynamic model for the development of "grey" system. Grey comprehensive evaluation is a comprehensive evaluation method based on grey system theory. This method is a multi-factor statistical analysis method based on relational degree analysis. The specific steps are as follows:

1. **Determine the reference sequence**

    Note i as the serial number of the i th evaluation object, i $= 1, 2,\ldots,$ m; k is the serial number of the v_{ik} evaluation index, k $= 1, 2,\ldots$ n, is the value of the v_{0k} evaluation index of the i th evaluation object. We take the best value of each index to form a reference series.

2. **Index value normalization processing**

    As the evaluation indicators usually have different dimensions and orders of magnitude, they cannot be compared directly. In order to ensure the comparability and reliability of the results, the original index values need to be normalized. The normalization formula is as follows:

$$X_{ik} = \frac{V_{ik} - \min_i V_{ik}}{\max_i V_{ik} - \min_i V_{ik}}$$

3. **Calculated correlation coefficient**

    We take the series of normalized constructions as the new reference series, $X_i = (x_{i1}, x_{i2}, \ldots, x_{in})$ as a comparative series. The correlation coefficient is calculated as follows:

$$\xi_{ik} = \frac{\min_i \max_k |X_{0k} - X_{ik}| + \rho \max_i \max_k |X_{0k} - X_{ik}|}{|X_{0k} - X_{ik}| + \rho \max_i \max_k |X_{0k} - X_{ik}|}$$

4. **The comprehensive evaluation results were calculated**

    Grey comprehensive evaluation is mainly based on grey weighted correlation degree to establish grey correlation degree, that is, R $=$ E $\times$ W. According to the size of grey weighted correlation degree, the correlation order of evaluation objects can be established.

3 Empirical Analysis

This study selects 4 clubs from a college and makes an empirical analysis based on multi-level grey system evaluation model.

1. Club activity effectiveness indicators include 4 first-level indicators, 12 s-level indicators and 27 third-level indicators, as shown in Table 2. For determining the weight of each indicator, 12 experts were selected, including 6 university supervision experts, 3 club leaders, 2 instructors and 1 university management officer. According to the contents of the indicators, the experts respectively compared and scored the relative importance of each indicator, constructed a judgment matrix, and tested the consistency of the judgment matrix, as shown in Table 4. According to the calculation of the judgment matrix, the weight of each influence factor on the effectiveness of military academy club activities can be obtained, as shown in Table 5.

Table 4. Each judgment matrix and consistency test at each level

Index	O	a_1	a_2	a_3	a_4	b_1	b_2	b_3	b_4	b_6	b_9	b_{11}	b_{12}
λ_{max}	4.215	3.000	3.018	4.220	2.000	2.000	4.156	4.220	3.094	2.000	2.000	3.054	3.065
CI	0.072	0.000	0.009	0.073	0.000	0.000	0.052	0.073	0.047	0.000	0.000	0.027	0.032
RI	0.9	0.580	0.580	0.900	0.000	0.000	0.900	0.9	0.580	0.000	0.000	0.580	0.580
CR	0.080	0.000	0.016	0.081	0.000	0.000	0.058	0.081	0.081	0.000	0.000	0.046	0.056

Note: CR is less than 0.1, passing the consistency test

Table 5. Weight of club activity effectiveness evaluation index system

Primary index	Secondary index	Three-level index	Total sort weight
Club management a1 (0.45)	Rules and regulations b_1(0.4)	Club constitution f_1(0.25)	0.0450
		Club system f_2(0.75)	0.1350
	Organization construction b2 (0.2)	Organizational structure f_3(0.4)	0.0360
		Member recruitment f_4(0.11)	0.0099
		Change of cadres f_5(0.1)	0.0090
		Personnel structure f_6(0.39)	0.0351
	Operation mechanism b_3(0.4)	Plan summary f_7(0.51)	0.0918
		Club meeting f_8(0.19)	0.0342
		Advisor f_9(0.18)	0.0324
		Member management f_{10}(0.12)	0.0216
Club activity a_2 (0.2)	Routine activity b_4(0.12)	Activity form f_{11}(0.36)	0.0086
		Number of activities f_{12}(0.1)	0.0024
		Activity quality f_{13}(0.54)	0.0130
	Characteristic activity b_5(0.56)	Characteristic activity f_{14}(1)	0.1120

(continued)

Table 5. (*continued*)

Primary index	Secondary index	Three-level index	Total sort weight
	External communication b6 (0.32)	Frequency of organization $f_{15}(0.2)$	0.0128
		Organizational effectiveness $f_{16}(0.8)$	0.0512
Activity effect a3 (0.11)	Population size $b_7(0.2)$	Number of club members $f_{17}(1)$	0.0220
	Propaganda effect b8 (0.07)	Propaganda effect $f_{18}(1)$	0.0077
	Awards and honors $b_9(0.46)$	Number of winners $f_{19}(0.2)$	0.0101
		Award level $f_{20}(0.8)$	0.0405
	Scientific research activity b10 (0.27)	Scientific research activity $f_{21}(1)$	0.0297
Club evaluation a4 (0.24)	Internal evaluation b_{11} (0.67)	Responsible person evaluation $f_{22}(0.33)$	0.0531
		Advisor evaluation $f_{23}(0.14)$	0.0225
		Club member evaluation $f_{24}(0.53)$	0.0852
	External evaluation $b_{12}(0.33)$	Management evaluation $f_{25}(0.19)$	0.0150
		Supervisory expert evaluation $f_{26}(0.73)$	0.0578
		Student evaluation $f_{27}(0.08)$	0.0063

2. As shown in Table 4, the overall ranking of C-level reflects the importance of each evaluation index in the military academy club evaluation system. In the entire evaluation system, the club system f2 in the rules and regulations is the index with the highest weight among all evaluation indicators. In the weight of O-A layer, club management is the highest weight of the four components, It can be seen that the normative construction of colleges and universities clubs is very important.
3. Questionnaires were formulated for each evaluation index reference point, a total of 120 questionnaires were issued, 117 were recovered and 117 were valid, and the data of each index were sorted out and normalized to obtain the reference series, as shown in Table 6.

Table 6. Values of club indexes and satisfaction values

	V_1	V_2	V_3	V_4	Reference column
f_1	0.900	0.807	1.000	0.797	1.000
f_2	0.666	0.939	0.620	0.785	0.939
f_3	0.944	0.909	0.790	0.934	0.944
f_4	0.624	0.250	0.396	0.332	0.624
f_5	0.472	1.000	0.690	0.598	1.000
f_6	0.694	0.426	0.550	0.000	0.694
f_7	0.834	0.953	0.620	0.864	0.953
f_8	0.644	0.439	0.790	0.348	0.790
f_9	0.490	0.530	0.610	1.000	1.000
f_{10}	0.658	0.709	0.910	0.434	0.910
f_{11}	0.954	0.480	0.870	0.316	0.954
f_{12}	0.532	0.588	0.900	0.117	0.900
f_{13}	0.888	0.588	0.560	0.066	0.888
f_{14}	0.522	0.000	1.000	0.266	1.000
f_{15}	0.550	0.044	1.000	0.383	1.000
f_{16}	0.900	0.527	0.970	0.566	0.970
f_{17}	0.900	0.774	0.960	0.481	0.960
f_{18}	0.850	0.456	0.550	0.316	0.850
f_{19}	1.000	0.706	1.000	0.367	1.000
f_{20}	0.800	0.706	1.000	1.000	1.000
f_{21}	0.000	0.030	0.000	0.684	0.684
f_{22}	0.960	0.875	1.000	0.747	1.000
f_{23}	0.906	0.733	0.924	0.649	0.924
f_{24}	0.900	0.720	0.990	0.782	0.990
f_{25}	0.922	0.814	0.978	0.756	0.978
f_{26}	0.862	0.858	0.934	0.930	0.934
f_{27}	0.796	0.557	0.840	0.560	0.840

4. Calculate the grey correlation coefficient. The correlation values between each index and the best values in the reference series can be obtained, as shown in Table 7.

Table 7. Related numerical table

	V_1	V_2	V_3	V_4
ξ_1	0.8333	0.7215	1.0000	0.7112
ξ_2	0.6468	1.0000	0.6105	0.7645
ξ_3	1.0000	0.9346	0.7645	0.9804
ξ_4	1.0000	0.5720	0.6868	0.6313
ξ_5	0.4864	1.0000	0.6173	0.5543
ξ_6	1.0000	0.6510	0.7764	0.4188
ξ_7	0.8078	1.0000	0.6002	0.8489
ξ_8	0.7740	0.5875	1.0000	0.5308
ξ_9	0.4950	0.5155	0.5618	1.0000
ξ_{10}	0.6649	0.7132	1.0000	0.5123
ξ_{11}	1.0000	0.5133	0.8562	0.4394
ξ_{12}	0.5760	0.6158	1.0000	0.3897
ξ_{13}	1.0000	0.6250	0.6039	0.3782
ξ_{14}	0.5112	0.3333	1.0000	0.4052
ξ_{15}	0.5263	0.3434	1.0000	0.4476
ξ_{16}	0.8772	0.5302	1.0000	0.5531
ξ_{17}	0.8929	0.7289	1.0000	0.5107
ξ_{18}	1.0000	0.5593	0.6250	0.4836
ξ_{19}	1.0000	0.6297	1.0000	0.4413
ξ_{20}	0.7143	0.6297	1.0000	1.0000
ξ_{21}	0.4223	0.4333	0.4223	1.0000
ξ_{22}	0.9259	0.8000	1.0000	0.6640
ξ_{23}	0.9653	0.7236	1.0000	0.6452
ξ_{24}	0.8475	0.6494	1.0000	0.7062
ξ_{25}	0.8993	0.7530	1.0000	0.6925
ξ_{26}	0.8741	0.8681	1.0000	0.9921
ξ_{27}	0.9191	0.6386	1.0000	0.6410

5. Comprehensive evaluation result.

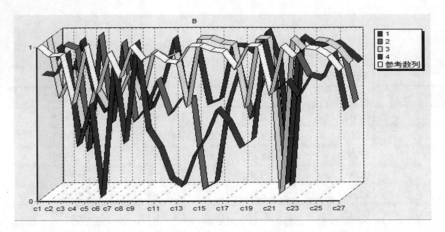

For the three-layer evaluation system of this study, the correlation coefficient of each index of layer F is synthesized to obtain the correlation degree of each index of layer B, and then the correlation degree of each index of layer A is synthesized using the correlation degree of layer B as the original data. By analogy, the correlation degree of club activity effectiveness index can be obtained, and the correlation degree of each index of each layer can be obtained:

$$R_{a1} = (0.7609\ 0.8531\ 0.7232\ 0.7487)$$
$$R_{a2} = (0.6594\ 0.4143\ 0.9681\ 0.4453)$$
$$R_{a3} = (0.7174\ 0.5916\ 0.8178\ 0.8146)$$
$$R_{a4} = (0.8875\ 0.7486\ 1.0000\ 0.7574)$$

Further, the correlation degree of the highest index O can be obtained:

$$R_O = (0.7662\ 0.7115\ 0.8490\ 0.6973)$$

According to the degree of correlation, the order of the results of the four clubs is as follows:

$$V3 > V1 > V2 > V4.$$

4   Conclusion

Based on the grey model, this study establishes a multi-level evaluation index system. In order to obtain accurate evaluation data, questionnaires and interviews are used to collect feedback information from participants and organizers. The comprehensive evaluation software is used to process the data, we build the weight matrix, and then obtain the comprehensive evaluation results through the grey correlation coefficient. The qualitative and quantitative methods are combined to ensure the scientific and rational evaluation results.

The research results show that the overall performance of the club activities is good, but according to the weight analysis of the evaluation system, the club activities have great room for improvement in the aspects of characteristic innovation and awards. As a result, the club can add more innovative elements in the future organization process to improve the activity experience, build more platforms to improve students' self-ability and organize higher-level competitions, so that students can get more opportunities to show, and then obtain better honors. Through the evaluation index system and grey correlation model established in this study, we can understand the club activities from various angles, find and solve problems in time, and continuously improve the quality of activities and participants' satisfaction.

References

1. Liu, F.: Development Status and Countermeasures of college student associations from the perspective of Quality Development. Nat. Higher Educ. **23**, 20–22 (2013)
2. Zhongyong, F.: Optimization path of college student association construction. School Party Build. Ideol. Educ. **20**, 66–67 (2019)
3. Kramer, G.L., et al.: Fortering Student Success in the Campus Cumunity. San Francisco:Jossey-Bass (2007)
4. El Karfa, A.: Open Classroom Communication and the Learning of Citizenship Values. English Teaching Forum.US Department of State Bureau of Educational and Cultural Affairs, Office of English Language Programs, SA-5, 2200 C Street NW 4th Floor, Washington, DC (2007). e-mail: etforum@state.gov; Website: http://americanenglish.state.gov/english-teaching-forum-0
5. Jia, C.: Research on the construction of evaluation index system of sports associations in colleges and universities —a case study of Jiangsu Province. Nanjing Inst. Phys. Educ. **3**(10), 18–23 (2016)
6. Chen, T.: Comprehensive evaluation of traditional mathematics teaching and multimedia teaching based on AHP. J. Higher Educ. (15) (2016)
7. Song, Y., Fang, F.: Research on the comprehensive evaluation system of college students' innovation and entrepreneurship associations. Juvenile J. **6**, 53–58 (2017)
8. Chen, W., Yang, X.: Research on the evaluation system of university students' associations. Qual. Educ. Western China **6**(13), 9–14 (2020)
9. Shao, J.: A summary of the research status quo on the cultivation of student association culture in vocational schools at home and abroad. Sci. Technol. Inf. **17**(4), 225–227 (2019)
10. Zhang, X.: Research on evaluation of college students' information literacy level based on analytic hierarchy process and ideal point method. Libr. J. **7**, 43–49 (2018)
11. Du, D., Yu, Q.: Modern Comprehensive Evaluation Methods and Case Selection. Tsinghua University Press (2005)

Research on Bitcoin Price Prediction Based on Text Analysis and Deep Learning

Ziying Liu[1], Xu Chen[1(✉)], and Xu-an Wang[2]

[1] Zhongnan University of Economics and Law, Wuhan, China
chenxu@whu.edu.cn
[2] Engineering University of Peoples Armed Police, Xi'an, China

Abstract. With the increasing popularity of Bitcoin, many investors are attracted to it. However, Bitcoin's decentralized and anonymous transaction features facilitate certain illegal activities, leading to significant price fluctuations and potential risks for investors. Therefore, predicting Bitcoin prices is crucial. This study analyzes factors influencing Bitcoin prices and trends by examining four years of closing prices. Data categories include trading, public, technical, macroeconomic factors, and global currency market indicators. The study involves LDA topic clustering, NLP sentiment analysis, data preprocessing, and the application of LSTM and CNN-LSTM deep learning models for price prediction. Machine learning methods are used to assess factors impacting Bitcoin prices, aiding investors and decision-makers in making more accurate predictions and improving investment decisions. The study suggests that integrating sentiment features enhances prediction accuracy, emphasizing the importance of public sentiment in Bitcoin price forecasting. By incorporating news headlines and public sentiment indicators, this research enriches cryptocurrency price prediction methods, offering new perspectives and theoretical support for digital currency research, potentially helping investors mitigate risks and supporting regulatory efforts in financial markets.

1 Introduction

Bitcoin, a popular cryptocurrency, has seen increasing investor confidence as the cryptocurrency market matures and regulations become more standardized. However, its decentralized and anonymous nature can lead to illegal activities and price volatility, making price prediction crucial. This research, combining theory and practice, uses text mining and machine learning to analyze price fluctuations and develop predictive models. It provides insights into the economic characteristics of digital currencies, helps investors make informed decisions, and offers guidance for regulatory bodies to govern trading markets, contributing to financial system stability. Its findings also have broader implications for the future of digital currencies, potentially impacting internet payment systems and monetary policies.

In the early stage of the study, domestic and foreign scholars mainly widely used econometric time series forecasting methods, and the mainstream literature used regression-based models, including vector error correction models, vector autoregressive models, and ARIMA models [1]. Munim uses the ARIMA model and the neural network autoregressive model to predict the next day's Bitcoin price [2]. Some studies have also looked at the impact of market sentiment on the price of Bitcoin, analyzing sentiment data such as social media and news reports to predict the direction of Bitcoin's price. Serafini and other scholars built the ARIMAX model based on the ARIMA model, combining Twitter tweet sentiment data and the daily price and trading volume of Bitcoin to predict the next day price of Bitcoin [3]. However, while econometric methods perform well with stationary data, it is difficult to capture the non-linear hidden patterns and the complex interconnections between the Bitcoin price and other economic factors in highly volatile Bitcoin data.

In order to solve the problem of complex correlation between traditional time series analysis methods in capturing nonlinear and non-stationary data machines, scholars at home and abroad have begun to study machine learning prediction models. Chen [3] used decision tree (DT) and support vector machine (SVM) techniques to predict the price trend of Bitcoin, and found that machine learning methods are more effective than traditional statistical methods in predicting Bitcoin prices. McNally [5] use LSTM and RNN based on Bayesian optimization to predict the price of Bitcoin, and believe that the prediction effect of LSTM network on the price of Bitcoin is better than that of the ARIMA model. For the verification of the effect of the LSTM model, some scholars used four machine learning models (ANN, SANN, SVM, LSTM) to compare short-term predictions, and found that the LSTM had the best performance [6–10]. Later, machine learning and deep learning algorithms at home and abroad have also innovated in prediction models, and the emergence of hybrid models has continuously improved the accuracy of predictions. At present, deep learning methods are an effective tool for predicting the price of bitcoin, because of its characteristics, it can learn more complex attribute information and the relationship between them, so as to make better predictions.

2 Data Selection and Preprocessing

2.1 Data Selection and Source

In this study, we analyze the daily closing prices of Bitcoin from January 1, 2020, to January 1, 2024, retrieved from the Investing.com financial website. Bitcoin is traded 24 h a day, and thus, we focus on the daily closing prices for our analysis.

Previous research suggests that effective predictive systems for Bitcoin prices can be constructed by incorporating factors such as trading metrics, public attention, cryptocurrency market conditions, and macroeconomic indicators. Consequently, our model includes five categories of influencing factors: trading factors, public factors, technical factors, macroeconomic factors, and global monetary market indicators.

Regarding trading factors, we include opening price, highest price, lowest price, and trading volume as influencing variables, all of which can be obtained from the Investing.com website.

For public factors, we consider the volume of tweets on Twitter and Bitcoin-related news headlines. The Twitter data on Bitcoin-related tweets was crawled from the bitinfocharts website, and the news headlines were retrieved from the Cryptonews.com cryptocurrency news site, both within the period from January 1, 2020, to January 1, 2024.

At the technical level, since blockchain is the underlying technology relied upon by the Bitcoin network, the generation of blocks directly impacts Bitcoin creation and trading, thereby affecting its supply and demand. We collect daily technical data such as average block size, difficulty, hash rate, and mining revenue from authoritative blockchain browsers (www.blockchain.com) and bitinfocharts. Technical indicators like mining difficulty, which adjusts to maintain a stable block generation time, directly affect Bitcoin supply and hence its pricing. In addition, we obtain two features related to Bitcoin network performance—block propagation delay and transaction propagation delay—from a Bitcoin network monitoring website (dsn.tm.kit.edu/bitcoin/) conducted by a project sponsored by the German Federal Ministry of Education, which provides insights into the network's operational status through multiple monitoring points in Germany.

In terms of macroeconomic and global monetary market factors, studies have confirmed the long-term impact of indicators such as the Dow Jones Index and inter-currency exchange rates on Bitcoin prices. Given Bitcoin's economic and monetary attributes, we also include indices like the Shanghai Stock Exchange Index and the US Dollar Index, which are related to global economic development. These and other 15 significant indicators such as the S&P 500 and major global exchange rates were gathered from the Choice financial terminal, which is a professional financial data platform under Eastmoney.com, offering convenient services for institutional investors and academic research.

The dataset for this study is as shown in Table 1.

2.2   Headline Text-Extraction

Sources and Crawling of News Headlines. This research extracted Bitcoin-related news headlines from the Cryptonews website between January 1, 2020, and January 1, 2024, amassing a total of 787 records. The data collection process involved utilizing the requests and lxml libraries to fetch web pages and parse content, with XPath used to extract headlines and publication dates, which were then stored in a CSV file. To enhance data quality, multiple stopword lists were incorporated and customized, followed by Chinese segmentation using jieba, and the results were saved in TXT format for subsequent LDA topic analysis.

Sentiment Analysis. For sentiment analysis, the ERNIE_SKEP model, proposed by Baidu, was employed for sentence-level sentiment classification. This model, grounded in knowledge enhancement, learns real-world semantic relationships through pre-training and supports transfer learning for task adaptation. Sentiment analysis of the text data yielded sentiment judgments and positive/negative sentiment scores.

LDA Topic Analysis. LDA analysis was implemented using the gensim library, with parameters set at three topics, 1000 documents per iteration, automatically learned α and η values, 400 iterations, and a minimum probability threshold of 0.001 for word-topic distributions. As shown in Fig. 1, the optimal number of topics was determined through estimating different clustering scenarios based on consistency and perplexity values, settling on three topics.

Topic-based Emotion Feature Generation. For each sub-topic data in the unit of trading day, the sum of the sentiment score and the sentiment polarity score of the news

Table 1. Five different categories of influencing factors: trading factors, public factors, technical factors, macroeconomic factors, and global money market indicators

Dependent Variable	Daily Closing Price of Bitcoin	
Influencing Factors		
Trading Factors	Opening Price, Highest Price, Lowest Price, Trading Volume	
Technical Factors	Blockchain Metrics	Average Daily Block Size, Difficulty Level, Hash Rate, Mining Reward (Including Block Reward and Transaction Fees), Average Transaction Fee (Excluding Mining Reward), Transaction Quantity, Market Capitalization
	Network Performance Indicators	Block Propagation Delay, Transaction Propagation Delay
Public Sentiment Factors	News Headlines, Twitter	
Macroeconomic Factors	Americas Market	S&P 500 Index, Dow Jones Industrial Average
	European Market	FTSE 100 Index
	Asian Pacific Market	Hang Seng Index, Nikkei 225 Index Greater ﹀
	China Market	Shanghai Stock Exchange Index, CSI 300 Index
Global Monetary Market Indicators	US Dollar Index, USD/JPY, EUR/USD, GBP/USD, USD/CHF, USD/SGD, USD/CNH	

text contained in it is calculated, so as to obtain the topic-based sentiment feature data required for training the prediction model in this paper.

The sentiment score characteristics of topic k on day t are as follows.

$$S_{k,t} = \frac{1}{n_{k,t}} \sum S_i \tag{1}$$

The emotion judgment score is calculated as above.

In this chapter, we complete the grouping of sentiment features of text based on topics by first performing sentiment analysis on Bitcoin news texts and then using the LDA model to extract topics. Then, through the above steps, all news text data is numerically converted into a 1*6 vector.

2.3 Data Preprocessing

After preliminary analysis of the acquired data, the following problems existed.

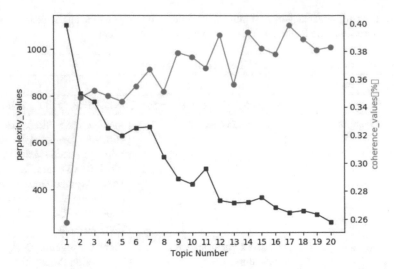

Fig. 1. The perplexity value and coherence value of LDA in different clustering cases.

For the macroeconomic data obtained from the Choice financial data terminal and the data related to global currencies, there is a problem that the weekend data is not available due to the weekend market closure. In this paper, the linear interpolation method is used to fill in the missing data at the weekend to reduce the error caused by the missing data.

For the text data of news headlines crawled from news websites, due to the human factor, there is a problem that the publisher publishes multiple news items in one day or only publishes one news news in multiple days. In this paper, we convert the news release time in minutes to daily and deduplicate the duplicate data. For its sentiment score data, the null values are filled in using the mean-filled method.

For the block propagation delay and transaction propagation delay data collected from the Bitcoin monitoring website, due to network reasons, there are problems of data missing, duplication and anomaly. The source data uses UNIX timestamps to record data in hours, and this article first converts UNIX timestamps into natural time in days, and arranges the data in natural time order.

Finally, pandas is used to merge all the collected data based on time, the data dimension is 67 before the merge processing, and the data dimension is 29 after the redundant data processing is merged and removed.

2.4 Descriptive Analysis and Feature Selection

All the collected data are summarized and descriptively analyzed by numerical data, data cleaning, completion, deduplication, etc., as shown in the following Table 2.

Through the above text numerization process, this paper merges the news text data with other numerical data to obtain a total of 29 data features. In order to eliminate redundant features and improve the interpretability of the model, this paper selects the top 8 weighted features through XGBOOST features: BTC market capitalization, closing

Table 2. The mean and standard deviation of the data used in the study

Data	Mean	Std	Data	Mean	Std
Closing	28888.260	15105.334	EUR_USD exchange rate	1.114	0.062
Opening	28863.236	15110.698	GBP_USD exchange rate	1.284974	0.0626
Highest price	29554.144	15520.258	USD_CHF exchange rate	9.26460	0.032228
The lowest price	28121.391	14535.169	USD_CNH exchange rate	6.7963	0.31114
Emotional Judgment	0.494	0.445	USD_JPY exchange rate	122.1705	15.6954
Sentiment score	0.499	0.235	USD_SGD exchange rate	1.3611	0.028531
The number of daily transactions	3.127494e+05	9.44990e+04	U.S. dollar index	98.9340	5.919308
Daily on-chain trading volume	8.039989e+05	8.8667e+05	Shanghai Index	3265.567	227.0144
Mining difficulty	2.983440e+13	1.47390e+13	CSI 300 Index	4362.544	562.194
Average daily transaction fee	0.000159	0.000165	Dow	3199.575	3502.036
Miner income	1046.918661	285.231165	UK FTSE Index	7062.058	616.9964
Hash rate	2.175097e+20	1.108614e+20	Nikkei 225	27359.811	3460.575
market value	5.47907e+11	2.8577e+11	Hang Seng Index	22959.017	3828.549
The average daily block size	1.3260e+06	2.8265e+05	U.S. S&P Index	3968.207	523.933
Number of tweets	119072.945	64740.009			

price_SPX, closing price_N225, closing price_FTSE, closing price_DJIA, closing price _GBPUSD, twitter, and BTC mining difficulty.

3 Experimental Process

3.1 Model Introduction

LSTM Model Long Short Term Memory (LSTM) is a special type of recurrent neural network. In short, when using a feedforward neural network, the neural network will think that what we input at time t is completely irrelevant to what we input at time t + 1, and for tasks such as natural language processing, it is obvious that the data at time t +

n can be used more rationally. In order to apply the information in the time dimension, the first model recursive neural network (RNN) appeared, and the following figure is a simple recursive neural network.

Since RNN is suitable for short-term dependence problems, the prediction effect of RNN will decrease with the increase of time span, and RNN is prone to two problems: gradient vanishing problem and gradient explosion problem. Therefore, LSTM networks were proposed in 1997 to solve the long-term dependency problem that is common in general recurrent neural networks, as well as the gradient vanishing or gradient explosion problem in RNNs.

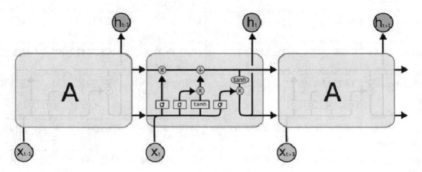

Fig. 2. The LSTM cell structure.

The LSTM cell structure is shown in Fig. 2 and contains two hidden states, which are the output h_t and the cell. The c_t key to LSTM standing out from the RNN is the line running through the cell in the above diagram – the hidden state of neurons, which we can simply understand as the "memory" of the recurrent neural network for the input data, which is represented by the cell state in c_t t + 1. A summary of all input information by the neural network before the moment.

CNN-LSTM. CNN is a feedforward neural network consisting of an input layer, a convolutional layer, a pooling layer, a fully connected layer, and an output layer. CNN networks refer to those neural networks that use convolution operations instead of general matrix multiplication operations at least one layer of the network. Because CNN adopts convolution operation in calculation, its operation speed is greatly improved compared with the general matrix operation, and the alternating use of convolutional layer and pooling layer can effectively extract local features of data and reduce the local feature dimension. Due to the weight sharing, the number of weights can be reduced and the complexity of the model can be reduced. The output of a one-dimensional convolution for feature extraction of a time series is:

$$Y = \sigma(W \cdot X + b) \tag{2}$$

Y represents the extracted features, σ is the activation function, W is the weight matrix, X is the time series, and b is the bias vector-LSTM.

The CNN-LSTM architecture for multi-feature sequence classification, as illustrated in Fig. 3, is characterized by its compositeness, incorporating a signal input layer, a CNN convolutional layer, a pooling layer, an LSTM layer, and a final classification output layer. The CNN-LSTM-based multi-input sequence classification process unfolds as follows: Initially, the input signal is standardized and passed through the CNN convolutional layer, which employs a wide convolutional kernel to adaptively extract features. Subsequently, the pooling layer, specifically the max pooling layer, is applied to the extracted features, dimensionality reduction is achieved, and the salient feature information is preserved. Thereafter, the reduced-dimensional feature data serves as the input for the LSTM layer, facilitating the training of the neural network and the autonomous learning of sequence-specific features. The training process is executed using the Adam algorithm, which propagates training errors backward and updates model parameters iteratively. Finally, the Softmax activation function is utilized to classify the signal features, thus completing the classification task of multi-feature input sequences.

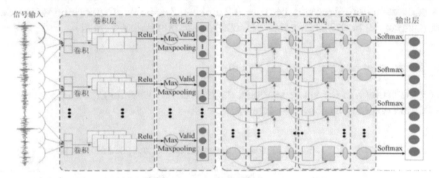

Fig. 3. The CNN-LSTM network structure.

3.2 Data Normalization and Data Set Division

Using MinMaxScaler method to normalize data, which is based on the principle of linearly transforming the data, mapping the original data to the range of [0, 1]. The specific formula is as follows:

$$X_{nom} = \frac{X - X_{min}}{X_{max} - X_{min}} \tag{3}$$

where X is the raw data, and X_{max}, X_{min} are maximum and minimum values of the data, respectively.

As for how to divide the dataset, the data time range in this article is from January 1, 2020 to January 1, 2024, with a total of 1463 records. In this paper, 1000 entries are divided into training data, and the other 463 entries are used as test data.

3.3 Experimental Setup

In this paper, three time steps of 7, 10, and 14 were selected for training of 1000 epochs. In LSTM, after data preprocessing and partitioning, a sequential model is established using tf.keras.models. It incorporates LSTM, Dropout, and dense layers, with an input shape of (samples, time steps, features). The model features three LSTM layers, each paired with a Dropout layer to enhance generalization and mitigate overfitting. LSTM parameters include units (neurons), activation functions, and 'return_sequences'. Dropout rates are set at 0.15, 0.05, and 0.01, respectively. A final fully connected layer, with neurons matching feature columns, uses a linear activation function. The model is compiled with mean squared error loss and the Adam optimizer.

The CNN-LSTM model starts with a CNN input layer and a reshape layer to treat data samples as 2D images with a window size 'x' and 8 features, followed by a 2D convolutional layer with 64 filters, ReLU activation, and same padding. A maxpooling layer and dropout layer are added for downsampling and overfitting prevention, respectively. Two LSTM layers are then connected for temporal sequence modeling, followed by a fully connected layer to output predictions. The model uses mean squared error for loss and Adam optimizer with beta1 = 0.9 and beta2 = 0.999. Other settings include a batch size of 32, base learning rate of 0.0001, and 1000 training epochs.

3.4   Evaluation of Experimental Results

Using the settings in the above sections, the prediction evaluation indicators of the two models are obtained as shown in Table 3.

Table 3. The mean and standard deviation of the data used in the study

Model	Sequence length	RMSE	MAE	MAPE
LSTM	7	1050.02163	1050.02162	0.0329037
	10	898.162750	641.564486	0.0249698
	14	952.605457	784.760434	0.0252689
LSTM-CNN	7	883.166770	706.967938	0.0274424
	10	813.743747	626.462203	0.0248897
	14	814.848941	622.272807	0.0249912

From Table 3, it can be seen that:

(1) The RMSE, MAE and MAPE of the CNN-LSTM model are all smaller than those of the LSTM model, indicating that the CNN-LSTM model will have a better effect on Bitcoin price prediction, that is, the LSTM can further capture features after the introduction of the CNN model to improve the performance of the model.
(2) By adjusting the step size of the two models, it can be found that for the daily data of Bitcoin price, the step size should not be too long or too short, and in this experiment, the step size of 10 is the best for both LSTM and CNN-LSTM models.

Through the three evaluation indicators in the two cases, it can also be seen that the model with emotional features will perform better than the model without emotional features (as shown in Figs. 4 and 5). It shows that public sentiment is a relatively important influencing factor in the study of Bitcoin price prediction.

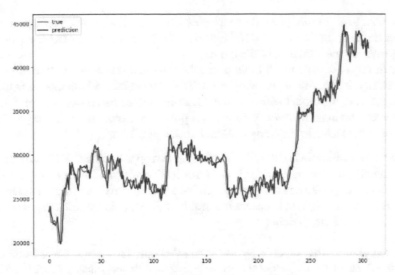

Fig. 4. Forecast renderings without public factor.

Fig. 5. Forecast renderings with public factor

4 Conclusion

This paper selects and collects five different categories of data that may have an impact on the price of Bitcoin, including trading factors, public factors, technical factors, macroeconomic factors and global money market indicators, and conducts LDA topic clustering and ERINE NLP pre-trained model sentiment analysis on text-based data, and introduces deep learning model LSTM and its variant CNN-LSTM model for learning and prediction. Comparative studies have found that:

(1) The CNN-LSTM model will have a better effect on Bitcoin price prediction, that is, after the introduction of the CNN model, the LSTM can further capture features to improve the performance of the model;
(2) For the daily data of the Bitcoin price, the step size should not be too long or too short, in this experiment, the step size is 10, and the effect of both models is the best.
(3) The effect of the model with emotional features will be better than that of the model without emotional features. It shows that public sentiment is a relatively important influencing factor in the study of Bitcoin price prediction.

The future of Bitcoin is still full of challenges and opportunities. With the continuous advancement of technology and the improvement of regulation, Bitcoin is expected to be more widely recognized and adopted on a global scale. However, investors and regulators still need to remain vigilant to ensure the health of the Bitcoin market and guard against potential risks and wrongdoing.

Acknowledgments. This work is supported by "the Fundamental Research Funds for the Central Universities", Zhongnan University of Economics and Law, Project Name (2722024BY023)).

References

1. Amjad, M., Shah, D.: Trading bitcoin and online time series prediction. In: NIPS 2016 Time Series Workshop, pp. 1–15 (2017)
2. Munim, Z.H., Shakil, M.H., Alon, I.: Next-day bitcoin price forecast. J. Risk Financ. Manag. **12**(2), 103 (2019)
3. Serafini, G., Yi, P., Zhang, Q., et al.: Sentiment-driven price prediction of the bitcoin based on statistical and deep learning approaches. In: 2020 International Joint Conference on Neural Networks (IJCNN), pp. 1–8. IEEE, Piscataway, NJ (2020)
4. Chen, T.H., Chen, M.Y., Du, G.T.: The determinants of bitcoin's price: utilization of GARCH and machine learning approaches. Comput. Econ. **57**, 267–280 (2010)
5. McNally, S., Roche, J., Caton, S.: Predicting the price of bitcoin using machine learning. In: 2018 26th Euromicro International Conference on Parallel, Distributed and Network-based Processing (PDP), pp. 339–343. IEEE, Piscataway, NJ (2018)
6. Khuntia, S., Pattanayak, J.K.: Adaptive market hypothesis and evolving predictability of bitcoin. Econ. Let. **167**, 26–28 (2018)
7. Zhang, W., Wang, P., Li, X., et al.: The inefficiency of cryptocurrency and its cross-correlation with Dow Jones Industrial Average. Phys. A: Stat. Mech. Appl. **510**, 658–670 (2018)
8. Jia, C.: Bitcoin price prediction based on news headline text mining and deep learning. Master's Thesis of Lanzhou University (2022)
9. Luo, Q.: Research on bitcoin price time series based on machine learning technology. Master's Thesis of Southwestern University of Finance and Economics (2020)
10. Erfanian, S.: Comparative analysis of machine learning algorithms for cryptocurrency price prediction. Doctoral Dissertation of Sichuan University (2021)

Effect of DoS Attack into LiDAR Ethernet

Yoonji Kim[1], Insu Oh[2], Jiung Hwang[2], Minchan Jeong[2], and Kangbin Yim[2(✉)]

[1] Department of Mobility Convergence Security, Soonchunhyang University,
Chungcheongnam-do, South Korea
`rladbsw17@sch.ac.kr`
[2] Department of Security Information Engineering, Soonchunhyang University,
Chungcheongnam-do, South Korea
`{iso,jiung5359,yim}@sch.ac.kr`

Abstract. For future fully autonomous vehicles, systems that utilize external sensors to recognize and judge the surrounding environment will become necessary. LiDAR plays an excellent role in autonomous driving because it provides accurate information by forming a 3D map. However, if the LiDAR sensor malfunctions and provides incorrect information, self-driving cars may be exposed to dangerous situations on the road. Many prior studies have recently been published demonstrating severe physical LiDAR spoofing attacks to induce obstacle misdetection. While existing research focuses on physically accessible spoofing attacks, this paper proposes an Ethernet data packet injection attack that can occur on a network using the Velodyne LiDAR VLP-16 (PUCK) sensor. The proposed method generated Ethernet data packets by replacing them with meaningless values based on analysis of Ethernet data packets on the network. A DoS attack was performed by injecting a large amount of data based on this. Additionally, based on these attacks, we would like to present a response methodology for LiDAR ethernet DoS attacks.

1 Introduction

Future autonomous driving requires a fully automated system equivalent to Level 5 that can autonomously monitor the surrounding environment without driver intervention. Modern cars are equipped with radar, ultrasonic, and camera sensors to provide drivers with an advanced driving assistance system. Still, advanced sensor technology is required to perform advanced assistance systems and automated decisions [1].

Current cars use a level 3 autonomous driving function using only cameras and radars, but cameras in cars function like human eyes, so in bad weather conditions, just as humans have difficulty recognizing objects, cameras also difficulty recognizing obstacles. Additionally, radar's ability to identify stationary objects is poor because it uses radio waves [2].

Light Detection and Ranging (LiDAR) sensors have been used in atmospheric observation equipment [3], agriculture [4], and portable electronic devices [5] in our lives, gradually expanding the scope of the application of LIDAR technology and have developed into a core technology of autonomous driving today. The LiDAR sensor's operating

L. Barolli (Ed.): IMIS 2024, LNDECT 214, pp. 69–78, 2024.
https://doi.org/10.1007/978-3-031-64766-6_8

mechanism includes firing a laser pulse from a transmitter based on a Time of Flight (ToF) method, and the pulse reaches an object within the sensor's measurement range. It is reflected to be detected by the photodiodes. Therefore, the distance to the object is calculated using this round-trip time. In addition, the LiDAR system can be classified into two main types, rotate LiDAR, and solid-state LiDAR depending on the beam steering mechanism. Rotate LiDAR provides a wide range of scanning capabilities essential for autonomous driving, while solid-state LiDAR eliminates mechanical wear and damage, providing a more compact and potentially more reliable solution [6].

A LiDAR sensor measures distance, direction, location, and the like using a laser and provides a high resolution, thereby requiring accurate distance measurement. Accordingly, the LiDAR may accurately and at high speed recognize the surrounding environment in real-time and show a difference from other autonomous sensors by generating a bird's eye view. In addition, despite the high cost, current high-end autonomous vehicles are used as part of vehicle sensor systems using LiDAR [2].

The detection range recognized by a 360 degree-rotating LiDAR sensor is shown in the following Fig. 1.

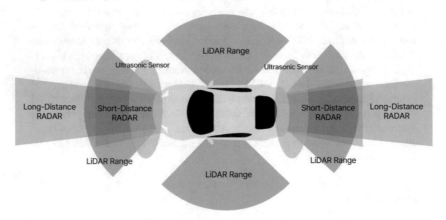

Fig. 1. Detection range recognized by a 360 degree-rotating LiDAR sensor

As LiDAR technology plays a crucial role in autonomous vehicles, security issues become very important. Suppose a spoofing or jamming attack occurs on the LiDAR sensor. In that case, an error may occur in the LiDAR sensor system, and it may not function properly, resulting in a fatal situation in autonomous driving. Therefore, analyzing and supplementing security threats for future autonomous driving technology is necessary.

Previous research has demonstrated spoofing and jamming attacks with threat models that manipulate point clouds by firing lasers at LiDAR sensors. However, in this paper, we examine the impact of LiDAR sensors by manipulating Ethernet packets with data collected using the Velodyne VLP-16 LiDAR sensor.

The rest of the paper is organized as follows: Sect. 2 describes related research targeting LiDAR sensors, and Sect. 3 explains the LiDAR sensor data collection environment and LiDAR sensor packet analysis. Sect. 4 explains LiDAR sensor packet manipulation,

Sect. 5 provides countermeasures against LiDAR DoS attacks, and Sect. 6 concludes with a conclusion.

2 Related Works

This section provides a brief overview of the existing literature on proposed attack techniques targeting LiDAR sensors (as summarized in Table 1).

Various physical attacks such as spoofing, jamming, and relay attacks can be carried out against LiDAR, and LiDAR attack models have been extensively studied. Yulong Cao et al. construct and evaluate two scenarios using a LiDAR spoofing attack that fakes obstacles close to the front as a threat model. Using Velodyne LiDAR, we uniformly sample 300 3D point cloud frames using 30 s of data collected on a real road and tracking accurate LiDAR sensor data released by Baidu Apollo. When an emergency brake attack was performed, it was confirmed that the speed immediately decreased within 1 s, and a spoofing attack was successful in which the attacker continued to attack and did not move even when the traffic signal turned green [7]. Zizhi Jin et al. aims to inject malicious points into a mechanical LiDAR with a physical laser attack and spoof 3D object detection to prevent it from recognizing existing objects or creating non-existent objects. We develop a laser transceiver capable of injecting up to 4200 spoofing points, improving the injection capability by 20 times, and design a PLA-LiDAR attack that injects a hostile point cloud into LiDAR with the correct shape and location. To verify the attack, they perform an extensive physical evaluation using the 16-channel Velodyne VLP-16 and 63-channel Robosense RS-16 [8].

Hochel Shin et al. present two types of attacks against LiDAR sensors, saturation attack, and spoofing attack, using saturation attack tool and spoofing attack tool. It succeeded in disabling the LiDAR sensor from detecting a specific direction and proved that the LiDAR sensor is vulnerable to exposure to light sources [9]. Jonathan Petit et al. demonstrate a relay attack, which aims to relay the original signal sent from LiDAR to another location to create a fake echo and make an object appear closer or farther than its actual location, and a spoofing attack, which is an attack that creates a fake object. In a relay attack, an attacker received a LiDAR signal from a vehicle, relayed it to another vehicle in a different location, and successfully created a fake echo to detect the location of the obstacle differently from the actual location. The spoofing attack shows a spoofing attack in which multiple copies of a point are created regularly by adjusting the delay to output multiple pulses [10]. Jiachen Sun and Yulong Cao propose Choi's Black Box LiDAR spoofing attack based on the identified vulnerabilities. A realistic physical attack was expressed using an actual data set (KITTI), and the effect of the attack was tested by injecting it into a point cloud 5–8 m in front of the victim's AV through a digital spoofing attack. As a result, the target model generally showed a success rate of 80%, and the CARLO method is proposed to completely detect spoofing attacks through the defense measures proposed and reduce the success rate to 5.5% [11].

Recent research proposes a threat model for spoofing or relay attacks targeting LiDAR through physical attacks. Although much research has been conducted on attack models for physical systems, there is a lack of research on identifying security threats by accessing LIDAR's internal system. LiDAR uses Ethernet packets to transmit large

Table 1. An overview of attack type on LiDAR

Paper	Attack Type	LiDAR Model	Description
[3]	Spoofing	HDL-64ES3	Evaluating scenarios that induce emergency braking and cause traffic jams by creating fake obstacles through LiDAR spoofing attacks
[4]	Spoofing	VLP-16	Design of PLA-LiDAR attack that spoofs LiDAR sensor by injecting 4200 malicious points
[5]	Spoofing, Saturation	VLP-16	Perform Spoofing by Relaying by creating saturation attack tools and spoofing attack tools
[6]	Spoofing	IBEO LUX 3	Realize a relay attack by manipulating the LiDAR sensor and manipulate the location of obstacles through a spoofing attack
[7]	Spoofing	VLP-16	Black box LiDAR spoofing attack based on LiDAR vulnerability

amounts of data at high speed and contains all 3D data such as the location of objects. Therefore, this paper proposes an attack risk using an Ethernet network-based approach.

3   Environment Setup

The LiDAR sensor used in this paper was Velodyne's VLP-16. VLP-16 uses 16 infrared lasers and fires approximately 18,000 times per second. Velodyne, a leader in the modern LiDAR field, is a core sensor in autonomous driving technology that is used for autonomous driving testing or when implementing actual autonomous vehicle functions.

Turing Drive, a self-driving bus in Taiwan, is a company that developed Taiwan's self-driving electric bus. It operated an autonomous shuttle service using the Velodyne VLP-16 (Puck) sensor [12] and was the first in Korea to pilot a late-night self-driving bus. In addition, since LiDAR usually generates large amounts of data in real-time with high-performance sensors and transmits data at high speeds, most consider using Ethernet. The types of LiDAR sensors released in the LiDAR market are listed in the Table 2 below.

Table 2. Various LiDAR manufacturers using Ethernet ports

Manufacturer	Name	State	Port
Velodyne [13]	VLP-16	Rotating	Ethernet
Livox [14]	TELE-15	Solid	Ethernet
Innoviz [15]	Innoviz TWO	Solid	Ethernet
Ouster [16]	Osdome	Rotating	Ethernet

The analysis was conducted targeting the Velodyne LiDAR sensor, which is widely used in the market, and the Ethernet packet injection experiment was conducted by constructing an environment as shown in Fig. 2. If a large amount of data is injected while collecting a large amount of Ethernet packets, there is a possibility that a collision may occur when proceeding to one port, and appropriate attack packets may not be generated. Therefore, a switch was installed between the LiDAR sensor and the PC to separate the packet injection port and collection port. Packet collection was carried out using Wireshark, a tool for analyzing network packets, and the LiDAR was fixed and operated in an experimental space of 95 m^2.

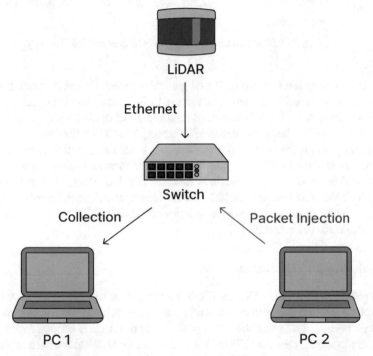

Fig. 2. Environment set up for packet injection and collection.

4 DoS Attack into LiDAR Ethernet

4.1 Packet Structure

VLP-16 LiDAR data is provided to an application program by processing data in real-time in the form of User Datagram Protocol (UDP) through an Ethernet communication network (as shown in Fig. 3). Since Velodyne LiDAR is a rotary LiDAR method, it rotates around a vertical axis and fires a laser. Therefore, the right-angled coordinates (X, Y, Z) are not directly included in the point cloud data, and each laser provides measurements of spherical coordinates (distance R, azimuth ω, altitude angle α) to calculate XYZ coordinates through azimuth calculation.

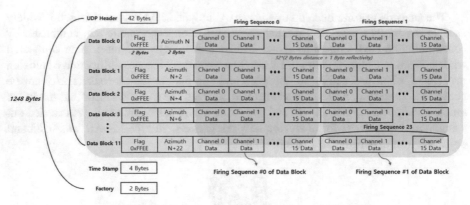

Fig. 3. LiDAR channels and packets in case of VLP-16

UDP communication is a protocol that generally transmits and receives high transmission speed and is used to transmit and receive LiDAR data. The data packet generated by the sensor includes a UDP header and consists of 12 data blocks, timestamps, and factory fields. The UDP header consists of Ethernet, IP, and UDP layers, each layer providing information on the source, destination, and port. The length of the LiDAR data packet consists of a total of 1248 bytes, two launch sequences consisting of 16 channel data exist, and only one azimuth angle is returned per data block. A data block with a value of a LiDAR point consists of 100 bytes of binary data, and the starting data of the LiDAR data block always starts with two bytes of '0xFF, 0xEE', and the azimuth of 2 bytes is calculated and entered [13].

4.2 Fabricating LiDAR Packets

The characteristic of a DoS (Denial of Service) attack is to maliciously attack a system, thereby depleting the system's resources and preventing it from being used for its intended purpose. A self-made Ethernet packet injection tool is needed to test the network system to identify vulnerabilities and check and understand the system's response in an actual attack situation. Therefore, in order to implement a DoS attack through the LiDAR sensor network, a specific tool was created to manipulate and inject packets, and python was used as the programming language.

```
from scapy.all import *

eth_layer=ether(dsf="ff:ff:ff:ff:ff:ff",src="60:76:88:38:8f:92")
ip_layer=IP(dst="255.255.255.255", src=192.168.1.201")
udp_layer=UDP(sport=2368, dport=2368)
data_block = 'ff ee' + ('00' * 98)
payload = data_block * 12
timestamp = '00 00 00 00'
factory = '37 22'
packet_data = bytes.fromhex(payload.replace(" ","") + times-
tamp.replace(" ", "") + factory.replace(" ", ""))
```

```
packet=eth_layer/ip_layer/udp_layer/Raw(load=packet_data)
while(1):
sendp(packet, iface="#Port_name")
```

Scapy, used in the injection tool, is a packet manipulation library written in Python that provides various functions to create, manipulate, capture, and analyze network packets. Based on the analysis of the LiDAR packet data previously, a string of 98 '0x00' is created following the packet '0xff, 0xee' that indicates the start of the data block, and this is repeated 12 times to create a data block. Then, we set the timestamp and code to be added to the end of the packet and combine all layers to create the entire packet.

4.3 Attack Result

We injected about 4,900 manipulated packets for 10 s each, and the visual image of LiDAR was confirmed using the veloview program. As a result, the 3D map formation was unstable because the correct point cloud was not captured, as shown in Fig. 4. In a normal state, data is constantly coming in, so it rotates 360° and receives data sequentially to form a normal 3D map. However, if incorrect data is injected due to a DoS attack, an intervention occurs in the middle of the data, and the frame is interrupted, so data does not come in normally. As a result of checking the LiDAR data while injecting the manipulated Ethernet packet, it was confirmed that the artificially created data packet was being transmitted.

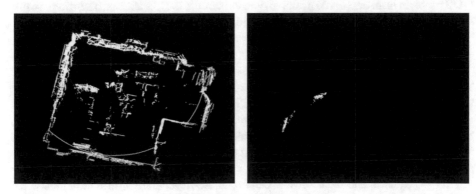

Fig. 4. Injection result from LiDAR

In addition, because of capturing data packets for 10 s, the number of data packets under normal conditions was about 8,900 on average, but the number of data packets under normal conditions was about 16,000 on average, resulting in delays in normal data and data disappearance. The amount of LiDAR data collected in normal and attack states is shown in Fig. 5.

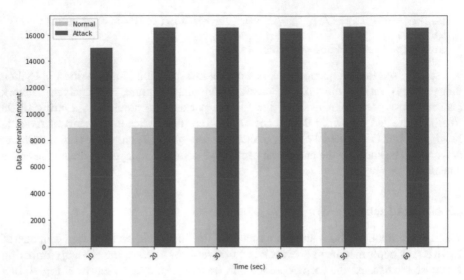

Fig. 5. Comparative analysis of normal and attack traffic by time interval

5 Countermeasures

In general, since LiDAR Ethernet data needs to be transmitted at high speeds, the UDP protocol, which allows relatively fast communication, is used. However, because the UDP protocol is an unreliable protocol, it is vulnerable to DoS attacks. Therefore, in this section, we propose countermeasures to mitigate DoS attacks performed in this paper.

5.1 Use of Sensor Fusion

To reduce the cause of errors and accidents in autonomous vehicle systems, reliability is increasing through the combination of various sensors in autonomous vehicle systems without relying on a single sensor. Therefore, LiDAR sensor fusion technology [17] can be used to integrate into one consistent model, and even if the LiDAR sensor becomes unusable through a physical attack or Ethernet packet DoS attack, the surrounding situation can be determined through a camera or radar sensor. In addition, the convergence of LiDAR, camera, and radar sensors can obtain more accurate information than when used individually and is efficient due to the high complementarity between sensors. Therefore, reducing the dependence on one sensor minimizes the impact of generating incorrect data due to physical damage and software errors.

5.2 Independent Network Separation for LiDAR

Using virtualization technology, access to necessary equipment and networks can be controlled through network separation and indiscriminate injection of UDP packets can be prevented. Network separation is the separation of networks to block illegal access and leakage of internal information through external Internet networks [18]. Therefore, it is used by general companies to block the inflow of malicious code and control personal information leakage. Attackers attempt to infiltrate the internal network from the outside to attack the autonomous driving system, including LiDAR sensors. However, if the network to which the LiDAR system is connected is separated and used as a separate network, access from the outside is restricted, so data cannot be injected unless physical access is made to the UDP port directly connected to the LiDAR. In addition, network separation can limit the impact of attacks and prevent them from spreading to other areas, thereby strengthening the stability and security of the entire system.

5.3 Entropy Based Ethernet IDS

Looking at the experiment results, a difference of approximately two times the amount of normal and attack data was observed. This result provides important hints for entropy-based intrusion detection systems (IDS). Entropy-based IDS is a system that detects abnormal behavior by analyzing the entropy of network traffic, and entropy increases when abnormal symptoms occur [19]. We expect to mitigate DoS attacks occurring on LiDAR Ethernet data based on a methodology that detects rapid increases in entropy.

6 Conclusion

The LiDAR sensor is essential to completing a level 5 autonomous vehicle because it compensates for the shortcomings of cameras and radars and recognizes the surrounding environment to provide information in a 3D map. LiDAR sensors transmit packets through an Ethernet communication network because they need to transmit data at high speeds with a variety of environmental information collected in large quantities. Therefore, in this paper, taking this into consideration, we proposed a DoS attack method that can occur on a LiDAR Ethernet network, and the attack causes a difference of about two times in the amount of data between normal and attack data and at the same time causes a delay in the 3D map. The phenomenon could be observed. Additionally, to prevent attacks against this, this paper shows a response plan to mitigate DoS attacks on LiDAR Ethernet data.

Acknowledgments. This work was supported by the National Research Foundation of Korea (NRF) grant funded by the Korea government (MSIT) (No. 2021R1A4A2001810) (NRF-2018R1A4A1025632), This work was supported by Institute for Information & communications Technology Planning & Evaluation (IITP) grant funded by the Korea government (MSIT) (No. 2022-0-01197, Convergence security core talent training business (Soon Chun Hyang University)).

References

1. Gomes, T., Roriz, R., Cunha, L., Ganal, A., Soares, N., Araújo, T., Monteiro, J.: Evaluation and testing system for automotive LiDAR sensors. Appl. Sci. **12**(24), 13003 (2022)
2. Campbell, S., O'Mahony, N., Krpalcova, L., Riordan, D., Walsh, J., Murphy, A., Ryan, C.: Sensor technology in autonomous vehicles: A review. In: 29th Irish Signals and Systems Conference (ISSC), pp. 1–4. IEEE (2018)

3. Comerón, A., Muñoz-Porcar, C., Rocadenbosch, F., Rodríguez-Gómez, A., Sicard, M.: Current research in lidar technology used for the remote sensing of atmospheric aerosols. Sensors **17**(6), 1450 (2017). https://doi.org/10.3390/s17061450
4. Debnath, S., Paul, M., Debnath, T.: Applications of LiDAR in agriculture and future research directions. J. Imaging **9**(3), 57 (2023)
5. Chan, J., Raghunath, A., Michaelsen, K.E., Gollakota, S.: Testing a drop of liquid using smartphone LiDAR. Proc. ACM on Interact. Mob. Wearable and Ubiquitous Technol. **6**(1), 1–27 (2022)
6. Khader, M., Cherian, S.: An introduction to automotive lidar. Texas Instruments (2020)
7. Cao, Y., Xiao, C., Cyr, B., Zhou, Y., Park, W., Rampazzi, S., Chen, Q. A., Fu, K., Mao, Z.M.: Adversarial sensor attack on lidar-based perception in autonomous driving. In: Proceedings of the 2019 ACM SIGSAC Conference on Computer and Communications Security, pp. 2267–2281 (2019)
8. Ji Jin, Z., Ji, X., Cheng, Y., Yang, B., Yan, C., Xu, W.: Pla-lidar: Physical laser attacks against lidar-based 3d object detection in autonomous vehicle. In: 2023 IEEE Symposium on Security and Privacy, pp. 1822–1839 (SP) (2023)
9. Shin, H., Kim, D., Kwon, Y., Kim, Y.: Illusion and dazzle: Adversarial optical channel exploits against lidars for automotive applications. In: Cryptographic Hardware and Embedded Systems–CHES 2017: 19th International Conference, pp. 445–467 (2017)
10. Petit, J., Stottelaar, B., Feiri, M., Kargl, F.: Remote attacks on automated vehicles sensors: experiments on camera and lidar. Black Hat Europe **11**, 995 (2015)
11. Sun, J., Cao, Y., Chen, Q.A., Mao, Z.M.: Towards robust {LiDAR-based} perception in autonomous driving: General black-box adversarial sensor attack and countermeasures. In: 29th USENIX Security Symposium (USENIX Security 20), pp. 877–894 (2020)
12. TURING Drive Taps Velodyne Lidar for Autonomous Driving Development (2020). https://velodynelidar.com/blog/turing-drive-taps-velodyne-lidar-for-autonomous-driving-development/
13. Velodyne: Velodyne User Manual (2019). https://velodynelidar.com/wp-content/uploads/2019/12/63-9243-Rev-E-VLP-16-User-Manual.pdf
14. Innoviz tehnologies, Designing a Level 3 LiDAR for Highway Driving, https://downloads.innoviz-tech.com/ (2022)
15. Livox: Livox Tele-15 user manual v1.2.pdf. https://www.livoxtech.com/downloads (2020)
16. Ouster: datasheet-rev7-v3p0-osdome.pdf (2023). https://data.ouster.io/downloads/datasheets/datasheet-rev7-v3p0-osdome.pdf
17. Kocić, J., Jovičić, N., Drndarević, V.: Sensors and sensor fusion in autonomous vehicles. In: 2018 26th Telecommunications Forum (TELFOR), pp. 420–425. IEEE (2018)
18. Hwang, S.: Network separation construction method using network virtualization. J. Korea Instit. Inform. Commun. Eng. (2020) 1071–1076
19. Wang, Q., Lu, Z., Qu, G.: An entropy analysis based intrusion detection system for controller area network in vehicles. In: 2018 31st IEEE International System-on-Chip Conference (SOCC), pp. 90–95 (2018)

Enhancing Road Safety with In-Vehicle Network Abnormal Driving Behavior Detection

Md Rezanur Islam[1], Kamronbek Yusupov[1], Munkhdelgerekh Batzorig[2], Insu Oh[2],
and Kangbin Yim[2(✉)]

[1] Department of Software Convergence, Soonchunhyang University, Asan, Korea
{arupreza,yuskamron}@sch.ac.kr
[2] Department of Information Security Engineering, Soonchunhyang University, Asan, Korea
{munkhdelgerekh,catalyst32,yim}@sch.ac.kr

Abstract. This study delves into leveraging Controller Area Network (CAN) data to detect and analyze abnormal driving patterns, underlining its significant role in bolstering road safety measures. By meticulously examining the comprehensive data supplied by the CAN system, which encapsulates real-time inputs from many vehicle sensors and mechanisms, this research marks a pivotal stride in the domain of vehicular safety and intelligent transport networks. The investigation elucidates on categorizing three specific types of unusual driving conduct, showcasing the accuracy and dependability of utilizing CAN data for such purposes. This methodology is a critical breakthrough in crafting instantaneous monitoring systems for erratic driving behavior, aiming to foster safer driving environments.

1 Introduction

Detecting abnormal driving behavior is critical in enhancing road safety and optimizing the efficiency of transportation systems. In recent years, advancements in vehicle technology have opened up new possibilities for analyzing driver behavior using in-vehicle network Controller Area Network (CAN) data [1]. This data, which includes information from various sensors and vehicle components, offers a rich source of insights into how drivers operate their vehicles [2, 3]. Abnormal driving behavior can manifest in various ways, including aggressive driving, distracted driving, reckless maneuvers, mistakenly keeping the door open, and belt off driving, all of which pose significant risks on the road. Detecting and mitigating these behaviors in real time is essential to prevent accidents and improve traffic flow. Understanding driver behavior at a granular level, from individual driving actions to broader patterns, is crucial for achieving these goals. Traditional traffic enforcement and monitoring methods often fail to catch up on detecting subtle or nuanced abnormal driving behaviors that may not immediately lead to accidents or violations [4]. Leveraging CAN data allows for a more comprehensive and nuanced analysis of driver behavior, enabling the identification of deviations from typical driving patterns. This study presents a novel approach to detecting abnormal driver behavior using in-vehicle network CAN data. CAN is a medium that connects Electronic Control Units (ECUs) [5], each responsible for various functions, and facilitates the exchange of data internally based on the driver's actions. The wealth of information provided by CAN data is leveraged to create a robust system capable of real-time detection and analysis of abnormal driving actions. The system identifies individual driving maneuvers and considers broader patterns and trends in driver behavior. By harnessing the power of in-vehicle network data, the aim is to contribute to developing more effective and data driven solutions for enhancing road safety. The approach offers the potential to create intelligent systems that can alert drivers, provide feedback, and prevent accidents caused by abnormal driving behavior. Through extensive experiments and analysis, the effectiveness and practicality of the method in real-world driving scenarios are demonstrated and presented in Fig. 1.

2 Related Work

Numerous approaches have been developed for detecting abnormal behavior, particularly in various applications such as driver monitoring, vehicle tracking, and safety. These approaches predominantly rely on external data sources, such as images of the driver, images of the moving vehicle, GPS data, physical attributes of the driver, and external vehicle data, including speed, throttle position, and brake pressure. This sector must still be a mountable system despite the many methods employed. In a paper by Wang et al. [6], a hierarchical deep learning approach is employed to classify driving maneuvers and identify drivers' behavior using large-scale GPS sensor data. This involves integrating Deep Neural Networks (DNN) and Long Short-Term Memory (LSTM) networks, preprocessing GPS data, constructing a joint-histogram feature map for regularization, and applying DNN. The model achieved over 94% accuracy in maneuver classification and 92% in driver identification, presenting significant advantages such as high accuracy and detailed characterization of driving behaviors. However, the approach suffers from complexities in implementation, computational intensity, and decreased LSTM performance with increased data scale. Kamaruddin et al. [7] employ speech emotion recognition for analyzing driver behaviors using Mel Frequency Cepstral Coefficients (MFCC) from datasets like the Real-time Speech Driving Dataset (RtSD) and the Berlin Emotional Speech Database (Emo-db). Using classifiers like Multi-Layer Perceptron (MLP), Adaptive Neuro-Fuzzy Inference System (ANFIS), and Generic Self-organizing Fuzzy Neural Network (GenSoFNN), they achieve varying success rates, notably detecting sleepiness with at least 65% accuracy across models. The approach holds potential for real-time applications in vehicle safety systems by alerting drivers of risky behaviors like sleepiness or aggression. However, the moderate overall accuracy, reliance on precise speech data, and complexity in implementation indicate the need for further refinement and research for practical application. In another study [8], the authors employ a deep learning technique, specifically Stacked Denoising Sparse Autoencoders (SdsAEs), to detect abnormal driving behaviors using driving behavior data features such as vehicle speed, throttle position, and brake pressure, normalized against a virtual driver model. The model achieves high micro-recall and micro-precision rates of 92.66% and 92.69%, respectively, demonstrating its effectiveness in distinguishing between everyday driving and four abnormal behaviors: drunk/fatigued, reckless, and phone use. However, the approach's complexity and the need for substantial and diverse data for training are significant challenges. Additionally, the model's reliance on specific features may limit its applicability across driving contexts or vehicles with varying sensor capabilities. Radtke et al. [9] discuss an algorithm for abnormal driving behavior detection using Adversarial Inverse Reinforcement Learning (AIRL), combining Generative Adversarial Networks (GAN) with Proximal Policy Optimization (PPO) in a multi-agent setting. This approach allows for interaction-aware prediction and more accurate driver behavior modeling. The highD dataset, a top view of vehicle movement collected from drones, is utilized for training and testing. The paper demonstrates that this method outperforms rule-based models and achieves comparable accuracy to state-of-the-art methods with significantly lower inference time. Finally, Shahverdy et al. [10] propose classifying driver behaviors into five distinct categories using Convolutional Neural Networks (CNN). The system collects vehicle data such as acceleration, gravity, RPM, speed, and throttle via a smartphone and an OBD-II adapter for ECU. The data is transformed into images using recurrence plots to capture spatial dependencies, which the CNN classifies into safe, aggressive, distracted, drowsy, or drunk driving behaviors. The approach effectively achieves high classification accuracy, offering an efficient and nonintrusive mechanism for behavior detection. However, it demands considerable computational resources for implementation and operation, relies heavily on the quality of input data, and might face challenges adapting to varying real-world driving scenarios. Despite these challenges, the method significantly advances in applying deep learning to improve road safety and intelligent transportation systems.

Many studies focus on detecting abnormal driver behavior using different data sources, but they often overlook the potential of using CAN data. However, CAN data has the potential to achieve greater precision and reliability in detecting abnormal behavior among drivers. The uniqueness of CAN data lies in its transmission mechanism. It is generated and transmitted based on the actions of ECUs within the vehicle, which govern the vehicle's internal functions that are influenced by the driver's actions. CAN data transmission is situation specific, delivering payload data tailored to specific driving scenarios facilitated by DBC (CAN Database) files [11]. This provides a more direct and fine-grained insight into driver behavior because it reflects the internal workings of the vehicle as influenced by the driver's inputs. Leveraging CAN data can enhance the precision and reliability of abnormal behavior detection, making it a valuable asset in improving road safety and driving behavior analysis.

3 Paper Preparation

3.1 Experimental Setup

In vehicle data collection, researchers typically utilize the On-Board Diagnostics II (OBD-II) system for diagnosis related data collection [12]. However, it should be noted that not all Electronic Control Units (ECUs) are connected to the OBD-II port [13], limiting the types of data that can be collected. To overcome this limitation, the research employs the ECU Direct Approach (EDA), in which data is collected through an internal gateway shown in Fig. 1. The method of line tapping tools is utilized to gather raw data, with an integrated central control unit (ICU) serving as the device that can access the CAN network within the vehicle [13]. The PEAK CAN system [PEAK] [14] is an interfacing device. For data analysis, Python Jupyter [15] and Pytorch for deep learning [16] are employed. The experiments are conducted on a system equipped with an Intel(R) Core (TM) i9-10900K CPU @ 3.70 GHz and an NVIDIA GeForce RTX 2080 super graphics card.

Fig. 1. Data Collection Through CAN Gateway [13]

3.2   Feature Extraction

The experimental process commences with data collection, involving the selection of a specific route. Data collection occurs in two stages: firstly, during normal and secure driving conditions, and secondly, during the collection of data containing abnormal activities. Subsequently, the data analysis phase is initiated. The data analysis technique hinges on identifying differences in the payload for entries sharing the same ID. Given that CAN data comprises eight-byte hexadecimal payloads, the analysis is performed by systematically examining each byte's position, one by one.

This Algorithm 1 is designed to compare two DataFrames, df1 and df2, with a list of specific IDs and divide the data into two DataFrames: one containing rows with matching IDs and another containing rows with non-matching IDs. The algorithm initializes two empty lists, match ids and non-match ids, to store the results. It iterates through each ID (id) in the given list of IDs. For each ID, it filters the rows in df1 and df2 based on the 'CAN ID' column to create df1 i and df2 i, which are subsets of the original DataFrames. Then, the algorithm calls the function Algorithm 2 to compare the data in df1 i and df2 i. This function should return two DataFrames: one with matching rows (match) and one with non-matching rows (non-match). The matching and non-matching rows are appended to match ids and non-match ids, respectively. After processing all the IDs, the algorithm converts match ids and non-match ids into DataFrames and removes any rows with missing values. Finally, it returns the two DataFrames: match ids containing rows with matching IDs and non-match ids containing rows with non-matching IDs. In summary, this algorithm systematically compares two DataFrames based on specific IDs, enabling the separation of data into matching and nonmatching categories for further analysis.

This Algorithm 2 is designed to compare two payload DataFrames, denoted as df1 and df2, and identify the differences in their byte values. It aims to find unique values in each column of df1 that are not present in the corresponding columns of df2. The

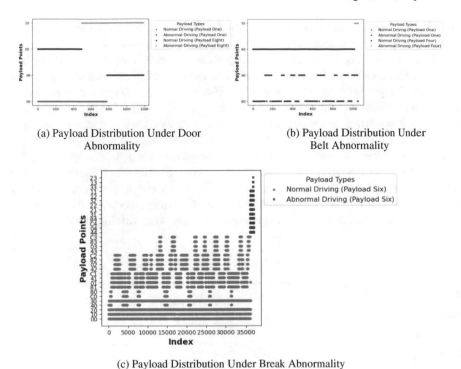

(a) Payload Distribution Under Door
Abnormality

(b) Payload Distribution Under
Belt Abnormality

(c) Payload Distribution Under Break Abnormality

Fig. 2. Payload Distributions Under Different Abnormalities

algorithm initializes an empty list called out to store the results. It then iterates through each column (col) in df1. For each column, it extracts the unique values (unique values df1) found in that column in df1, as well as the unique values (unique values df2) in the corresponding column of df2. Next, the algorithm enters a nested loop to examine each value (val) in unique values df1. It checks whether each val is present in unique values df2. If val is not found in unique values df2, it appends a tuple consisting of the column name (col) and the unique value (val) to the out list. Finally, the algorithm returns the out list, which contains all the column names and values from df1 that are not present in the corresponding columns of df2. In summary, this algorithm systematically identifies and collects the discrepancies between two DataFrames, providing a clear and structured approach to pinpointing where and what values differ between the two datasets.

In this study, the primary focus is on detecting three specific abnormal behaviors: door abnormalities, belt abnormalities, and brake abnormalities. The main emphasis is identifying brake abnormalities and investigating door and belt abnormalities, as they are readily available in current vehicles. This approach allows for the evaluation of the performance of the analysis algorithms (Algorithm 1 and 2).

Specific CAN IDs associated with these abnormal behaviors have been identified through the analysis. CAN ID 0x018 corresponds to door and belt functions, while CAN IDs 0x4B0 and 0x220 are responsible for brake related data. ECUs transmit data over the CAN network based on interdependent functions. For instance, the RPM, speed, and brake functions are interconnected. When the brake is engaged, the RPM decreases, and the speed also decreases. Consequently, a sequential impact is observed in the CAN data, particularly in the case of brake-related actions. However, door and belt functions exhibit different interdependencies. In the case of interdependent functions, substantial changes are noticeable in the data. Figure 2c demonstrates that under brake abnormality conditions, the number of unique payload values is higher compared to door and belt abnormalities shown in Fig. 2a and b. Only byte positions one and four are affected for door abnormalities, while for belt abnormalities, it is byte positions one and eight. In contrast, for brake abnormalities, except for byte positions two, four, six, and eight, the other positions remain relatively unchanged. Therefore, the data analysis is structured based on the payload values, considering these distinct characteristics. This approach ensures a comprehensive and nuanced examination of the data, enabling effective detection and differentiation between these abnormal driving behaviors.

Table 1. Deep Learning Hyperparameters.

Hyperparameters	Value
Batch size	64
Hidden size	64
Output size	4
Learning Rate	0.001
Loss Function	CrossEntropy
Optimizer	Adam
Early Stopping	10
Best Accuracy	0.9607
Counter	33

3.3 Deep Learning Model

In this study, a deep learning model was developed to classify abnormal driving behavior. The LSTM-based classifier was designed to analyze complex data sequences derived from various vehicle sensors. The model architecture comprises an LSTM layer and a fully connected layer for classification. The hyperparameters of the model as shown in Table 1, including batch size, hidden size, and learning rate, were tailored to optimize its performance. A CrossEntropy loss function and the Adam optimizer were also employed for training. An early stopping mechanism was incorporated to ensure efficient training and prevent overfitting, monitoring the model's accuracy over epochs. The experiments yielded promising results, with the best accuracy of 96.07% achieved after 33 training epochs. This research significantly contributes to the development of advanced systems for real-time detection of abnormal driving behavior, potentially enhancing road safety and traffic management.

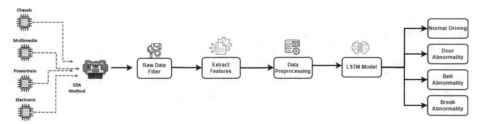

Fig. 3. Comprehensive Working Flow of Abnormal Behave Detection

4 Result Evaluation and Discussion

4.1 Working Flow

The workflow for a vehicle monitoring system that employs an EDA method to process and analyze data from various car subsystems, such as Chassis, Multimedia, Powertrain, and Electronics, is as follows: The data collected from these subsystems is initially passed through a 'Raw Data Filter' to eliminate noise and irrelevant information as shown in Fig. 3. Following this initial filtration, the data undergoes a 'Feature Extraction' process in which specific characteristics are identified and extracted for further analysis. After the relevant features have been extracted, the data is preprocessed to prepare it for the final analysis phase. This preprocessing may include normalization, scaling, or data transformation to ensure it is in a suitable format for machine learning models. The preprocessed data is then input into an LSTM model, a recurrent neural network (RNN) well-suited for time-series data. The LSTM model's responsibility is to identify patterns and abnormal behaviors in the data, which could indicate different vehicle operation states, such as 'Normal Driving,' 'Door Abnormality,' 'Belt Abnormality,' or 'Brake Abnormality.' Each of these outcomes is crucial for ensuring the vehicle's safe operation and providing actionable insights that can lead to maintenance or immediate corrective actions if necessary.

4.2 Result Evaluation

In the result evaluation section, the performance of the classification model was analyzed through various metrics. The confusion matrix offers a detailed view of the model's predictions, showing a high accuracy level in identifying 'Normal', 'Door Abnormality', 'Belt Abnormality' classes with no false positives or negatives. 'Break Abnormality' exhibited some misclassifications with 'Normal', highlighted by an 11.93% error rate in this category. The model achieved a notable 96.33% as shown in Table 2, accuracy in predicting the 'Normal' class and a perfect 100% in detecting 'Door' and 'Belt' abnormalities.

The Receiver Operating Characteristic (ROC) curve reinforces these results, showing Area Under the Curve (AUC) scores near perfection for 'Door Abnormality' and 'Belt Abnormality' at 1.00 and slightly lower for 'Normal' and 'Break Abnormality' at 0.97 and 0.94 respectively. This indicates a remarkable discriminative capability of the model across all classes. The classification report reveals high precision and recall for all classes, resulting in F1 scores demonstrating a well-balanced model. The 'Door Abnormality' and 'Belt Abnormality' classes achieved perfect precision and recall, achieving an F1-score of 1.00. The 'Normal' and 'Break Abnormality' classes also showed high precision and recall, with F1 scores of 0.96 and 0.90, respectively.

The model's overall accuracy stands at an impressive 97%, with macro and weighted averages across precision, recall, and F1-score of 0.97, signifying the model's robustness. The error rate is low at approximately 3.46%, and the false alarm rate is non-existent, indicating the model's dependability in practical scenarios. The ROC AUC score of 0.9768 further confirms the strong performance of the model across all classes as shown in Figs. 4 and 5. These metrics collectively validate the classification model's effectiveness in differentiating between classes with high reliability and minimal error, making it a valuable tool for its intended application domain.

Table 2. Classification Report and Performance Metrics

Class	Precision	Recall	F1-score
Normal	0.96	0.97	0.96
Door Abnormality	1.00	1.00	1.00
Belt Abnormality	1.00	1.00	1.00
Break Abnormality	0.91	0.89	0.90
Macro Avg	0.97	0.96	0.97
Weighted Avg	0.97	0.97	0.97
Error Rate	0.0346		
False Alarm Rate	0.0000		
ROC AUC Score	0.9768		
Accuracy	0.97		

4.3   Discussion and Limitation

The research presented here highlights CAN data's significant yet underutilized potential in analyzing abnormal driving behaviors. CAN data is renowned for its accuracy and reliability and is an optimal solution in this sector. However, its complexity poses a challenge. The intricate nature of in-vehicle networks, characterized by numerous ECUs and the voluminous generation of data, coupled with the dynamic characteristics of CAN data, makes analysis intricate. In this study, three types of abnormal behaviors have been successfully classified. However, identifying and categorizing various types of abnormal driving behaviors to develop a comprehensive abnormal behavior detection system requires an extensive survey. A pivotal in this process is the CAN DBC file, which serves as the blueprint for CAN network communications. Utilizing the CAN DBC could significantly enhance the accuracy and reliability of the system. Nevertheless, the proprietary nature of CAN DBC files, typically held confidential by manufacturers, presents a barrier. The future objective involves adopting a reverse engineering approach to the CAN DBC. This strategy aims at extracting relevant features in conjunction with a detailed survey of abnormal driving behaviors. By doing so, the driving safety field can be advanced, implementing an effective system for detecting abnormal driving behaviors. This endeavor could lead to more secure and safer driving environments.

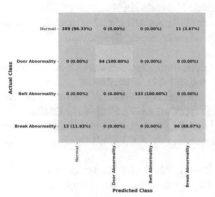

Fig. 4. Confusion Matrix in the Classification Report for Abnormal Driving Behavior Detection

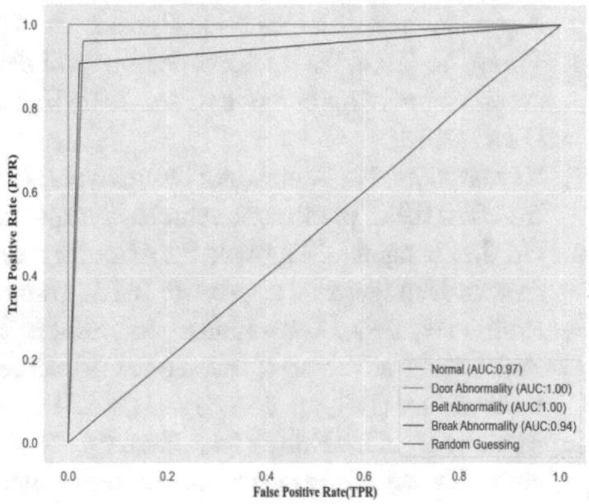

Fig. 5. ROC Curve for Abnormal Driving Behavior Detection

5 Conclusion

The study highlights the potential of utilizing CAN data to pinpoint abnormal driving behaviors. The researchers successfully classified three distinct behaviors, showing the effectiveness and precision of using CAN data. These findings emphasize the valuable insights that CAN data provides into the intricate workings of vehicles influenced by driver actions. Consequently, this approach represents a highly promising avenue to enhance road safety. This research contributes significantly to intelligent transportation systems, laying the groundwork for developing more advanced methods to detect abnormal driving behaviors. Ultimately, these advancements aim to create safer roads for everyone.

Acknowledgments. This work was supported by Institute for Information and communications Technology Planning and Evaluation (IITP) grant funded by the Korea government (MSIT) (No. 2022-0-01197, Convergence security core talent training business (Soon Chun Hyang University)) and supported by the National Research Foundation of Korea (NRF) grant funded by the Korea government (MSIT) (No. 2021R1A4A2001810).

References

1. Azadani, M.N., Boukerche, A.: Driving behavior analysis guidelines for intelligent transportation systems. IEEE Trans. Intell. Transp. Syst. **23**(7), 6027–6045 (2021)

2. Abdennour, N., Ouni, T., Amor, N.B.: Driver identification using only the CAN-Bus vehicle data through an RCN deep learning approach. Robot. Auton. Syst. **136**, 103707 (2021)
3. Gazdag, A., Lestyán, S., Remeli, M., Ács, G., Holczer, T., Biczók, G.: Privacy pitfalls of releasing in-vehicle network data. Veh. Commun. **39**, 100565 (2023)
4. Khan, K., Zaidi, S.B., Ali, A.: Evaluating the nature of distractive driving factors towards road traffic accident. Civ. Eng. J. **6**(8), 1555–1580 (2020)
5. Islam, M.R., Sahlabadi, M., Kim, K., Kim, Y., Yim, K.: CF-AIDS: Comprehensive Frequency-Agnostic Intrusion Detection System on In-Vehicle Network. IEEE Access (2023)
6. Wang, Y., Ho, I.W.-H.: Joint deep neural network modelling and statistical analysis on characterizing driving behaviors. In: 2018 IEEE Intelligent Vehicles Symposium (IV), pp. 1–6. IEEE (2018)
7. Kamaruddin, N., Wahab, A.: Driver behavior analysis through speech emotion understanding. In: 2010 IEEE Intelligent vehicles symposium, pp. 238–243. IEEE (2010)
8. Hu, J., Zhang, X., Maybank, S.: Abnormal driving detection with normalized driving behavior data: a deep learning approach. IEEE Trans. Veh. Technol. **69**(7), 6943–6951 (2020)
9. Radtke, H., Bey, H., Sackmann, M., Schön, T.: Predicting driver behavior on the highway with multi-agent adversarial inverse reinforcement learning. In: 2023 IEEE Intelligent Vehicles Symposium (IV), pp. 1–8. IEEE (2023)
10. Shahverdy, M., Fathy, M., Berangi, R., Sabokrou, M.: Driver behavior detection and classification using deep convolutional neural networks. Expert Syst. Appl. **149**, 113240 (2020)
11. Hoang, T.N., Islam, M.R., Yim, K., Kim, D.: CANPerFL: improve in-vehicle intrusion detection performance by sharing knowledge. Appl. Sci. **13**(11), 6369 (2023)
12. Rimpas, D., Papadakis, A., Samarakou, M.: OBD-II sensor diagnostics for monitoring vehicle operation and consumption. Energy Rep. **6**, 55–63 (2020)
13. Koh, Y., Kim, S., Kim, Y., Oh, I., Yim, K.: Efficient CAN dataset collection method for accurate security threat analysis on vehicle internal network. In: Barolli, L. (ed.) Innovative Mobile and Internet Services in Ubiquitous Computing: Proceedings of the 16th International Conference on Innovative Mobile and Internet Services in Ubiquitous Computing (IMIS-2022), pp. 97–107. Springer International Publishing, Cham (2022). https://doi.org/10.1007/978-3-031-08819-3_10
14. Plšičík, R., Danko, M.: API for data transfer using USB to CAN converter. In: 2023 IEEE 32nd International Symposium on Industrial Electronics (ISIE), pp. 1–6. IEEE (2023)
15. Islam, M.R., Oh, I., Yim, K.: CANTool an in-vehicle network data analyzer. In 2022 International Conference on Information Technology Systems and Innovation (ICITSI), pp. 252–257. IEEE (2022)
16. Ketkar, N., Moolayil, J.: Introduction to pytorch. In: Ketkar, N., Moolayil, J. (eds.) Deep learning with python: learn best practices of deep learning models with PyTorch, pp. 27–91. Apress, Berkeley, CA (2021). https://doi.org/10.1007/978-1-4842-5364-9_2

A Security Transaction Scheme of Internet of Vehicles System Based on Dual Blockchain and SM9 Technology

Lili Jiao[✉]

Police Officer College of the Chinese People, Chengdu 610213, China
253427977@qq.com

Abstract. With the rapid development of intelligent vehicle technology, the number of vehicles is showing an exponential growth and generating massive amounts of data. Traditional vehicle network systems have shortcomings in data protection, such as key management, data storage, and other aspects. The development of blockchain technology provides a good solution for solving such problems. This paper proposes an identity authentication chain and data storage chain based on SM9 algorithm, it is effective to solve the problems of identity authentication and data storage in vehicle networking systems. This scheme can not only improve the system security performance, but also effectively prevent single point of failure, identity forgery, data tampering, privacy disclosure and other problems caused by malicious attacks on the system. It can also meet the real-time and stability of high-speed vehicle data transmission, and have good practicability and progressiveness.

1 Introduction

Internet of Vehicles (IoV) is a network where vehicles equipped with intelligent in vehicle systems use specific network communication protocols to store and interact data between vehicles and data platforms. Through the new generation of wireless communication protocols, road vehicles can directly interact with road equipment and vehicles, achieving multiple functions such as road condition warning, information protection, and AI driving [1]. In recent years, as an important part of the Internet of Things and intelligent transportation, the Internet of Vehicles has provided mobile high-speed data transmission solutions for mobile edge computing (MEC) self-organizing communication networks [2], meeting the requirements of the Internet of Vehicles for real-time computing and big data analysis and storage. The mobile self-organizing vehicle network based on MEC connects vehicles and roads, data computing units, and data storage units, and customizes the mobile vehicle management system, allowing members of the vehicle network to route and share data through the mobile network [3].

L. Barolli (Ed.): IMIS 2024, LNDECT 214, pp. 89–99, 2024.
https://doi.org/10.1007/978-3-031-64766-6_10

2 Background Knowledge

2.1 SM9 Algorithm

The SM9 algorithm, as a public key cryptography algorithm, became a national commercial cryptography industry standard (GM/T 0044-2016) in 2016 and was included in the international standard in November 2018. Its main feature is that the public key data can use identify number, mobile phone number, e-mail and other information representing the user's identity, so it is called identity based public key cryptography algorithm [4]. As an identity identification encryption algorithm, State Secret SM9 has a key length of 256 bits, and the public and private keys are calculated by the Key Generation Center (KGC) using the device identity, the system's master public key, master private key, and public parameters, reducing the pressure of system key management and improving the overall security performance of the system [5]. Therefore, it is widely used in scenarios such as network identity authentication, email, document circulation, the Internet of Things, and cloud data platforms. Meanwhile, the SM9 algorithm is also very suitable for regulated application scenarios, as it has significant advantages in controlling massive interconnected devices. Therefore, this paper adopts the SM9 algorithm to ensure the secure transmission of the vehicle network system.

2.2 Blockchain

Blockchain is a distributed ledger that is based on cryptographic principles rather than credit, allowing any two parties to reach an agreement to trade directly without the involvement of a trusted third-party intermediary. The blockchain collects and verifies transaction data sent across the entire network at regular intervals, packaging legitimate data into a data block called a block. The data blocks are linked by hash values to form a data chain, which is called a ledger. Therefore, blockchain is also known as a distributed ledger database, where all data blocks correspond to the ledger, and each page of the ledger corresponds to each data block as shown in Fig. 1 [6].

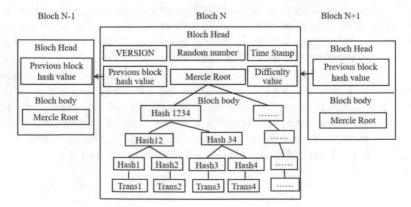

Fig. 1. Data structure of blocks.

Blockchain is a chain formed by combining data blocks in a chained data structure in chronological order, generating and updating data through a distributed consensus mechanism, and then using cryptographic algorithms to ensure the trustworthiness of network member nodes and the immutability and unforgeability of data blocks. Therefore, it has the key advantage of decentralization, enabling reliable peer-to-peer transactions among nodes in a distributed system that does not require mutual trust, making it very suitable for vehicle networking application scenarios.

The paper [7] proposes a vehicle networking data sharing and storage scheme based on alliance chain and zero knowledge proof, achieving the goal of decentralized data storage and improving system security. The paper [8] proposes a blockchain based identity authentication and password management scheme for the Internet of Vehicles, which realizes key negotiation and data exchange in distributed links and improves the effectiveness of system identity authentication. On the basis of the above research, in order to achieve efficient identity authentication and data security storage in the Internet of Vehicles system, this paper proposes a dual chain system architecture based on identity authentication blockchain and data storage chain, which can achieve vehicle identity authentication, key generation, and data exchange in an environment without important authorities, making the system have better security performance.

3 Framework of IoV System Based on Double-layer Chain

Due to the high mobility of vehicles, the architecture relies solely on a blockchain to achieve vehicle identity authentication, key management, and data synchronization interaction, which places high demands on the storage and computing power of onboard devices. Therefore, in order to solve this problem, we divide the communication network of the vehicle network according to regions, so that vehicle nodes and road side unit (RSU) nodes in the same region form a blockchain communication network. The RSU is equipped with identity chain (IC) and data chain (DC), which can not only achieve vehicle identity authentication, key generation, and data exchange, but also achieve data synchronization and persistence between the two chains, providing a data source for IPFS point-to-point distributed file systems [9]. The system architecture is shown in Fig. 2.

In this architecture, the onboard utilizes vehicle sensors to receive real-time information such as drivers, road information, and traffic conditions, and utilizes V2V network protocol to achieve wireless communication between vehicles and workshops, and V2I network protocol to achieve wireless communication between vehicles and RSU. Install Identity Authentication Chain (IC) and Data Storage Chain (DC) in the RSU road testing unit to solve the data storage and computing power of the onboard system. Among them, IC is responsible for managing the identity information of vehicles, including vehicle key generation, identity verification, and information change; The DC is responsible for accessing vehicle data and off chain storage, and storing the data obtained from uploading contracts in IPFS. After successful storage, a transaction containing IPFS metadata information is broadcasted in the DC. Among them, IPFS provides stable data persistence and high-speed access services for dual chain vehicle networking systems.

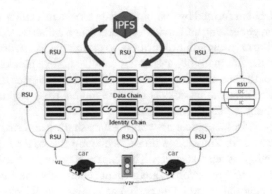

Fig. 2. Framework of dual chain vehicle networking system.

4 Design of Secure Transaction Scheme of the IoV System

The secure transaction scheme proposed in this paper is a decentralized vehicle networking system based on Ethereal block-chain and smart contract technology. It mainly consists of five parts: system initialization, RUS key generation, vehicle registration, identity authentication, and data access [10].

4.1 System Initialize

The system provides a safety parameter λ, and the KGC selects a Bilinear Group $BP = (G_1, G_2, G_T, e, N), N > 2^\lambda$. Then the system selects a generator P_1 of additive group G_1 and a generator P_2 of additive group G_2, and $P_1 = \psi(P_2)$. The system selects two password hash functions $H_1 : \{0, 1\} * \times Z_N^* \to Z_N^*, H_2 : \{0, 1\} * \times Z_N^* \to Z_N^*$. KGC generates a random number $s \in [1, N - 1]$ as the system master key, then calculates $P_{pub} = s \cdot P_1$ as the main public key of the system, and the system selects another byte hid to represent the private key generation function identifier. Last, the system publishes the public parameters $param = \{BP, P_1, P_2, P_{pub}, H_1, H_2, hid\}$.

4.2 Key Generation

After installing successfully an RSU device on the Ethernet client, the system generates a unique identifier ID_{RSU} for every RSU device. Then, the system calls the deployed authentication contract to generate a key for each RSU device. The authentication contract is to calculate $t_1 = H_1(ID_{RSU}\|hid, N) + s, t_2 = s \cdot t_1^{-1}, sk_{RSU} = [t_2]P_1$ over a finite field through KGC, and sk_{RSU} is as the private key of the RSU, to calculate $PK_{RSU} = sk_{RSU} \cdot P_2$ as the public key of the RSU. Then to generate initialization transaction records $T_{RSU} = \{ID_{RSU}, PK_{RSU}, timestap\}$, the $timestap$ is to represent the system time limit and add transaction information to the IC.

4.3 Vehicle Registration

After the vehicle is connected to the vehicle network, the sensor obtains the attributes of the vehicle equipment and establishes a connection between the new node and the attached RSU node. The V2I communication sends a registration request to the RSU. After successful registration, generate a transaction on the IC that includes the vehicle's identity certificate. The specific plan is as follows:

① The vehicle is connected to the vehicle network, and the RSU checks whether the onboard equipment of the vehicle has the attribute of safe driving. If not available, the service will be refused.

② Establish a connection between the new node and the RSU node of the attachment, and use sensors to obtain vehicle identity information M and attribute list S to access the public key PK_{RSU} of the RSU node on the IC chain, to generate a message $msg = M\|S$, and randomly select a number $r \in [1, N-1]$, and calculate $Q_{RSU} = H_1(ID_{RSU}\|hid, N) \cdot P_1 + P_{pub}, C_1 = [r]Q_{RSU}, g = e(P_2, P_{pub}), w = g^r, K = KGC(C_1\|w\|ID_{RSU}), C_2 = msg \oplus K$, then the system generates message authentication code $C_3 = MAC(K_2, C_2)$. Last, send the message $C = C_1\|C_2\|C_3$.

③ After RSU receives reply message C, the RSU calculates $w' = e(C_1, sk_{RSU}), k' = KGC(C_1\|w'\|ID_{RSU}), msg' = C_2 \oplus k', u = MAC(k', C_2)$. If $u = C_3$, then the RSU obtains the information M and S from msg' and verifies the validity of M and S. If the test is passed, access permissions will be generated and written back to the S list. Then the RSU will generate the vehicle identity ID_v and calculate $PK_v = H_1(ID_v\|T)$ as the partial public key of the vehicle, $sk_v = (PK_v + s)^{-1} \cdot P_{pub}$ as the partial private key of the vehicle. Among them, T is the system time when registering identity. The vehicle stores the public key PK_v in the M list, and generate a transaction $T_x = \{ID_v, PK_v, M, T\}$, and add the transaction to the IC. When the transaction enters the transaction pool, it is considered registered successfully. Finally, the information $\{PK_v, sk_v\}$ is sent to the vehicle user through a secure channel.

④ The vehicle user randomly generates $x_v \in [1, N-1]$ as the secret value and calculates $P_v = x_v \cdot P_2$ as the public key. Therefore, the complete public-private key pair of the vehicle user is $\{(sk_v, x_v), P_v\}$.

4.4 Identity Authentication

After the vehicle is connected to the network, an identity authentication request is sent, and the RSU retrieves blocks from the IC to verify the validity of the identity certificate and whether the certificate has expired. After successful verification, negotiate the session key for subsequent business operations. The detailed steps are as follows:

① The vehicle terminal selects a random number $r \in [1, N-1]$, and calculates $U = r \cdot P_2, h = (M\|PK_v\|T_1\|hid) \oplus H_2(r \cdot e(P_{pub}, P_1)), N = (r-h) \cdot sk_v, I = PK_v \oplus H_2(r \cdot PK_{RSU})$, among them, T_1 is the current time. And then sends $M_1 = (N, h, T_1, U, I)$ to the nearest RSU.

② After receiving M_1, the RSU first determines that $T_1 - T$ is greater than *timestap*, indicating that the authentication request has timed out and the authentication is terminated. If $T_1 - T <timestap$, then the RSU calculates $PK_v = H_2(sk_{RSU} \cdot U) \oplus I$ by using its own private sk_{RSU}, and retrieves blocks containing PK_v from the IC to obtain identity credentials for M, and then calculate $u = e(N, PK_v \cdot P_1 + P_{pub}), t = [h] \cdot e(P_{pub}, P_1)$, and then $h' = (M\|PK_v\| T_1\|hid) \oplus H_2(u \cdot t)$. If $h' = h$, it indicates that the vehicle is legal, so that the identity authentication is successful, otherwise the authentication fails.

③ The RSU selects a random number $y \in [1, N-1]$, and calculates $Y = y \cdot P_2$, $g_y = [y] \cdot e(P_{pub}, P_1), v = H_1(ID_{RSU} \| T_3 \| hid) \oplus H_2(g_y), W = (y-v) \cdot sk_{RSU}$, $K = PK_{RSU} \oplus H_2(y \cdot P_V)$, T_3 is current time, and generates the session $key1 = H_1(PK_v \| PK_{RSU} \| H_2(sk_{RSU} \cdot P_v))$, then send the messages $M_2 = (W, v, K, Y)$ to vehicle terminal.

④ After the vehicle terminal receives message M_2, it calculates $PK_{RSU} = H_2(x_v \cdot Y) \oplus K$, $m = e(W, P_1 + P_{pub})$, $n = [v] \cdot e(P_{pub}, P_2)$, $v' = H_1(PK_{RSU} \| T_3 \| hid) \oplus H_2(m \cdot n)$ by using the private x_v. If $v = v'$ then the identity of the RSU is legal, and then exports the session key $key1' = H_1(PK_v \| PK_{RSU} \| H_2(sk_{RSU} \cdot P_v))$. After authentication, the vehicle communicates with the RSU of the attachment using the session key $key1'$.

4.5 Data Access Contract

The vehicle collects real-time information such as driver, road information, and traffic conditions through sensors during driving, and continuously exchanges data with data access contracts. The data access contract should ensure the reliability of each access, and the working process is divided into five parts: data preprocessing, asynchronous transmissions, data persistence, upload voucher generation, and data retrieval. The variables and method names involved are shown in Table 1.

Table 1. S Attribute List Fields

Field	Type	Meaning
ID_V	Unit	Vehicle terminal unique identification
Role	Unit	The identity of the uploader
Keyset	Mapping	Keyword set of data
Access Set	Mapping	Access permission control set

(1) The vehicle terminal uses sensors to collect real-time driving data F and device attribute list S, and after preprocessing, searches for the public key PK_{RSU} of the attachment RSU on the IC chain. Then generates a random number $\alpha \in [1, N-1]$ to calculate $C_1 = [\alpha] \cdot PK_{RSU}$, $U_1 = e(P_2, P_{pub})^\alpha$, $K_1 = KGC(C_1 \| U_1 \| ID_{RSU})$, $C_2 = F \oplus K_1$ for getting a list of uploaded files List<File>=< ID_{RSU}, PK_v, $S, C_1, C_2, Hash\_SM3(F) >$.

(2) After receiving a data transmission request, the RSU connects to the data security transmission channel and receives a list of uploaded files List<File>, and calculates $U_1' = e(C_1, sk_{RSU})$, $K_1' = KGC(C_1 \| U_1' \| ID_{RSU})$, $F' = C_2 \oplus K_1'$, $Hash\_SM3(F')$. Then to verify if $Hash\_SM3(F')$ match $Hash\_SM3(F)$ in List<File>. To ensure the accuracy and completeness of data. After the verification is completed, the RSU disconnects the upload connection from the onboard.

(3) After RSU verification is passed, a random number is randomly generated $\beta \in [1, N-1]$ and to calculate, $\ell = H_1(F \| keyset \| AccessSet \| hid, N)$, $\varepsilon =$

$(\beta - \ell) \bmod N$, $U_2 = e(P_2, P_{pub})^{\beta} C = [\varepsilon] \cdot sk_{RSU}$, and to get $T = \{ID_{RSU}, \ell, C, Hash\_SM3(F)\}$. The RSU establishes a secure data transmission channel with the IPFS node, concurrently and asynchronously transmit the message T to IPFS, and to query the RSU of the uploaded data through the ID_{RSU} for subsequent data statistics.

(4) After the IPFS upload is successful, a response message is returned to the RSU, which includes the storage address *addr* of the file. Based on this storage address, the upload contract is called to construct an upload transaction. The upload transaction includes the encrypted storage address, feature set, and access control set. The feature set contains keywords to accelerate retrieval, and the access control set is used for permission control. When calling the query contract, if the public key is not in the access control set, access is denied.

(5) The query contract quickly matches the fields contained in the query conditions on the DC, finds all transactions in the transaction entity keyset, and obtains the storage address and hash value of each transaction data in IPFS. In order to accelerate the retrieval speed, a time field is set in the query conditions, and only the blocks that meet the time range are queried based on the timestamp of the block header, avoiding traversing all transaction blocks. This significantly improves the retrieval efficiency of the system when the transaction volume is too large.

5 Analysis of the Scheme

5.1 Validity

In identity authentication, RSU needs to verify whether the calculated h' is equal to h which is sent by the vehicle terminal. It is equivalent to verify $u \cdot t = [r] \cdot e(P_{pub}, P_1)$. The verification process is as follows:

$$
\begin{aligned}
u \cdot t &= e(N, PK_v \cdot P_1 + P_{pub}) \cdot [h] \cdot e(P_{pub}, P_1) \\
&= e((r - h) \cdot sk_v, PK_v \cdot P_1 + s \cdot P_1) \cdot e(P_{pub}, P_1)^h \\
&= e\left(\frac{P_{pub}}{PK_v + s}, PK_v \cdot P_1 + s \cdot P_1\right)^{(r-h)} \cdot e(P_{pub}, P_1)^h \\
&= e(P_{pub}, P_1)^{(PK_v+s)^{-1} \cdot (PK_v+s) \cdot (r-h)} \cdot e(P_{pub}, P_1)^h \\
&= e(P_{pub}, P_1)^r \\
&= [r] \cdot e(P_{pub}, P_1)
\end{aligned}
$$

In session key negotiation, for verifying session key consistency $key = key1'$, it is equivalent to verify $H_2(sk_{RSU} \cdot P_v) = H_2(x_v \cdot PK_{RSU})$. That is to verify $sk_{RSU} \cdot P_v = sk_{RSU} \cdot x_v \cdot P_2 = x_v \cdot sk_{RSU} \cdot P_2 = x_v \cdot PK_{RSU}$. So that the correctness of the scheme is confirmed.

5.2 Safety Analysis

(1) Key Security

Key security has always been a core issue faced by the system. In this scheme, the key mainly involves the system's master public key and private key, RSU's private key, and partial public key and private key for each vehicle. The main public key and private key of the system are generated by the system during initialization and do not require a dedicated key management center to manage them. The system only needs to manage a portion of the public and private key pairs $\{PK_v, sk_v\}$ for all vehicle users, and store them in the blockchain of each RSU. The real key used for signature authentication is the secret value x_v generated by the user, which is saved by the user. Its public key is obtained by calculating $P_v = x_v \cdot P_2$. So in the blockchain of the entire network, there is no need to save the complete public and private key pairs of users, which not only protects user privacy information but also ensures the security of system keys.

(2) Data Security

This solution is based on the distributed processing mechanism of Ethereum EVM, which encrypts and stores all transaction data through smart contracts. There is no possibility of bypassing smart contracts to obtain plaintext. In addition, transaction records are stored on each RSU. Based on the characteristics of blockchain distributed ledgers, if an RSU node is attacked to attempt to tamper with transactions, it must have more than 51% of the network's computing power. This attack not only requires a large amount of computational cost, but also consumes the time cost for attackers to repackage transactions through PoW. When the network size is large enough, the data on the DC can hardly be tampered with. At the same time, transactions and raw data are backed up at multiple nodes, and there is no risk of data loss due to a single point of failure [11].

(3) Preventing Forgery

This scheme has enforceability of messages in an adaptive environment. In vehicle identity authentication, the RSU receives the core information $(M\|PK_v\|T_1\|hid) \oplus H_2(r \cdot e(P_{pub}, P_1))$, $(r - h) \cdot sk_v$ are formed by the vehicle user using their private key sk_v to sign personal vehicle identity information M, and then encrypting it through RSU public key. After receiving the message, the RSU first needs to decrypt the message with its own private key to obtain the public key $PK_v = H_2(x_{RSU} \cdot U) \oplus I$, and retrieve the block containing PK_v in the IC. The M corresponding to PK_v in the block is used to verify the signature of the vehicle's identity information, ensuring bidirectional authentication of identity. In this scheme, the attacker does not forge a signature due to the inability to obtain the private key of the vehicle user. It is also impossible to obtain the private key of the RSU and encrypt the authentication request. So that the entire scheme can effectively prevent third-party forgery attacks while ensuring the security of bidirectional identity authentication.

5.3 Performance Testing

(1) Storage overhead

The simulation experiment of this scheme adopts the Truffle framework and management contract, while the Ganache architecture manages blockchain accounts and keys. According to the IEEE 1609.2 trial standard [12], the communication message length is 100 bits, the SM9 algorithm key length is 256 bits, the vehicle identity information length is 100 bits, the device information length is 100 bits, and the real-time road condition information length of the vehicle is 256 bits with a timestamp of 16 bits. The experiment takes the average value as the experimental result. Analyzing the execution status of EVM smart contracts at each node, it was found that the block header occupied 256B, and the average transaction on the IC chain was estimated to be 100B, with an average of 20 records per block. At the same scale of 1 million vehicles as the papers [7, 8], the proposed solution in this paper increases the storage space required for RSU, but significantly reduces the requirements for storage devices at the vehicle. As shown in Table 2.

Table 2. Storage space consumption

Scheme	Vehicle terminal/MB	RSU/MB
Scheme [7]	1.2	132
Scheme [8]	2.1	45
This scheme	0.5	195

(2) Transaction throughput

At intersections with concentrated traffic, RSU will face a large number of concurrent requests and generate massive transaction records. Therefore, the throughput of transactions is of great significance for the stability and processing speed of the dual chain vehicle network system. The block on the DC is approximately 1.3 MB, and the RSU performs an average of one packet on chain operation per second for transactions. Fig. 3 shows the difference in transaction throughput between the proposed solution and the traditional single chain architecture. The results showed that when the transaction concurrency reached 600 transactions per second, the processing capacity of the single chain architecture was only 345 transactions, while the processing capacity of the double chain architecture reached 378 transactions, an increase of about 15%. After exceeding 700 transactions per second, both the double chain and single chain reached the concurrency bottleneck in the simulated environment, and the maximum throughput of the double chain processing was about 380 transactions per second. As shown in Fig. 3.

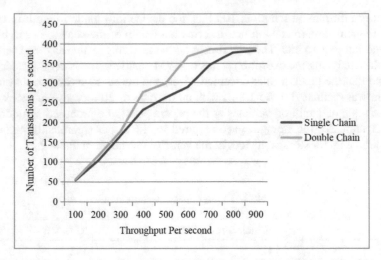

Fig. 3. Throughput Comparison

6 Conclusion

This paper proposes a dual chain vehicle networking model based on identity authentication chain and data access chain, which can achieve vehicle user identity authentication in a distributed environment while solving the problem of fast data access during high-speed vehicle movement, achieving decentralized security and stability requirements. In order to improve the efficiency of data warehouse access, the introduction of IPFS avoids the risk of single point of failure and achieves multi-party data sharing and traceability. The experimental results indicate that this scheme has good application value in environments with rapidly increasing vehicles. The next step will focus on researching the cross-domain authentication problem of the Internet of Vehicles and further improving the efficiency of data interaction in large-scale blockchain networks.

References

1. Wang, X., liu, J., An, X.: A review of information security research in vehicle networking. Transport. Technol. Manag. (2), 123-126 (2021)
2. Zhang, H., Cheng, Y., Liu, K.: Mobility management strategies for integrating mobile edge computing and content delivery networks in vehicular networking. J. Electron. Inform. **42**(6), 1444–1451 (2020)
3. Feanc, O., Cicirell, I., Antoni, O., et al.: Edge computing and social internet of things for large-scale smart environments development. IEEE Internet Things J. **5**(4), 2557–2571 (2017)
4. Lu, Y., Liu, S., Li, J.: Application of SM9 in bluetooth security protection of smart energy charging pile. Comput. Appl. Softw. **39**(11), 324–336 (2022)
5. Zhang, C., Peng, C., Ding, H., Xu, D.: Searchable encryption scheme based on china state cryptography standard SM9. Comput. Eng. **48**(7), 159–166 (2022)
6. Wang, S., Shi, S., Wen, Y., Luo, M.: Scheme to share equipment data based on blockchain. National Defense Technol. **42**(5), 52–57 (2021)

7. Zhang, X., Chen, X.: Data security sharing and storage based on a consortium blockchain in a vehicular ad-hoc network. IEEE **7**, 58241–58254 (2019)
8. Kouicem, D.E., Bouabdallah, A., Lakhlef, H.: An efficient and anonymous blockchain-based data sharing scheme for vehicular networks. In: IEEE 2020 IEEE Symposium on Computers and Communications (ISCC), pp. 1–6. New York IEEE (2020)
9. Hu, Y., She, K.: Blockchain and smart contract based dual-chain internet of vehicles system. Netinfo Secur. **8**(11), 26–35 (2022)
10. Tang, F., Gan, N., Yang, X., Wang, J.: Anti malicious KGC certificateless signature scheme based on blockchain and domestic cryptographic SM9. Chin. J. Netw. Inform. Secur. **8**(6), 9–19 (2022)
11. Tang, F., Gan, N.: Anti malicious KGC certificate less signature scheme based on block-chain and domestic cryptographic SM9. Chin. J. Netw. Inform. Secur. **8**(6), 9–18 (2022)
12. An, T., Ma, W., Liu, X.: Aggregate signature scheme based on SM9 cryptographic algorithm in VANET. Comput. Appl. Softw. **37**(12), 280–284 (2020)

Swap and Carry Strategy for Utilizing Spare Batteries as an Emergency Power Supply on Battery Swapping EV

Mayu Hatamoto and Tetsuya Shigeyasu$^{(\boxtimes)}$

Prefectural University of Hiroshima, Hiroshima, Japan
r422005gq@ed.pu-hiroshima.ac.jp, sigeyasu@pu-hiroshima.ac.jp

Abstract. Electric Vehicles (EVs) have attracted attention to reduce greenhouse gas. Battery swapping EVs make it possible to drive longer distance than plug-in charging EVs, because battery swapping EVs can swap batteries in a few minutes. However these services need a lot of initial costs to be distributed widely. One of the reasons is that a number of spare batteries need to be installed in Battery Swapping Stations (BSSs). Then we will propose to utilize spare batteries for other purposes except for driving EVs, like an emergency power supply at the time of disaster. For the purpose, spare batteries must be kept in a certain level. But batteries in a BSS which located in an overcrowded area tend to be low because the BSS is used frequently, and the batteries cannot supply enough power. Therefore, this paper focuses on the issues by battery transportation, and proposes a method that can equalize the amount of electricity levels of spare battery in BSS by transporting a battery from the BSS in which the average state of charge (SOC) of spare batteries is high to the one is low.

1 Introduction

Recently, EVs have drawn attention for reducing to emit greenhouse gas. EVs can effectively contribute in global warming, because it does not use fossil fuel while driving, different from petrol cars.

Battery electric vehicles (BEVs) need no fossil fuel or hydrogen, and only use electricity. There are two types of BEVs, one is plug-in charging EV, the other is battery swapping EV. The former has several weaknesses, for example, long charging time, short driving distance, and so on. While the latter can overcome these weak points, and it is expected to contribute increase both of EV operating time and traveling distance. But, battery swapping EV has a demerit that the initial cost tends to be high when the service is introduced. The reason is that the service require a number of spare batteries in accordance with the increasing number of EVs, therefore, utilizing the spare batteries for other usages except for EVs, effectively reduces the economic burden, and it realizes the widely spreading the EV service. Some similar ideas are already under consideration which make use of electricity in EVs to other life scenes like V2H (Vehicle to Home) [1].

A service for battery swapping EV can use more electricity than plug-in charging EV by adding spare batteries. So, if the capacity is used effectively, it will become

L. Barolli (Ed.): IMIS 2024, LNDECT 214, pp. 100–110, 2024.
https://doi.org/10.1007/978-3-031-64766-6_11

possible to supply electricity generated from less dependence on methods which emit CO_2.

Therefore, this study considers utilization of spare batteries as an emergency power supply when power failures have been occurred by disasters. For the purpose, new system securing enough electricity with spare batteries in BSS, is proposed.

However, if traffic is biased among driving areas, a gap in the average SOC of spare batteries are generated between BSSs. For instance, in a heavy traffic area, it is expected that the average SOC becomes low, since EVs frequently use the BSS. Consequently, the insufficient electricity amount can not relieve emergency condition.

Therefore, to cope with the problem, in this paper, we develop the method to realize equalizing the average SOC of spare batteries between BSSs, by carrying a battery from the BSS having a larger amount of electricity to other BSS. In order to evaluate the proposed method, we use evaluation model with the open source traffic simulator SUMO (Simulation of Urban MObility) [2], and clarify the effect of battery transportation.

The remainder of this paper is organized as follows. Section 2 describes the current situation surrounding battery swapping EV. In Sect. 3, the evaluation model and the characteristics are explained. Section 4 proposes new battery swapping algorithm. Finally, the conclusion and the future work are described in Sect. 5.

2   Background of a Service for Battery Swapping EV

It has been reported that a recent transportation system accounts for more than 30% of the world CO_2 emissions [3]. According to the situation, EU has already decided to limit to sell new cars emitting CO_2 (like petrol and diesel-driven cars) from 2035 [4], and, the similar trend will spread globally. Incidentally, new car sales of BEVs and plug-in hybrid electric vehicles (PHEVs) account for 14% of total new car sales all around the world as of 2022 [5].

By the way, some battery swapping services has already started. For instance, Gogoro [6] in Taiwan has launched the service for electric motorcycles since 2015 as one of the pioneers in the field of battery swapping mobility.

In EV, Ample [7] in America reports that battery swapping EV can swap its battery with in 5 min at a BSS they developed. Their BSS can be installed on an area for only two parking spaces. They also provide a mechanism swapping batteries at the drive-thru.

NIO [8] in China has spread its service widely not only in China but also overseas [9]. EV users can get fully charged battery automatically at a NIO's swap station by only 3 min shorter than Ample's station.

In Japan, Gachaco [11] providing electric share motorcycles service, considers various possibilities of battery utilization. For examples, they utilize a battery as a home-use battery, emergency power supply at a disaster, and so on. Besides, degraded batteries by frequently swapped, will be reused as for different purposes, such as, stationary battery other than mobility application [12].

By the way, a service for battery swapping EV has several issues in order to deploy. First, the costs will be increased for launching a service, because a number of spare batteries need to be installed in a BSS. In addition, saving a space to swap batteries is an another problem. Needless to say, EV services require a space both for storing spare batteries and Swapping batteries. Moreover, an EV needs to swap a battery periodically to keep driving. Therefore, suitable number of BSSs needs to be built avoiding waste of time with waiting for swaps.

3   Development of Evaluation Platform

3.1   Outline of Evaluation Model

Figure 1 shows the field model used for evaluating a service of battery swapping EV. As the figure shows, the road network consists of mesh, and each road is 5 km long and four-lane (i.e., two lanes in each direction). Parameters used for the evaluation is shown in Table 1.

Table 1. Simulation Parameters.

Items	Values
POI	6 [locations]
BSS	2 [locations]
# of Spare battery in BSS	50
Ev	50 [cars]
Battery capacity	40 [kWh]
Electricity efficient	8 [km/kWh]
Threshold for judging battery swapping	0.3
Work time to swap batteries at BSS	300 [sec]
Charging speed of spare battery in BSS	3 [kW/h]
Stop time at POI	1 [sec]

We explain Ev behavior as follows. On the evaluation, an Ev first selects one POI randomly, and move to the POI with the shortest route. After the Ev arrives at the selected POI, it newly selects an another one randomly and restarts to travel. In addition, each POI is selected with an equal probability. If an Ev battery SOC falls below the predefined threshold while driving, the Ev heads to the nearest BSS. When the Ev gets to the BSS, it swaps own battery for a spare battery most charged in the BSS. An Ev repeats these behavior.

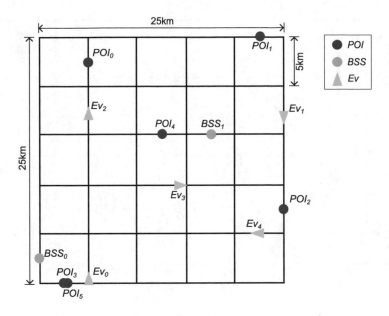

Fig. 1. Evaluation model.

3.2   Detail of Evaluation Model

We explain about Table 1 in detail. Each BSS has multiple spare batteries and all batteries will be charged simultaneously whenever their SOC are below 100%. BSSs charge each battery with the speed 3 [kW/h]. Then, it takes about 13 h to complete battery charging from an empty state.

We assume that only one Ev can swap its battery at a time in a BSS. So, if the number of cars over the capacity arrive at BSSs, they must wait its turn to swap its battery on a road.

We set on Ev maximum speed is 50 [km/h]. An Ev can drive 320 [km] since its battery capacity is 40 [kW], if the battery is full. Usage time on POI means the time from an Ev arriving at POI until starting to move to next POI. In evaluation model, we assume it is 1.0 [sec] ideally (the shortest), but in real situation, the length depends on use situation.

3.3   Characteristics of Evaluation Model

As shown in the Fig. 1, due to the biased POI locations, driving area of Evs are also biased. Under the condition, if we increase the number of POIs around the BSS_1, the number of used for battery swapping of BSS_1 is also increased while the BSS_0 is not changed. Hence, the average battery charge ratio of spare batteries of the BSS_1 becomes lower than the BSS_0.

Figure 2 shows the characteristics of the average battery charge ratio of spare batteries, and Table 2 are the results of the overall average of the average ratio in simulation and the number of battery swaps. As the Fig. 2 shows that there is a difference between

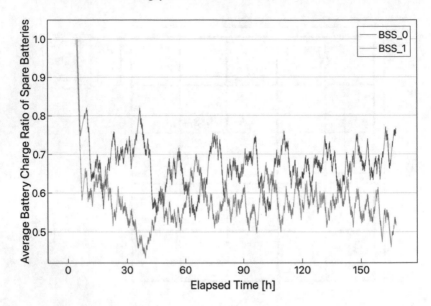

Fig. 2. Characteristics of in the average battery charge ratio of spare batteries of each BSS.

Table 2. The overall average of the average battery charge ratio of spare batteries of BSS and the number of battery swaps.

Items	BSS_0	BSS_1
Battery level	0.6817	0.5849
# of battery swaps	707	953

the two BSSs. This results confirms that the overall average value is $BSS_0 > BSS_1$ and this phenomenon is also confirmed by the Table 2. In addition, since the number of battery swaps is $BSS_0 < BSS_1$, it is thought that the number of used of each BSS influences to deteriorate an average SOC of spare batteries.

4   Equalization of the Amount of Electricity Stored Among BSSs

In this chapter, we report the effect of equalizing the average SOC of spare batteries by battery transportation among BSSs.

4.1   Effect of the Equalization by Simple Battery Transportation Based on Fixed Criterion Values

In this section, we report an effect of the equalization battery transportation based on a fixed criterion.

When an Ev travels to different area than current area, the Ev is likely to carry battery. Figure 3 shows area division in the evaluation model. In detail, POI_3, POI_5, and

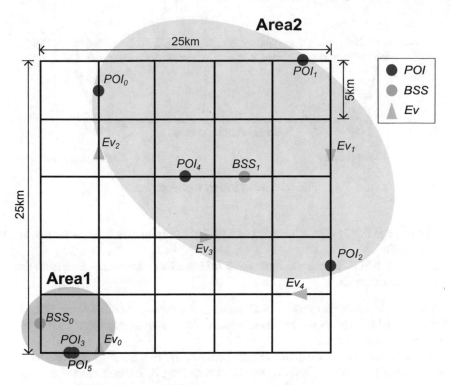

Fig. 3. Area divisions in the evaluation model.

BSS_0 belong to Area1. While, POI_0, POI_1, POI_2, POI_4, and BSS_1 belong to Area2. Figure 4 illustrates an image of battery transportation.

Specifically, when an Ev travels to a next destination POI located in a different area than a current area, it judges whether battery carrying or not based on a fixed criterion as below. First, the Ev swaps own battery to a spare battery which is the highest ratio in the BSS in a current area. Next, the Ev drops the battery at a BSS in a destination area. At this time the Ev swaps own battery to a spare battery whose level is the threshold or more and the lowest in the BSS.

The two fixed criteria to judge whether to carry battery are decided according to the overall average of the average SOC of spare batteries in simulation as follow. One value named OverTh is 0.68. The other named UnderTh is 0.58. If following conditions are met, Ev carry battery:

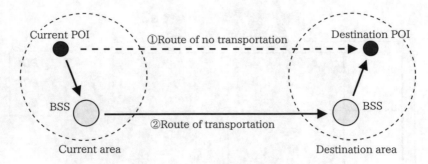

Fig. 4. Image of battery transportation.

1. The average SOC of spare batteries of a BSS at a current area is larger than or equals to OverTh.
2. The average SOC of spare batteries of a BSS at a destination area is smaller than or equals to UnderTh.

Figure 5 illustrates the results of changes in the average SOC of spare batteries of a BSS, and Table 3 shows the results calculated the overall average of the average ratio of the simulation results.

These results confirmed that the gap between the two BSSs is reduced by transporting batteries compared with no transportation (Fig. 2, Table 2). So we can say that the battery transportation is effective to equalize the average SOC of spare batteries between BSSs. Incidentally, the number of battery swaps is 297.

But, the method using fixed criteria cannot consider applicability to social implementation, because the fixed criteria do not reflect the changes of road condition.

Table 3. The overall averages of the average SOC of spare batteries in BSSs.

Items	BSS_0	BSS_1
Battery level	0.7085	0.6535

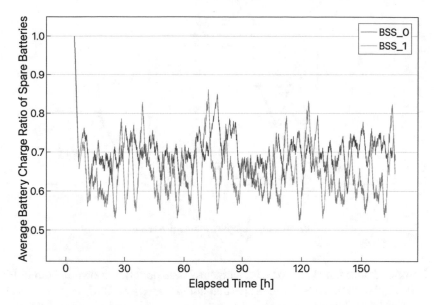

Fig. 5. The changes in the average SOC of spare batteries in BSSs (fixed criteria).

4.2   Effect of Judgement of Battery Transportation Considering Realtime Traffic Condiotion

This section evaluates the proposed method using battery carrying threshold changing in real time according to road condition.

To reflect real-time traffic condition, value of a threshold is derived from moving average. Concretely, the moving average of the average SOC defined as the following formula. In the formula, the average battery charge ratio, time, and window size is defined as x, t [sec], a [sec], respectively.

It also let the moving average at time t be A_t. Figure 6 is the image of moving average.

$$A_t = \frac{1}{a} \sum_{m=t-a+1}^{t} x_m \tag{1}$$

We suppose A_t is derived from the moving average of the average SOC of spare batteries of each BSS. An Ev transports a battery if the moving average value of a BSS in a destination area is lower than a BSS in a current area. Figure 7 shows the changes in the average SOC of each BSS, and Table 4 are the results of the overall average of the average SOC in simulation and the number of battery swaps. For the both results are conducted under the condition where the value of a is set as 3,600 [sec].

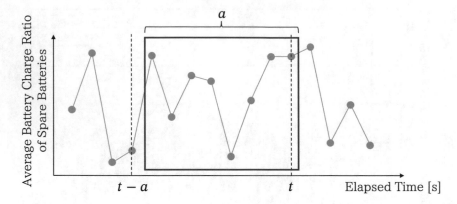

Fig. 6. An example of moving average.

Compared with the results of without battery transportation shown in both of Fig. 2 and Table 2, results of the Fig. 7 indicates that difference among BSS_0 and BSS_1 became small. In addition, Table 4 quantitatively confirms that by introducing the battery transportation the difference of overall average SOC among two BSSs became small, also. From these results, we can conclude that our proposed battery transportation for equalizing average SOC of spare batteries among BSSs effectively works even if the threshold for judging transportation is adaptively charged based on the real time road conditions.

In addition, battery transportation raises the level of the average SOC of spare batteries as a whole as shown in Tables 2, 3, 4.

On the other hand, battery transportation strains Ev users due to the increase in the number of battery swaps by battery transportation, as can be seen Tables 2, 4.

Table 4. The overall average of the average SOC of spare batteries of BSSs and the number of battery swaps.

Items	BSS_0	BSS_1
Battery level	0.7285	0.6853
# of battery swaps	1086	1509

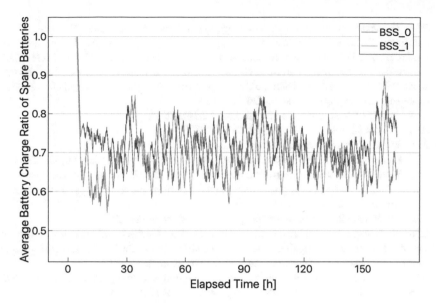

Fig. 7. The changes in the average SOC of spare batteries of BSSs (moving average).

5   Conclusion and Future Work

This paper reveals battery transportation effectively equalizes the average SOC of spare batteries among BSSs. Transporting batteries with adaptively changing judging threshold based on the average SOC changing in real time, worked effectively. In addition, it is also clarified that our proposed battery swapping also increases the total electricity charged in spare batteries.

In order to clarify the relationship between window size a and degree of changing road conditions (ex. day time and night time) in terms of battery transporting effect, further investigations with changing various parameters will be conducted. In additions, weak points of our proposed battery transportation, such as increasing number of battery swappings and increasing total travel distance of Evs, must be addressed in our future works.

Acknowledgement. This work is partially supported by grant-in-aid for scientific research of Electric Technology Research Foundation of Chugoku.

References

1. Ohira, Y., Hirashima, K.: Powering homes with EVs: Panasonic, Omron set to enter Japan's V2H market, NIKKEI Asia (2023). https://asia.nikkei.com/Business/Technology/Powering-homes-with-EVs-Panasonic-Omron-set-to-enter-Japan-s-V2H-market. Accessed 27 Feb 2024
2. Lopez, P.A., et al.: Microscopic traffic simulation using SUMO. In: 2018 21st International Conference on Intelligent Transportation Systems (ITSC), Maui, HI, USA (2018). https://doi.org/10.1109/ITSC.2018.8569938

3. IEA: Transport. https://www.iea.org/energy-system/transport. Accessed 27 Feb 2024
4. European Parliament: EU ban on the sale of new petrol and diesel cars from 2035 explained (2023). https://www.europarl.europa.eu/topics/en/article/20221019STO44572/eu-ban-on-sale-of-new-petrol-and-diesel-cars-from-2035-explained. Accessed 26 Feb 2024
5. IEA: Global EV Data Explorer (2023). https://www.iea.org/data-and-statistics/data-tools/global-ev-data-explorer. Accessed 26 Feb 2024
6. Gogoro. https://www.gogoro.com/. Accessed 26 Feb 2024
7. Ample. https://ample.com/. Accessed 26 Feb 2024
8. NIO. https://www.nio.com/. Accessed 26 Feb 2024
9. NIO Press Release: NIO Announces Details of its Expansion into German, Dutch, Danish and Swedish Markets at European Launch Event (2022). https://www.nio.com/news/nio-announces-expansion-german-dutch-danish-swedish-markets-european-launch-event. Accessed 26 Feb 2024
10. NIO Power. https://www.nio.com/nio-power. Accessed 26 Feb 2024
11. ENEOS Newsroom: Establishment of Gachaco, Inc. Gachaco will provide sharing service of standardized swappable batteries for electric motorcycle (2022). https://www.eneos-innovation.co.jp/english/newsroom/20220330. Accessed 26 Feb 2024
12. EXEO: Launch of Collaboration for Popularizing and Spreading Gachaco Station (2022). https://www.exeo.co.jp/en/news/5276.html. Accessed 26 Feb 2024

Development of DTN Buffer Management for Rapid Grasping of Disaster Situations While Minimizing AoI

Fuka Isayama[1(✉)] and Tetsuya Shigeyasu[2]

[1] Graduate School of Comprehensive Scientific Research, Prefectural University of Hiroshima, 1-1-71 Ujinahigashi, Minamiku, Hiroshimashi, Hiroshima-ken 734-8558, Japan
`r422001vf@ed.pu-hiroshima.ac.jp`
[2] Prefectural University of Hiroshima, 1-1-71 Ujinahigashi, Minamiku, Hiroshimashi, Hiroshima-ken 734-8558, Japan
`sigeyasu@pu-hiroshima.ac.jp`

Abstract. At a disaster situation, it is important to gather information both of victims and disaster stricken area. In addition, the importance of fast data gathering, grows rapidly if such information frequently changes with time. Many systems have been proposed to gather disaster information at disaster affected area based on DTN (Delay/Disruption Tolerant Networking) working effectively regardless the damages of telecommunications infrastructure. However, most of them being developed to use evaluation indicators such as message arrival rate and message delay. Those indicators, however, do not well represent the characteristics of the performance for gathering frequently changing information. Therefore, this paper proposes to use AoI (Age of Information) as an indicator. According to the performance evaluations that, firstly we clarify that transmission order management according to the message priority reduces AoI of most important messages, approx. 50% compared with AoI of conventional method when the number of senders lager than 20. In addition, we also clarify that buffer allocation on the basis of the priority level further reduces unneeded AoI of low priority messages, approx. 60% compared with AoI without buffer management when the number of senders lager than 20.

1 Introduction

When a natural disaster occurs, it is necessary to collect disaster information for utilizing at a disaster relief activities handling by headquarters.

In order to collect these information, many systems have been proposed to use DTN (Delay/Disruption Tolerant Networking) [1], working effectively regardless the damages of communication infrastructure by enabling end-to-end communications even when just after the disaster [2–4].

In these literatures, the proposed systems have been evaluated by indicators such as message arrival rate, massage delay time, and the number of message relays. However, these evaluation indicators do not well represent the gathering performance if the target information changes with time progress.

L. Barolli (Ed.): IMIS 2024, LNDECT 214, pp. 111–123, 2024.
https://doi.org/10.1007/978-3-031-64766-6_12

Therefore, this study evaluates DTN based disaster information collection system by AoI (Age of Information) [5,6] as an indicator that essentially indicates the information freshness.

Incidentally, information collected in the event of a disaster can be classified into two groups. One is the information relating to a target should be dealt with quickly. Injury or illness information and information generated in hazardous areas are included in this group. And the other is the information relating to a target having a time to cope with. Social infrastructures information, family contact information, and information generated in nonhazardous areas, are included in this group. The former group must be forwarded quickly faster than the latter information. In this paper, we clarify transmission order management based on the priority reduces AoI of the specified messages effectively.

In addition, buffer allocation for each message priority avoids undesired AoI degradation of low priority messages.

Rests of this paper are organized as follows. Section 2 considers related work of DTN and AoI. In Sect. 3, the proposed method to improve AoI is discussed. The performance evaluations are in Sect. 4. Finally, the Sect. 5 concludes the paper.

2   Related Work

2.1   DTN

DTN is a network technology enabling end-to-end communication even under unstable conditions of existing communication infrastructures by disaster. DTN consists of sender node generating original messages, relay nodes forwarding a received message, and destination nodes receiving messages finally.

In a DTN, nodes exchange messages in a manner of SCF (Store, Carry and Forward). According to the routing DTN method, a node receiving a new message forwards the message to the other next node within its communication range. If there is no node in its range, the node carries the message until it encounters other new node. Typical routing methods of DTN are Epidemic Routing [7] and Spray and Wait [8]. Epidemic Routing transfers a message whenever a node holding messages encounters another node. It is reported that, the epidemic routing can improve the message delivery ratio under the low node density network. However, a risk of network congestion is also increased by increasing node density and message generation rate.

In contrast, the Spray and Wait reduces consumption bath of bandwidth and buffer, by restricting number of duplications of a message. This method is effectively works. However, there is a difficulty to select the adequate restriction, otherwise, excessive restriction induces low message reachability and long delay.

Hence, this paper uses the Epidemic Routing to develop new DTN system.

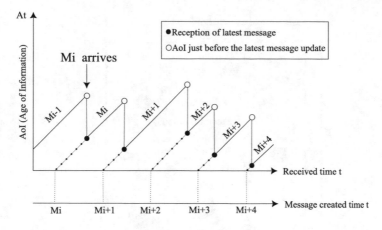

Fig. 1. The image of AoI.

2.2 AoI

AoI represents information freshness under changing the target condition. Then, AoI is a suitable indicator for investigating the performance of gathering victim's vital data, information of ongoing spreading of disaster damage.

AoI is the elapsed time of latest message among it received. Then the value increased successively as the time passes the newer message is received. If the latest message is received, a value of AoI is updated.

Suppose η_t is the generation time of the latest message among it received at time t, the following equation is the definition of AoI(A_t).

$$A_t = t - \eta_t \tag{1}$$

Figure 1 shows an example of AoI changes over time. As the figure shows, the value of AoI fluctuates by the receptions of latest messages. At the receptions, the values drastically reduced, then, the AoI generates like a sawtooth wave.

For evaluating system performance by AoI, average value of AoI is used commonly. The definition of average value of AoI is as follow.

$$m_A := \lim_{T \to \infty} \frac{1}{T} \int_0^T A_t \, dt \tag{2}$$

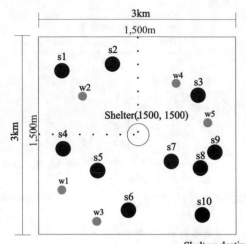

Shelter: destination node
si   : sender node
wi   : relay node

Fig. 2. Evaluation topology.

3   Proposed Method to Improve AoI

In Sect. 3.2, we propose buffer management for improvement of AoI. And we propose transmission order management based on the message priority in Sect. 3.3. However, this method transmits high priority messages frequently, while the lower priority message are not. Therefore, we further propose an algorithm to prevent degradation of AoI of lower priority message by buffer assignment every the level of priority in Sect. 3.4.

3.1   Network Model of Disaster Affected Area

Figure 2 shows the network model of disaster information system in this paper. As the figure shows, the field is a square with a side of 3 km. In the center of the field, there is a destination node as a disaster headquarters or an evacuation center. The sender nodes $s1-s10, s20, s30$ create and transmit original massages. The relay nodes $w1-w100$ relay messages. In addition, the sender nodes and the relay nodes move according to RWP (Random Way Point).

3.2   Buffer Management for Improvement of AoI

In the conventional method, when a message is received, a free buffer is allocated for a new message regardless of the message content. For improving AoI, in the literature [9], we have proposed to discard older generated message from its buffer when the newly received message carries same target and newly generated according to the nature of AoI definition. For the comparison, our method refers the content of messages (target of information, time of generation).

3.3 Transmission Order Management Based on the Message Priority

In this paper, we introduce transmission order management based on the message priority for effectively reducing the value of AoI of specified messages. In the following, this process call RANK method. In this method, higher rank message is allocated small value. In the case of a message forwarding, sender node transmits buffer messages in ascending order of rank value (high priority first). On the contrary, in the case of a message drop by buffer overflow, messages will be discarded in descending order of rank value (low priority first).

3.4 Buffer Assignment Every the Level of Priority

In the literature [9], we have reported that our RANK method degrades AoI of lower rank messages undesirably while the higher message's AoI is effectively improved because of the method reduced the opportunity of lower rank messages significantly. In order to cope with the issue, this paper proposes new buffer assignment for each rank to avoid undesired degradation of lower rank message's AoI.

This method contributes for avoiding monopoly of buffer by high priority message. Concretely, a node received the message with RANK i, stores it with in the buffer allocated for RANK i. If the buffer is overflowed, the node discards one of buffered messages according to the Eq. 3. The equation derives the value of each buffered messages.

Here, d and t are the parameters of a distance and a time.

$$value = (d_2/d_1)\alpha + (t_2/t_1)\beta \tag{3}$$

d_1 is an initial distance between sender and destination (in the case of this paper, a disaster headquarters and sending node). d_2 is a distance between current remaining distance to a destination (in the case of this paper, a disaster headquarters and current node location t). t_1 is an initial TTL (Time To Live) of the message. t_2 is an elapsed time since the message was generated.

Figure 3 shows an example of buffer allocation according to the equation. When a reception of a new message, a receiver node checks the remaining buffer of the rank corresponding to the message. At a reception of RANK1 messages as shown in the Fig. 3, the receiver node calculates the value of messages in the RANK1 buffer, because of the buffer is full. After the calculation of the values, the receiver drops the message having the largest value, then, the received message is stored.

$$RANK1 : RANK2 : RANK3 = 50\% : 30\% : 20\% \tag{4}$$

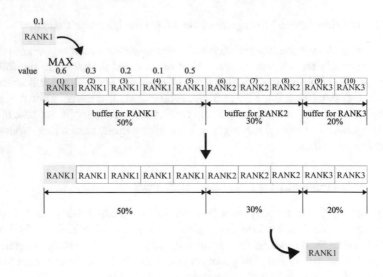

Fig. 3. An examples of buffer allocation

4   Performance Evaluations

For the evaluations, we used the network simulator One (The Opportunistic Network Environment Simulator) [10].

Table 1 shows parameters used by simulation. In this paper, we evaluate AoI in case of the number of sender nodes is 10, 20 and 30. The method where all messages are send as the same rank, is named as unranked, the method where messages are categorized into RANK1–3, is names as ranked.

As described before, nodes transfer message in the order rank. In the case of multiple messages same higher rank, nodes transfer message in the order latest generation message. In addition, at the occasion of a buffer overflow at receiver node, the node drops a stored message in the order of rank. In the case of there are multiple lowest ranked messages, oldest message is dropped. In addition, if there are some buffered messages originally sent from the same node, all messages are dropped except the one selected by the procedure described in the Sect. 3.2.

Table 1. Simulation parameters

item	value		
Field size	3×3 [km^2]		
Number of sender nodes	10 ($1 \leq si \leq 10$)	RANK1	$s9, s10$
		RANK2	$s6–s8$
		RANK3	$s1–s5$
	20 ($1 \leq si \leq 20$)	RANK1	$s17–s20$
		RANK2	$s11–s16$
		RANK3	$s1–s10$
	30 ($1 \leq si \leq 30$)	RANK1	$s25–s30$
		RANK2	$s16–s24$
		RANK3	$s1–s15$
Number of destination nodes (fixed)	1		
Number of relay nodes	100		
Simulation period	43,200 [sec]		
Communication range	100 [m]		
Transmission speed	2 [Mbps]		
Message Size	1 [Mbyte]		
Buffer size	10 [Mbyte]		
Message generation interval	100–150 [sec]		
Movement model	RandomWaypoint		
Movement speed of node	3.0–4.0 [m/sec]		
α	0.5		
β	0.5		

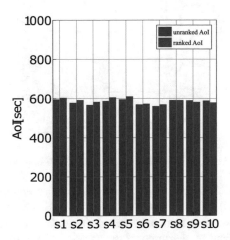

Fig. 4. AoIs when the number of sending node is 10.

4.1 The Effects of Transmission Order Management Based on the Message Priority

Figures 4, 5 and 6 show AoI of both of unranked and ranked method. In the case of ranked method, all messages are sent as rank1–3 according to the original sender.

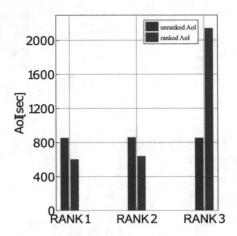

Fig. 5. Average AoIs when the number of sending node is 20.

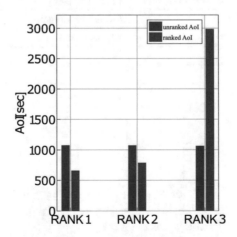

Fig. 6. Average AoIs when the number of sending node is 30.

Figure 4 shows the AoI when the number of sending node is 10. Because of the buffer size and message size are set to 10 MB and 1 MB, all relay nodes equally store and carry up to 10 messages. Hence, as described, each relay node can store 10 messages of different original sender node. If the number of sending nodes is less than 10, all messages can be stored in a buffer regardless the message rank. Then, no buffer overflow occurs and there is not difference between the AoI of the two methods.

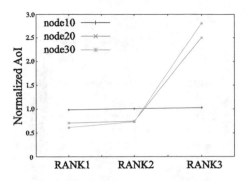

Fig. 7. Normalized AoI.

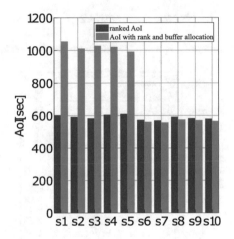

Fig. 8. AoIs when the number of sending node is 10.

Figures 5 and 6 show the AoI when the number of sending node is 20 and 30. These figures indicate the average AoI of nodes having the same rank.

From the results of both figures, we can confirm that AoI of highest rank, RANK1 is improved, approx. 50% compared with unranked AoI. This effect is due to the order management according to RANK method. From these results, due to transmission order management, different of AoI between highest and lowest rank becomes large more than necessary.

Figure 7 shows the Normalized AoI. In this figure, each point represents a value of ranked AoI divided by a value of unranked AoI. In the case of the number of sending node is 10, difference between ranked AoI and unranked AoI is not recognized. (all points are almost 1.0.) However, the number of senders becomes larger than 20, the AoIs of RANK1 and RANK2 improved. While the AoI of RANK3 degraded significantly, approx. 250% compared with unranked AoI.

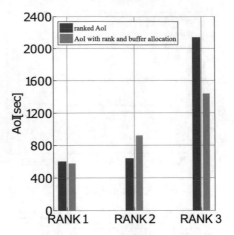

Fig. 9. Average AoIs when the number of sending node is 20.

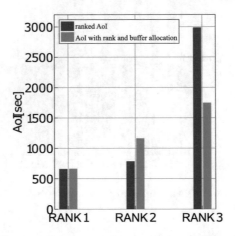

Fig. 10. Average AoIs when the number of sending node is 30.

4.2   The Effects of Buffer Assignment Every the Level of Priority

Figures 8, 9 and 10 show the AoIs of RANK methods of w/ and w/o buffer allocation described in Sect. 3.4. In these results, buffer allocation for RANK1, 2, and 3 are set to 50%, 30%, 20%.

Figure 8 shows the AoI when the number of sending node is 10. These results confirm that RANK3 AoIs of the method w/ buffer allocation (s1–s5) is largely degraded, while the AoIs both of RANK1 and RANK2 kept almost same value.

Figures 9 and 10 show the AoIs when the number of sending node is 20 and 30. These results confirmed that RANK3 AoIs of the method w/ buffer management is drastically improved, approx. 60% compared with the method w/o buffer allocation while the RANK2 AoI degraded slightly, approx. 150% compared with w/o. This phenomenon can be thought that buffer allocation works effectively by avoiding buffer

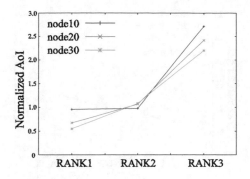

Fig. 11. Normalized AoI under buffer allocation (RANK1:RANK2:RANK3 = 60%:30%:10%)

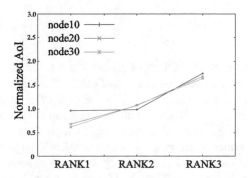

Fig. 12. Normalized AoI under buffer allocation (RANK1:RANK2:RANK3 = 50%:30%:20%)

domination under relatively larger number of message senders than buffer size. The proposed buffer allocation, however, degraded by unneeded over restriction under the relatively smaller number of senders than buffer size.

4.3 The Normalized AoI of Various Types of Buffer Allocation

Figures 11, 12 and 13 shows the normalized AoIs under the changing buffer allocations for each RANK. From these results, we can confirm that the degree of deteriorations of RANK3, reduced by increasing the amount of buffer allocation for RANK3.

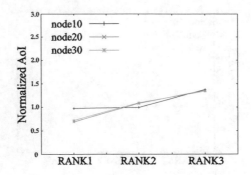

Fig. 13. Normalized AoI under buffer allocation (RANK1:RANK2:RANK3 = 40%:30%:30%)

5 Conclusion

In this paper, we discussed the effects of AoIs by the methods both of transmission order management for each sending node, and the buffer allocation according to the message rank. From the evaluation results, in the case of no room in buffer, simple message transfer based only by RANK method, significantly degrades the AoI of lower rank message due to the buffer domination by higher rank message. On the other hand, in the case of the number of message sender is relatively large than physical buffer size, our proposed buffer allocation effectively works to avoid buffer domination and reduced deterioration of low RANK AoIs. Moreover, we will examine the proposed method under the more realistic conditions.

Acknowledgement. This work was supported by JSPS KAKENHI Grant Number JP21K11851.

References

1. Farrell, S., Cahill, V.: Delay and Disruption Tolerant Networking. Artech House, London (2006)
2. Darmani, M.Y., Karimi, S.: Message overhead control using P-epidemic routing method in resource-constrained heterogeneous DTN. In: 2021 29th Iranian Conference on Electrical Engineering, 18–20 May 2021, Tehran, Iran, pp. 498–502 (2021)
3. Farrell, S., Cahill, V., Geraghty, D., Humphreys, I., McDonald, P.: When TCP breaks: delay- and disruption-tolerant networking. IEEE Internet Comput. **10**(4), 72–78 (2006)
4. Fall, K.: A delay-tolerant network architecture for challenged internets. In: ACM SIGCOMM 2003 Conference on Applications, Technologies, Architectures, and Protocols for Computer Communication, pp. 27–34 (2003)
5. Talak, R., Karaman, S., Modiano, E.: Improving age of information in wireless networks with perfect channel state information. IEEE/ACM Trans. Netw. **28**, 1765–1778 (2020)
6. Inoue, Y., Takine, T.: Age of information: basic concepts and its theoretical analysis. IEICE Fundam. Rev. **13**(3), 197–208 (2020)
7. Vahdat, A., Becker, D.: Epidemic routing for partially connected ad hoc networks. Technical report, CS-2000-06, Duke University (2000)

8. Spyropoulos, T., Psounis, K., Raghavendra, C.S.: Efficient routing in intermittently connected mobile networks: the multiple-copy case. IEEE/ACM Trans. Netw. **16**(1), 77–90 (2008)
9. Isayama, F., Shigeyasu, T.: A new DTN routing strategy considering age of information. In: The 18th International Conference on Broadband and Wireless Computing, Communication and Applications, pp. 263–272 (2023)
10. Ari Keranen: The ONE – The Opportunistic Network Environment simulator. https://akeranen.github.io/the-one/

A Study on Detecting Damaged Building Based on Results of Wi-Fi RTT Measurements

Natsumi Hiramoto[✉], Tetsuya Shigeyasu, and Chunxiang Chen

Prefectural University of Hiroshima, Hiroshima, Japan
r322002ni@ed.pu-hiroshimaz.ac.jp

Abstract. Recently, earthquakes became a serious social problem due to affecting many aspects. In particular, it is difficult to forecast earthquakes, and it is important to figure out the situation at the disaster affect quickly immediately after an earthquake. Human safety is dependent on accurate and rapid assessment of the situation inside buildings. This study discuses a system detecting damaged building by smartphones and wireless access points to support activities at the disaster affected area safety. The proposed system that inspects the safety of the building in real time uses wireless access points employing Wi-Fi RTT (Round Trip Time), installed in a building before a disaster occurs. The effectiveness of the proposed study is evaluated by simulation, and the results confirm that if the proposed study can learn enough data from daily activities, it can warn people to avoid entering unsafe buildings, and it contributes to prevent secondary damage in the event of an anomaly.

1 Introduction

In recent years, natural disasters have been occurring frequently on a global scale [1,2], and earthquakes have became a particularly serious social problem. Earthquakes have a significant impact on the social life in our country.

One month after the Noto Peninsula Earthquake on January 1, 2024, more than 10,000 people were still forced to live in evacuation centers [3]. Furthermore, Japan is expected to experience a Nankai Trough Earthquake in the future, the problem how to prepare for earthquakes is a pressing issue we are now facing. However, since the technology to predict earthquakes in advance has not yet been established, what we can do now is to respond quickly immediately after an earthquake. This is very important in terms of preventing secondary damage by understanding the situation.

Slope failures and landslides, which are included in secondary damage, are relatively easy to figure out, and residents in the affected area can leaving them. Similarly, evacuees can avoid to get near to the damaged buildings even if it is hard to recognize from the outside due to the small size of the damage. Otherwise, if aftershocks occur while the evacuee entered the building, and the building collapses due to extended damage, it could lead to the spread of secondary damage.

Therefore, it is necessary to detect damage to buildings with damage that cannot be seen from the outside by some means, and to notify the evacuees when entering the building or immediately after entering the building.

Many studies on automatic detection of building damage have been reported in literatures. The literature [4] has proposed a system that automatically detects building

L. Barolli (Ed.): IMIS 2024, LNDECT 214, pp. 124–134, 2024.
https://doi.org/10.1007/978-3-031-64766-6_13

damage by learning in advance during normal conditions. Specifically, the seismometer are installed in a building to remotely and immediately assess the level of earthquake damage and whether the damaged building can continue to be used. This method can remotely assess the condition of a building immediately after an earthquake without having to send someone to the site. A method using LiDAR (Light Detection and Ranging) technology has also been proposed [5]. However, these systems requires specialized sensors to be installed on the building in advance, and it makes additional costs.

In this paper, we discuss the system for automatically detecting building anomalies immediately after a natural disaster without requiring diagnosis by personnel with special expertise or the installation of special equipments. The information on the relative distances among wireless access points in a building obtained by Wi-Fi RTT [6,7], which measures the distance between wireless access points and devices implemented in the IEEE802.11mc wireless LAN (Local Area Network) standard, is automatically learned by a smartphone application at a daily activities. When a disaster occurs, the change of wireless access points due to a disaster is automatically detected by comparing the learned data with the observed data in advance the disaster, and the alert warning is sent to a person who enters the building if the damage is detected. By the system, the second damage can be expected to be reduced. In order to clarify the effectiveness of the proposed system, we report the results of evaluations of the detection accuracy of the system.

2 Related Research

There have been many studies on the assessment of the situation when a building collapses. This section describes various related studies which have been proposed using satellites and drones.

2.1 Studies Using Satellites

The remote sensing technology to perform from satellites and aircraft is used to assess damage caused by natural disasters. Recent improvements in sensor performance, such as high resolution optical sensors with a resolution of 1[m] or less and all weather type Synthetic Aperture Radar (SAR), have contributed greatly to the widespread use of remote sensing technology. SAR is a system in which microwaves are irradiated to the ground from an antenna mounted on a platform. The literature [8] has proposed the system to detect earthquake damage to buildings. This method determines damaged building by comparing SAR data from different observation modes over time. Methods utilizing SAR data and AI technology have also been developed [9]. Some studies have assessed building damage using high-resolution satellite images before and after the Turkish earthquake [10]. The urgency of rescue operations is determined by investigating collapsed buildings and areas with high population density. However, these methods evaluate building damage from the exterior of the building, so it is difficult to detect minimal damage and are not suitable for understanding indoor conditions.

2.2 Studies Using UAV (Unmanned Aerial Vehicle)

There are a growing number of actual cases where drones equipped with cameras are used to capture images from the sky over the disaster site to assess the disaster situation.

As a part of study on understanding the situation of flood damage caused by typhoons and heavy rains, a method has been proposed to detect flooded houses and vehicles that are inundated in the image and estimate their submerged water level from the overhead image taken [11]. During disasters, drones can operate quickly in disaster areas more quickly than other manned aircraft. However, it is necessary to check many images and videos taken by drones to determine and record the damage. AR (Augmented Reality) solutions by integrating BIM (Building Information Modeling) and UAV have also been proposed [12]. These methods can check indoor conditions. However, it is difficult to fly drones for long periods of time, and there are many issues such as the need for drone operation skills.

3   Distance Measurement Using Wi-Fi RTT Implemented in IEEE802.11mc

Wi-Fi RTT is defined in IEEE802.11mc [7] and is a function to measure the distance between a AP (Access Point) and a smartphone. This function measures the distance between the two devices by measuring the round trip time.

Figure 1 shows how Wi-Fi RTT is used to estimate distance. t is a round trip propagation delay between the two devices, and a time required for transmission and reception is calculated from the time stamp recorded in the header of the packet. The distance d between the two devices is calculated from the t and the radio wave speed V. In addition, speed V of the radio wave is the fixed number and is 2.998×10^8 m/s.

$$t = (t_4 - t_1) - (t_3 - t_2) \tag{1}$$

$$d = \frac{V \cdot t}{2} \tag{2}$$

$$Distance = \frac{Time \cdot Speed}{2}$$

Fig. 1. Measurement of distance between devices using IEEE802.11mc

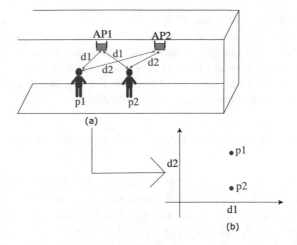

Fig. 2. Distance measurement with APs and distance relationship graphs

4 A Method for Automatically Detecting Building Damage Without Any Special Equipment

This section describes the basic idea for detecting building damage without any special equipment.

4.1 Gathering Relative Distances at Daily Activities

The method can automatically measures normal data, relative distance by smartphone application inside a building. The application gathers those relative distances between

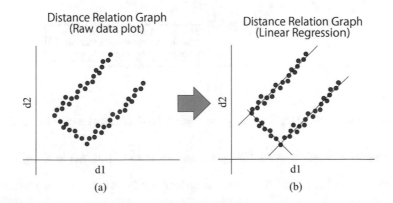

Fig. 3. Creation of distance relationship graphs and regression lines

pre-installed Wi-Fi AP in place and device of user. Gathered data will be used for detection whether a building is damaged by disaster. Figure 2 shows a measurement image of the relative positions between APs at the normal conditions. As shown in the figure, the proposed system automatically measures two distances (as shown as d1, d2) between any one of two APs and the user's smartphone in the building (Fig. 2(a)), the system and plots the measurement results on a graph (Fig. 2(b)).

4.2   Making Distance Relation Graphs Based on Measured Data at Daily Activities

As explained in the previous section, the distance data of APs is automatically gathered at a daily activities while users stay in the building. So, enough number of data can be gathered, and this helps to detect the damage of building at a disaster. Gathered data will be plotted on a graph, named Distance Relation Graph.

Figure 3(a) shows an example of a distance relationship graph. As shown in the figure, distance relation graph becomes the U-shape which consists of three straight lines when the enough number of measured data is gathered. Therefore, by dividing the measurement results into three parts and deriving a regression line for each part, the results shown in Fig. 3(b) can be obtained.

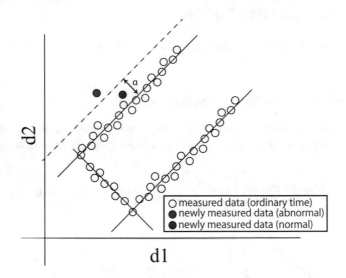

Fig. 4. Abnormality judgment using regression lines of distance relationship graphs

4.3   Anomaly Detection of Buildings Using Distance Relationship Graphs

This section describes a method for detecting damaged building using the distance relationship graphs described in the previous sections. For detecting anomaly, a regression lines obtained from the distance relation graph shown in Fig. 3(b) are used. Due to a regression line is derived from the distance relation graph, if the location of the APs to

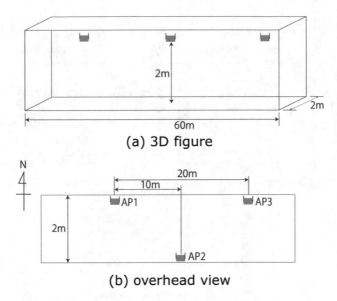

(a) 3D figure

(b) overhead view

Fig. 5. Validation model for anomaly detection by distance relationship graphs

be measured in same condition without any damage, the measurement data will be plotted close to this regression line. Our proposed system judges a building has damaged if the newly measured data is plotted at a place a certain distant from regression line on a distance relation graph.

Figure 4 illustrates the method in detail. As shown in the figure, the distance between a newly added point and a nearest regression line is calculated, and if the distance is greater than a pre-determined threshold value α, it is judged that the building is damaged. In this figure, if the point indicated by the red circle is measured, it is judged that an abnormal. On the other hand in the case of the blue circle, it is judged that there is no abnormal due to the distances is smaller than α.

4.4   Validation of the Building Anomaly Detection Method

This section validates the building anomaly detection using the distance relation graph proposed in the previous section. The model of the building used in the discussion is shown in Fig. 5. As shown in the figure, this section assumes that a AP is installed in a corridor in a building. In the assumed model, three APs are installed in a corridor that is 2[m] wide and 60[m] long. All three APs are installed at a 2[m] in height. The APs are installed at every 10[m], but each AP is installed alternately on the north side (top) and the south side (bottom) side of the corridor.

In the validation process described below, we consider a case where the position of the AP changes for emulating the effects of an earthquake in a situation where the creation of a distance relation graph including the derivation of the regression line has been already completed. The smartphones used to measure the distance are assumed to be held by residents of the affected area who enter the building. In addition, the smart phones are at a 1[m] in height from the floor.

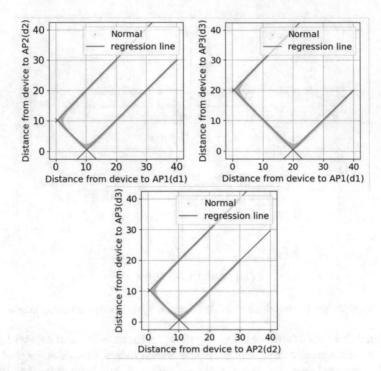

Fig. 6. Distance relation graph by the assumed model (normal condition)

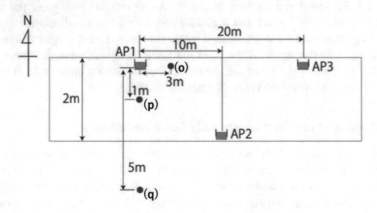

Fig. 7. Location change of AP1 due to anomaly

4.4.1 Distance Relation Graph of Normal Condition

In the assumed model, automatic distance measurement is performed for the three APs. Figure 6 shows the distance relation graph based on the model.

In the figure, the light blue dots indicate the relative distance to the AP from a randomly selected location on the corridor, and the blue line is the regression line derived from those plotted data.

4.4.2   Measurement Results When APs Are Moved Due to Earthquakes

Followings discuss the results under the condition that one of three APs changes its location. Patterns of the location change used for the evaluation shown in Fig. 7.

Case(o) AP1 Moved 3[m] to the East

As shown in Fig. 7(o), AP1 moved 3[m] to the east, and then the distances between random positions in the corridor and each three APs was measured.

Figure 8 shows the measured distance data. In this figure, the graph of the distance between AP1–AP2 and AP1–AP3 plots the measured results when moving 3[m] away from the normal data. Both results contain the data of AP1 that changed the location. On the other hand, in the results of distance AP2–AP3, then, our system can judge that the

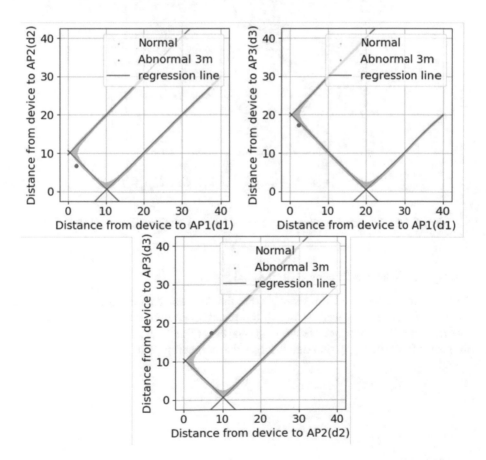

Fig. 8. Distance relationship graph in the assumed model (3[m] parallel movement of AP1)

AP1 has changed its location due to the disaster. Where no AP changed the location. From these graphs, we can confirm that only two graphs, AP1–AP2 and AP1–AP3 indicated that plotted point is distant from regression line.

Case(p) AP1 Moved 1[m] to the South

Here we consider the case where AP1 is moved 1[m] to the south (Fig. 7(p)).

Figure 9 shows the results the green points in this figure show the results taken at random positions after AP1 changed its location. As the figure shows it can be seen that there is no obvious difference from the result under normal conditions. Therefore, this result confirms that it is difficult detect the location change of AP1 in this case.

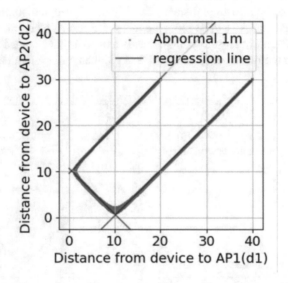

Fig. 9. Distance relationship graph when AP1 moved 1[m] to the south (AP1–AP2)

Case(q) AP1 Moved AP1 Moved 5[m] to the South

Figure 10 shows the results of the case that the AP1 changed its location 5[m] to the south. In the figure, as well as Fig. 9, the results measured from randomly selected positions in the corridor are shown in green. From this figure, it can be seen that the difference from the regression line is larger than when it is 1[m].

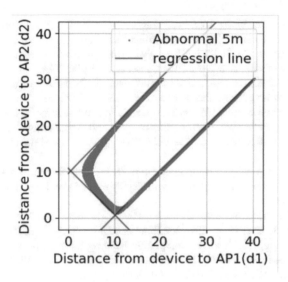

Fig. 10. Distance relationship graph when AP1 moved 5[m] to the south (AP1–AP2)

5 Conclusion

This paper discussed the system for detecting abnormality of the building affected by the disaster. For the case that the damage of the disaster is not obviously recognized from the outside, but it has a risk of secondary damage, it is hard to recognized need to avoid using the building by the evacuees. Hence, this paper proposed the system for automatically warning victims to the damage of the building. By the evaluation using simulation, results confirmed that our proposed system can detect the abnormality of the building in some cases. However, it is also clarified the there is a case that could not be detected. Hence, in the future, we plan to develop additional method for solving the weakness of our proposed system.

References

1. Institute for Economics & Peace: The Trend in Natural Disasters. Ecological Threat Register 2020, 49–50 (2020)
2. Mustafa, T.H.: The impact of natural disasters on education. In: International Conference on Social Science Studies, pp. 86–94 (2003)
3. Editor of the Japan Times: Two months on from Noto quake, 11,400 evacuees still in shelters. The Japan Times (2024)
4. Alcantara, E.A.M., Bong, M.D., Saito, T.: Structural response prediction for damage identification using wavelet spectra in convolutional neural network. Sensors **21**(20), 6795 (2021). https://doi.org/10.3390/s21206795
5. Kaartinen, E., Dunphy, K., Sadhu, A.: LiDAR-based structural health monitoring: applications in civil infrastructure systems. Int. J. Environ. Res. Public Health **22**(12), 4610 (2022). https://doi.org/10.3390/ijerph19010237

6. Android Developer: Wi-Fi location: ranging with RTT. https://developer.android.com/develop/connectivity/wifi/wifi-rtt of subordinate document. Accessed 25 Mar 2024
7. Android Developer: Wi-Fi RTT (IEEE 802.11mc). https://source.android.com/docs/core/connect/wifi-rtt of subordinate document. Accessed 25 Mar 2024
8. Kim, M., Park, S.-E., Lee, S.-J.: Detection of damaged buildings using temporal SAR data with different observation modes. Remote Sens. **15**, 1–13 (2023)
9. Miyamoto, T., Yamamoto, Y.: Using 3-D convolution and multimodal architecture for earthquake damage detection based on satellite imagery and digital urban data. IEEE J. Sel. Top. Appl. Earth Obs. Remote Sens. **14**, 8606–8613 (2021). https://doi.org/10.1109/JSTARS.2021.3102701
10. Haobin, X., et al.: A deep learning application for building damage assessment using ultra-high-resolution remote sensing imagery in Turkey earthquake. Int. J. Disaster Risk Sci. **14**, 947–962 (2023)
11. Rizk, H., Nishimur, Y., Yamaguchi, H., Higashino, T.: Drone-based water level detection in flood disasters. Sensors **19**(1), 237 (2022). https://doi.org/10.3390/s22124610
12. Xietian, X., Junjie, C., Shuai, L.: Integrating building information model and augmented reality for drone-based building inspection. J. Comput. Civ. Eng. **35**, 1–17 (2021). https://doi.org/10.1061/(ASCE)CP.1943-5487.0000958

Extended Gerber-Shiu Expected Discounted Penalty Functions in Risk Model Perturbed by Diffusion and Application

Zhimin Wang$^{(\boxtimes)}$, Haibo Zhang, Xiang Ma, and Xuan Wang

Engineering University of The Chinese People's Armed Police Force, Xian, Shanxi, China
18700599850@139.com

Abstract. An extension of the traditional risk surplus model is presented, integrating a diffusion process into the compound Poisson framework to enrich the model's descriptive power. This integration leads to the development of the Extended Gerber-Shiu (EGS) function, which introduces a premium for surpassing the initial capital level and takes into account two distinct stopping time scenarios. The problem formulation is derived using a retrospective differential analysis, resulting in a coupled integro-differential equation that the EGS function must satisfy. Addressing this formulation becomes the primary focus. The martingale approach is then applied, revealing that the EGS function can be partitioned into two simpler, independent problems: the fundamental Gerber-Shiu function and a first passage problem. The resolution of the latter is essential for the risk surplus model under study. Utilizing martingale measure transformation and Laplace transformation under various discount factors, the results that define the EGS function and its solution are determined. This work contributes to the field by enhancing the understanding and application of the EGS function in actuarial science and risk management, providing a foundation for further research and practical implementation.

1 Introduction

In the fields of actuarial science and probabilistic applications, the domain of risk theory (also known as collective risk theory) employs mathematical modeling to assess the susceptibility of insurance companies to bankruptcy or financial ruin. Consequently, the development of these risk models plays a pivotal role in advancing the study and understanding of risk theory.

In the research of risk theory, two methods are widely applied. One of the methods is the renewal technique introduced by Feller [7] and the other one is the martingale approach introduced by Gerber [9]. Up to now, the two previous mentioned methods are still main methods for research the ruin problems. In order to figure out how ruin happens, Gerber et al. [6, 10, 11] have introduced another two $r.v.$, the surpluses immediately before U_{τ_0-} and the deficits at ruin $\left| U_{\tau_0} \right|$, which are also insurance company concerned about. In those articles, the respective probability distributions are considered. From the perspective of pure mathematics, in the three $r.v.$, we know that the probability distribution function of $r.v.$ U_{τ_0-} is even more critical.

© The Author(s), under exclusive license to Springer Nature Switzerland AG 2024
L. Barolli (Ed.): IMIS 2024, LNDECT 214, pp. 135–147, 2024.
https://doi.org/10.1007/978-3-031-64766-6_14

For a comprehensive approach, it is essential to establish a consistent framework for examining the aforementioned quantities. This involves the analysis of the Gerber-Shiu expected discounted penalty function at the point of ruin, presently recognized as the Gerber-Shiu function. The function is mathematically expressed as

$$\mathbb{E}_u\big[e^{-\delta \tau_0} w\big(U_{\tau_0-}, |U_{\tau_0}|\big); \tau_0 < \infty\big].$$

Here $\delta \geq 0$ represents the discount rate, and $w(y, z)$ is a positive function ensuring the finiteness of the expected value.

Ruin probability and Gerber-Shiu function only consider the quantity related to ruin time. However, in risk management with life time, it is necessary to consider two kinds of stopping time, that is the ruin time and the time that the surplus firstly reach a pre-assumed high level. In this case, the ruin problem become a kind of two side exit problem. Motivated by this point, the Gerber-Shiu function is publicized with an additional reward for reaching a level over the fresh capital. Schmidli [15] studied an extended Gerber-Shiu function (EGS function for short in the rest of this paper) under traditional risk control models with fixed interest rates.

2   Model and Problem Formulation

This part, we based on the classical risk surplus model to make further promotion. Our claim and premium income may be from the influence of uncertain factors such as economic capital markets. The uncertainty of the market and other factors we use Brownian motion (Wiener or diffusion process) to describe.

2.1   Model of Risk Surplus Process that is Perturbed by Diffusion

The risk surplus process of policyholders is determined by

$$U_t = u + ct - \sum_{i=1}^{N_t} Y_i = u + ct - S(t) + W(t). \tag{2.1}$$

where $u \geq 0$ is the insurer's initial surplus; The premiums is charged continuously at a constant rate, $c \geq 0$, per unit time; Arrival times T_i; $\{N_t, t \geq 0\}$, represents the claim frequency of arrival time t, following the Poisson parameter λ Poisson process; The claim sizes $\{Y_i, i \geq 1\}$ are i.i.d. random variables with common distribution function $G(y)$ and $G(0) = 0$, and the average claim size is $\mu = \mathbb{E}[Y_1]$; $\{Y_i\}$ are also independent of $\{N_t\}$. The Wiener process is characterized by an infinitesimal drift of $S(t) = \sum_{i=1}^{N_t} Y_i$ and an infinitesimal variance of $2D > 0$. Consequently, for all $t > 0$, the random variable W(t) follows a normal distribution with a mean of zero and a variance of 2Dt. Additionally, it is postulated that the processes $\{S(t)\}$ and $\{W(t)\}$ operate independently of each other.

The diffusion term in model (2.1) represents the realistic randomness of the total claim amount; That is to say, the new model increases the randomness of premium income. This uncertainty may be caused by the environment of Perfectly Competitive Market.

2.2 Problem Formulation of the EGS Function

Throughout the subsequent analysis, we operate within a measure space (Ω, F). Unless specified otherwise, let $(\Omega, \mathrm{F}, \mathrm{P})$ denote a complete probability space. When conditioning on $U_0 = u$, we denote this by $\mathbb{P}_u$. The information structure is defined by the filtration $\mathbf{F} = \{\mathcal{F}_t\}$, which is assumed to be the smallest right-continuous filtration rendering $\{U_t\}$ adapted. It is important to note that the filtration is not completed, as we intend to modify the measure in subsequent steps.

Assume that $\rho > 0$, where the relative safe load can be expressed as

$$\rho \hat{=} \frac{c - \lambda\mu}{\lambda\mu} \, (\text{i.e., } c > \lambda\mu).$$

Let $a \geq u$ with $a > 0$. Let $\tau_0 = \inf\{t : U_t < 0\}$ $(\inf \emptyset = \infty)$ understand as the time of bankruptcy. Define $\tau^a = \inf\{t : U_t > a\}$ and $\tau^* = \tau_0 \wedge \tau^a$, the first exit time from the interval $[0, a]$. Note that $\tau^* < \infty$ a.s. In this article, the focus is on the object (the EGS function) of discussion with initial surplus u is

$$\zeta(u) = E\big[e^{-\delta\tau_0}w\big(U_{\tau_0-}, |U_{\tau_0}|\big)I_{\{\tau^a > \tau_0\}} \mid U_0 = u\big] + E\big[e^{-\delta\tau^a}\zeta_a I_{\{\tau^a < \tau_0\}} \mid U_0 = u\big] \tag{2.2}$$

for some given constant ζ_a. We could here choose $\zeta_a \leq 0$ to model a reward for reaching a before ruin. Note that $U_{\tau^a-} = U_{\tau^a} = a$, so the capital prior to leaving the interval through a is known.

The initial expected value in the description of $\zeta(u)$ bears resemblance to the limited-time Gerber-Shiu function introduced by Kuznetsov and Morales [13]. However, in this context, the fixed time horizon is substituted with a stochastic stopping time. The subsequent expected value resembles the first occurrence of Laplace transform when the surplus process reaches the critical value a, with the process halted upon ruin. This specific transformation (where $\zeta_a = 1$) is of particular interest in the current study. It will be demonstrated that $\zeta(u)$ is derivable using the Gerber-Shiu function at ($\zeta_a = 0$) and the Laplace transformation of τ^a on the condition that $\{\tau^a < \tau_0\}$ and $w(y, z) = 0$.

In particular, the following initial conditions are trivial,

$$\zeta(0) = 0, q_L - a.s.. \tag{2.3}$$

The function w(x,y), where x $\geq$ 0 and y $\geq$ 0, is posited as nonnegative and unbroken throughout this work. It is presumed that $\zeta(u)$ is a defined entity. In this study, w(x,y) is considered to be capped, meaning that the supremum over and y for w(x,y) is finite and represented by M, a positive constant, ensuring that $\zeta(u)$ remains bounded. It is also assumed that G(y) exhibits continuity, a simplification for the subsequent analysis and to guarantee that $\zeta(u)$ is differentiable without interruption.

Nonetheless, the assumptions can be relaxed. Should continuity be abandoned, it must be recognized that $\zeta(u)$ may encounter points of non-differentiability. Nevertheless, the function would maintain absolute continuity, rendering our methodology applicable, and $\zeta(u)$ would still represent a form of density. Focusing solely on positive functions is not restrictive, as w(y, z) can invariably be decomposed into its positive and negative components.

We prove that the following it:

Theorem 2.1. $\forall u \in [0, a]$ with $a > 0$, $\zeta(u)$ is the solutions to the following integral-differential equation.

$$D\zeta'(u) + c\zeta(u) - D\zeta'(0) - \delta \int_0^u \zeta(t)dt$$
$$-\lambda \int_0^u \zeta(u-y)[1 - G(y)]dy + \lambda \int_0^u \int_t^\infty w(t, y-t)dG(y)dt = 0. \quad (2.4)$$

Proof. Let $0 \le u < a$. Let further $\gamma(t) = u + ct + W(t)$ denote the capital at time t if no claim occurs in $[0, t]$. Choose Δt such that $\gamma(\Delta t) < a$.

Suppose that $U_0 = u$. Applied in Cai [1], consider the number of claims in a very short time interval $(0, \Delta t)$; there are three cases:

(i) no claim arrives with probability $(1 - \lambda \Delta t) + o(\Delta t)$. Then initial surplus increased from u to $U_{\Delta t} = \gamma(\Delta t) = u + c\Delta t + W(\Delta t)$;
(ii) one claim occurs with probability $\lambda \Delta t + o(\Delta t)$, claim amount is $y > 0$. Then initial surplus increased from u to $U_{\Delta t} = \gamma(\Delta t) - Y_1$.

In this situation, we should further consider whether initiating a claim would lead to bankruptcy:
(a) the claim cause ruin, i.e. $\tau_0 = \Delta t, \tau^a > \Delta t, y > \gamma(\Delta t)$. Then $\zeta(u)$ may be taken as

$$\int_{\gamma(\Delta t)}^\infty e^{-\delta \Delta t} w(\gamma(\Delta t), y - \gamma(\Delta t))dG(y);$$

(b) the claim don't cause ruin, i.e. $\tau_0 > \Delta t, \tau^a > \Delta t, y \le \gamma(\Delta t)$. Then $\zeta(u)$ may be taken as

$$\int_0^{\gamma(\Delta t)} e^{-\delta \Delta t} \zeta(\gamma(\Delta t) - y)dG(y);$$

(iii) more than one claim happen with probability $o(\Delta t)$.

By the Markov property of process U_t and Law of total expectation we have

$$\zeta(u) = \mathbb{E}[\zeta(u)|U_{\Delta t}] = \mathbb{E}\left[e^{-\delta\Delta t}\zeta(\gamma(\Delta t)N(\Delta t) = 0\right] \cdot \mathbb{P}[N(\Delta t) = 0]$$

$$+ \mathbb{E}\left[e^{-\delta\Delta t}\zeta(\gamma(\Delta t)N(\Delta t) = 1\right] \cdot \mathbb{P}[N(\Delta t) = 1] + o(\Delta t)$$

$$= (1 - \lambda\Delta t)e^{-\delta\Delta t}\mathbb{E}[\zeta(\gamma(\Delta t))]$$

$$+ \lambda\Delta t e^{-\delta\Delta t}\left[\int_0^{\gamma(\Delta t)} \zeta(\gamma(\Delta t) - y)dG(y) + \int_{\gamma(\Delta t)}^{\infty} w(\gamma(\Delta t), y - \gamma(\Delta t))dG(y)\right] + o(\Delta t). \quad (2.5)$$

Letting Δt tend to zero Eq. (2.5) shows that $\zeta(u)$ is right continuous and differentiable from the right. We substitute

$$\mathbb{E}[\zeta(\gamma(\Delta t))] = \mathbb{E}[\zeta(u + c\Delta t + W(\Delta t))] = \zeta(u) + c\Delta t\zeta'(u) + D\Delta t\zeta''(u)$$

$$\zeta(u) = (1 - \lambda\Delta t)(1 - \delta\Delta t)\left(\zeta(u) + c\Delta t\zeta'(u) + D\Delta t\zeta''(u)\right)$$

$$+ \lambda\Delta t(1 - \delta\Delta t)\left[\int_0^{\gamma(\Delta t)} \zeta(\gamma(\Delta t) - y)dG(y) + \int_{\gamma(\Delta t)}^{\infty} w(\gamma(\Delta t), y - \gamma(\Delta t))dG(y)\right]$$

$$+ o(\Delta t).$$

$$(2.6)$$

Rearranging Eq. (2.6), dividing by Δt and letting $\Delta t \to 0$.

$$D\zeta''(u) + c\zeta'(u) - (\lambda + \delta)\zeta(u)$$

$$+ \lambda\left[\int_0^u \zeta(u - y)dG(y) + \int_u^{\infty} w(u, y - u)dG(y)\right] = 0. \quad (2.7)$$

For $0 < u \leq a$ chose the initial capital $u + c\Delta t + W(\Delta t) \geq 0$ and repeat the above steps. This gives that $\zeta(u)$ is left continuous and differentiable from the left, and (2.7) holds.

Now we integrate (2.8) over t (say from 0 to u) that

$$\int_0^u D\zeta''(t)dt + \int_0^u c\zeta'(t)dt - \int_0^u (\lambda + \delta)\zeta(t)dt$$

$$+ \lambda\left[\int_0^u\int_0^t \zeta(t - y)dG(y)dt + \int_0^u\int_t^{\infty} w(u, y - t)dG(y)dt\right] = 0. \quad (2.8)$$

Since initial conditions (2.3), we get

$$D\zeta'(u) + c\zeta(u) - D\zeta'(0) - \delta\int_0^u \zeta(t)dt$$

$$- \lambda\left[\int_0^u \zeta(t)dt - \int_0^u\int_0^t \zeta(t - y)dG(y)dt\right] + \lambda\int_0^u\int_t^{\infty} w(u, y - t)dG(y)dt \quad (2.9)$$

and note that

$$\int_0^u \zeta(t)dt - \int_0^u \int_0^t \zeta(t-y)dG(y)dt$$

$$= \int_0^u \zeta(t)dt + \int_0^u \int_0^t \zeta(t-y)d\big[1-G(y)\big]dt$$

$$= \int_0^u \zeta(t)dt + \int_0^u \left\{ \zeta(t-y)\big[1-G(y)\big]\big|_{y=0}^t + \int_0^t \zeta'(t-y)\big[1-G(y)\big]dy \right\}dt$$

$$= \int_0^u [1-G(t)]dt + \int_0^u \int_y^u \zeta'(t-y)\big[1-G(y)\big]dtdy$$

$$= \int_0^u [1-G(t)]dt - \int_0^u \big[1-G(y)\big]dy + \int_0^u \zeta(u-y)\big[1-G(y)\big]dy$$

$$= \int_0^u \zeta(u-y)\big[1-G(y)\big]dy.$$

This yields

$$D\zeta'(u) + c\zeta(u) - D\zeta'(0) - \delta\int_0^u \zeta(t)dt$$

$$-\lambda\int_0^u \zeta(u-y)\big[1-G(y)\big]dy + \lambda\int_0^u \int_t^\infty w(t, y-t)dG(y)dt = 0.$$

This completes the proof.

We obtain problem formulation (2.4) such as the EGS function (2.2). In paper, the following, we will discuss around its as a key task.

3 Martingale Approach to Problem Formulation

Now way to solve the EGS function $\zeta(u)$ through the Eq. (2.4). In this section, We deal with its mainly by the martingale approach. We can split our $\zeta(u)$ in two subproblems. Using the method of inverting Laplace transformation, the subproblem provides a numerical method for solving the corresponding Laplace-transform.

Let $\mathcal{F}_t^N = \sigma\{N_s, 0 \le s \le t\}$, $\mathcal{F}_t^W = \sigma\{W_s, 0 \le s \le t\}$, $\mathcal{F}_t^U = \sigma\{U_t, 0 \le s \le t\}$ and $\mathcal{F}_t = \mathcal{F}_t^N \vee \mathcal{F}_t^W \vee \mathcal{F}_t^U = \sigma\{(N_s, W_s, U_s), 0 \le s \le t\}$. In this paper we shall use Lemma in Dynkin [5], we cite it here and prove it:

Lemma 3.1. (Dynkin's theorem).

Let $Y = \{Y(t); t \ge 0\}$ be a stochastic process. The possible values $\mathbb{S} \subseteq \mathbb{R}$ of $Y(t)$ is called the state space. Let Y be a Markov process with generator A and v a function in the domain of A such that $Av \equiv 0$. Then M, defined by $M(t) = v(Y(t))$, is an $\mathbf{F}^Y$-martingale.

First of all, we prove a verification theorem.

Theorem 3.2. Suppose we have a solution that $\xi(u)$ on $[0, a]$ to (2.4) with $\xi(a) = \zeta_a$. Then $\xi(u) = \zeta(u)$.

Proof. Note that as a continuous function on a compact interval $\xi(u)$ must be bounded.

The Dynkin's theorem claims that the process

$$
M_t = \begin{cases}
\xi(U_t)e^{-\delta t}, & \text{if } \tau^* > t \\
w(U_{\tau_0 -}, |U_{\tau_0}|)e^{-\delta \tau_0}, & \text{if } \tau^* = \tau_0 \leq t, \\
\zeta_a e^{-\delta \tau^a}, & \text{if } \tau^* = \tau^a \leq t,
\end{cases}
$$

is a martingale, see for example Davis [4]. By the Martingale property,

$$
\xi(u) = \mathbb{E}_u[M_t].
$$

Because $\{M_t\}$ is bounded, this yields

$$
\xi(u) = \lim_{t \to \infty} \mathbb{E}_u[M_t] = \mathbb{E}_u\left[\lim_{t \to \infty} M_t\right] = \mathbb{E}_u[M_\infty] = \zeta(u).
$$

The special case with $a = +\infty$ is the ordinary Gerber-Shiu function. Denote it by $\zeta^*(u)$. Suppose we found $\zeta^*(u)$. The article derived the expression of the function ζ.

Theorem 3.3. The function ζ can be expressed as

$$
\zeta(u) = \zeta^*(u) + (\zeta_a - \zeta^*(a))\mathbb{E}_u\left[e^{-\delta \tau^a} I_{\{\tau^a < \tau_0\}}\right]. \tag{3.1}
$$

Proof. The function $\zeta^*(u)$ is the solution to (2.4) with $\zeta(a) = \zeta^*(a)$.

Then $\xi(u) = \zeta(u) - \zeta^*(u)$ solves

$$
D\xi'(u) + c\xi(u) - D\xi'(0) - \delta \int_0^u \xi(t)dt - \lambda \int_0^u \xi(u-y)[1 - G(y)]dy = 0. \tag{3.2}
$$

This is the special circumstances of the Gerber-Shiu function with $w = 0$ and $\xi(a) = \zeta_a - \zeta^*(a)$. By our considerations, the function

$$
\widetilde{\zeta}(u) = \mathbb{E}_u\left[e^{-\delta \tau^a}(\zeta_a - \zeta^*(a))I_{\{\tau^a < \tau_0\}}\right]
$$

is a solution to (3.2) with the desired value at a. The result follows from Theorem 3.2,

$$
\widetilde{\zeta}(u) = \xi(u)\big(= \zeta(u) - \zeta^*(u)\big) = (\zeta_a - \zeta^*(a))\mathbb{E}_u\left[e^{-\delta \tau^a} I_{\{\tau^a < \tau_0\}}\right].
$$

With this, the proof is concluded.

Accordingly, our issue can be decomposed into a pair of sub-problems:

(1) Determine the Gerber-Shiu function;
(2) Address the case where w(y, z) = 0 and $\zeta_a = 1$.

Given the extensive treatment of the Gerber-Shiu function in existing literature, the focus of our investigation will be on the second sub-problem.

Remark 1. From (3.1) we see that.

$$\mathbb{E}_u\Big[e^{-\delta\tau_0}w\big(U_{\tau_0-},\,|U_{\tau_0}|\big)I_{\{\tau^a>\tau_0\}}\Big] = \zeta^*(u) - \zeta^*(a)\mathbb{E}_u\Big[e^{-\delta\tau^a}I_{\{\tau^a<\tau_0\}}\Big].$$

Consequently, the halted Gerber-Shiu function can be derived utilizing the Gerber-Shiu function for an indefinite time frame, coupled with the resolution of the aforementioned second issue. Essentially, the expression above corresponds to the bounded-time Gerber-Shiu function proposed by Kuznetsov and Morales [13], which integrates the time frame up to the initial breach of the threshold (a, ∞).

Remark 2. In their work, Gerber et al. [12] examined the Gerber-Shiu function under the scenario of a dividend barrier. Specifically, when the surplus reaches a level b > 0, the premium earnings are distributed as dividends. This setting mirrors the issue at hand, but precludes the explicit knowledge of ζ_b. In contrast, the boundary condition is defined $\zeta\prime(b) = 1$. The approach to obtaining the sought-after ζ_b value involves differentiating the expression in Eq. (3.1).

We start consider the model (2.2) with equation

$$cr + Dr^2 + \lambda h(r) = \delta, P - \text{a.s.} \tag{3.3}$$

where $h(r) \triangleq \mathbb{E}\big[e^{-rY_1} - 1\big] = \int_0^\infty e^{-ry}dG(y) - 1$, the left hand side is convex, vanishes in zero and tends to infinity as $r \to \infty$. So there exists a unique positive solution ϱ to the equation if $\delta > 0$ and if $c > \lambda\mu$. If $\delta = 0$, we let $a = +\infty$. We will consider the case $\delta = 0$ later.

3.1 $\delta > 0$

First, we consider the case $\delta > 0$. Meanwhile, letting $c > \lambda\mu$, the above equation exists a unique positive solution ϱ.

Proposition 1. When the above Eq. (3.3) exists a unique positive solution ϱ, the new process L, given by.

$$\{L_t = \exp\{\varrho(U_t - u) - \delta t\}\},$$

is an **F**-martingale with mean value one where the filtration **F** is given by $\mathcal{F}_t = \mathcal{F}_t^N \vee \mathcal{F}_t^W \vee \mathcal{F}_t^L = \sigma\{(N_s, W_s, L_s), 0 \le s \le t\}$.

Proof. The fact that S has independent increments relative to $\mathcal{F}_t^N$ is equivalent to that $S_t - S_s$, for $s \le t$, independent of $\mathcal{F}_s^N$. Since.

$$\mathbb{E}\Big[e^{-\varrho S(1)}\,\big|\,\mathcal{F}_t^N\Big] = \mathbb{E}\Big[e^{-\varrho\sum_{k=1}^{N(1)}Y_k}\,\big|\,\mathcal{F}_t^N\Big] = \int_{\cup_{i=0}^\infty\{N(1)=i\}} e^{-\varrho\sum_{k=1}^{N(1)}Y_k}dp$$

$$= \sum_{i=0}^\infty \int_{\{N(1)=i\}} e^{-\varrho\sum_{k=1}^{N(1)}Y_k}dp = \sum_{i=0}^\infty \int_\Omega I_{\{N(1)=i\}}e^{-\varrho\sum_{k=1}^{N(1)}Y_k}dp$$

$$= \sum_{i=0}^\infty \int_\Omega I_{\{N(1)=i\}}e^{-\varrho\sum_{k=1}^i Y_k}dp = \sum_{i=0}^\infty P_{\{N(1)-N(0)=i\}}\Big(\mathbb{E}e^{-\varrho Y_1}\Big)^i dp$$

$$= \sum_{i=0}^{\infty} \frac{\lambda^i}{i!} e^{-\lambda} \left(\mathbb{E} e^{-\varrho Y_1} \right)^i dp = \sum_{i=0}^{\infty} \frac{\left(\lambda \mathbb{E} e^{-\varrho Y_1} \right)^i}{i!} e^{-\lambda} dp = e^{\lambda \left(\mathbb{E} e^{-\varrho Y_1} - 1 \right)} = e^{\lambda h(\varrho)},$$

$\{W_t\}$ has also stationary and independent increments. Since

$$\mathbb{E}\left[e^{\varrho W(1)} \mid \mathcal{F}_t^W \right] = \int_{-\infty}^{+\infty} e^{\varrho x} \frac{1}{\sqrt{2\pi 2D}} e^{-\frac{x^2}{2 \cdot 2D}} dx = \int_{-\infty}^{+\infty} \frac{1}{2\sqrt{\pi D}} e^{-\frac{x^2}{4D} + \varrho x} dx$$

$$= \frac{1}{2\sqrt{\pi D}} \int_{-\infty}^{+\infty} e^{-\left(\frac{x - 2D\varrho}{2\sqrt{D}} \right)^2 + D\varrho^2} dx \overset{y = \frac{x - 2D\varrho}{2\sqrt{D}}}{=} \frac{1}{2\sqrt{\pi D}} e^{D\varrho^2} \int_{-\infty}^{+\infty} e^{-y^2} \cdot 2\sqrt{D} dy$$

$$= \frac{1}{\sqrt{\pi}} e^{D\varrho^2} \int_{-\infty}^{+\infty} e^{-y^2} dy = e^{D\varrho^2}. \forall t \ge s \ge 0, \mathbb{E}[L_t \mid \mathcal{F}_s] = \mathbb{E}[L_t - L_s + L_s \mid \mathcal{F}_s]$$

$$= L_s \mathbb{E}[L_t - L_s \mid \mathcal{F}_s] = L_s \mathbb{E}\left[e^{-\delta(t-s) + \varrho c(t-s) - \varrho(S(t) - S(s)) + \varrho(W(t) - W(s))} \mid \mathcal{F}_s \right]$$

$$= L_s e^{-\delta(t-s) + \varrho c(t-s)} \mathbb{E}\left[e^{-\varrho S(t-s) + \varrho W(t-s)} \mid \mathcal{F}_s \right] \overset{t_1 = t-s}{=} L_s e^{-\delta t_1 + \varrho c t_1} \mathbb{E}\left[e^{-\varrho S(t_1) + \varrho W(t_1)} \mid \mathcal{F}_s \right]$$

$$= L_s e^{-\delta t_1 + \varrho c t_1} \left(\mathbb{E}\left[e^{-\varrho S(1)} \right] \right)^{t_1} \left(\mathbb{E}\left[e^{\varrho W(1)} \right] \right)^{t_1} = L_s e^{-\delta t_1 + \varrho c t_1 + \lambda h(\varrho) t_1 + D\varrho^2 t_1} = L_s e^{(-\delta + \varrho c + \lambda h(\varrho) + D\varrho^2) t_1}$$

$$= L_s, \mathbb{P} - a.s. \text{ for } t \ge s.$$

$\{L_t\}$ is an **F**-martingale, therefore $\mathbb{E} L_t = \mathbb{E} L_0 = 1$.

This completes that proof.

Consider on $\mathcal{F}_t$ the measure $\mathbb{P}^\varrho[A] = \mathbb{E}_u[L_t; A]$.

This measure can be extended to $\mathcal{F}$. Under $\mathbb{P}^\varrho$, the process $\{U_t\}$ is still a Poisson risk surplus model with claim intensity $\lambda^\varrho = \lambda \mathbb{E}\left[e^{-\varrho Y_1} \right]$.

and claim size distribution $G^\varrho(y) = \frac{\int_0^y e^{-\varrho z} dG(z)}{\mathbb{E}[e^{-\varrho Y_1}]}$.

The analysis confirms that the condition for a positive net profit is maintained, hence

$$\mathbb{P}_u^\varrho[\tau_0 < \infty] < 1.$$

For a foundational understanding of measure transformation techniques, one may consult the work of Rolski et al. [14] as a starting point.

Consider now the quantity $\zeta(u) = \mathbb{E}_u\left[e^{-\delta \tau^a} I_{\{\tau^a < \tau_0\}} \right]$ we are interested in. Using the measure $\mathbb{P}^\varrho$, this is

$$\zeta(u) = \mathbb{E}_u^\varrho\left[\exp\{ -\varrho(U_{\tau^a} - u) + \delta \tau^a \} e^{-\delta \tau^a} I_{\{\tau^a < \tau_0\}} \right] = e^{-\varrho(a-u)} \mathbb{P}_u^\varrho[\tau^a < \tau_0],$$

where we used that $U_{\tau^a} = a$.

Theorem 3.4. If $w(y, z) = 0$ and $\zeta_\alpha = 1$, if $\delta > 0$ and $c > \lambda\mu$, then, then the function $\psi^\varrho(u)$ is given by:

$$\zeta(u) = \frac{1 - \psi^\varrho(u)}{1 - \psi^\varrho(a)},$$

where $\psi^\varrho(u)$ denotes the probability of ruin under the probability measure $\mathbb{P}^\varrho$.

Proof. The proof involves computing $\mathbb{P}_u^\varrho[\tau^a < \tau_0]$. Given that the safety loading under $\mathbb{P}^\varrho$ is affirmative, it follows that $\psi^\varrho(u) < 1$. By Dynkin's theorem, the stochastic process $\{1 - \psi^\varrho(U_{\tau_0 \wedge t}^0)\}$ is identified as a bounded martingale under $\mathbb{P}^\varrho$. The conclusion is reached upon stopping at $\tau^a \wedge \tau_0$, as this stopping time is almost surely finite.

3.2 $\delta = 0$

It remains to consider the case $\delta = 0$. In this case, if we look for $\mathbb{P}_u[\tau^a < \tau_0]$, then we would have to choose $\varrho = 0$, but under $\mathbb{P}$, $\psi(u) = 1$. So, the above approach does not work; we need to talk about it again. Indeed, We have to choose $a = +\infty$ under the case $\delta = 0$.

Let $\xi(u)$ be the solution to

$$D\xi'(u) + c\xi(u) - D\xi'(0) - \lambda \int_0^u \xi(u - y)[1 - G(y)]dy = 0. \tag{3.4}$$

with initial condition $\xi(0) = 0$, see (3.2).

Then $\zeta(u) = \xi(u)/\xi(a)$. Assuming that $\xi(\infty) = 1$ and $\xi'(\infty) = 0$, for $u \to \infty$ this yields the equation $c = D\xi'(0) - \lambda\mu$, from which it follows that

$$\xi'(0) = \frac{c - \lambda\mu}{D} = \rho\kappa, \tag{3.5}$$

with the notation $\kappa = c/D$. Hence the final form of (3.4) is

$$\xi'(u) + \kappa\xi(u) - \rho\kappa - \frac{\lambda}{D} \int_0^u \xi(u - y)[1 - G(y)]dy = 0. \tag{3.6}$$

Within the framework of Gerber's definitions [8], the formulation presented constitutes an advanced form of a defective renewal equation. Building upon this concept, the objective is to progress further by formulating a renewal equation in its more conventional understanding. Initially, Eq. (3.6) is manipulated by incorporating the integrating factor, followed by an integration process across a specified interval, for instance from zero to an unspecified upper limit.

$$\int_0^u e^{\kappa t}\xi'(t)dt - \int_0^u \kappa e^{\kappa t}(\rho - \xi(t))dt - \frac{\lambda}{D} \int_0^u \int_0^t e^{\kappa t}\xi(t - y)[1 - G(y)]dy = 0,$$

and get

$$e^{\kappa u}\xi(u) = \rho(e^{\kappa u} - 1) + \frac{\lambda}{D} \int_0^u \int_0^t e^{\kappa t}\xi(t - y)[1 - G(y)]dy. \tag{3.7}$$

We define that

$$h_1(x) = \kappa e^{-\kappa x}, h_2(x) = \frac{1}{\mu}[1 - G(x)], x > 0. \int_0^\infty h_i(x)dx = 1, i = 1,2.$$

So, $h_i(x)$, $i = 1,2$ are the probability density functions and denote by $H_i(x)$, $i = 1, 2$ the corresponding distribution functions. Multiplying $e^{-\kappa u}$ on both sides, then (3.7) can be written in the following more appealing form:

$$\xi(u) = \rho H_1(u) + (1 - \rho) \int_0^u \xi(u)h_1 * h_2(u - z)dz. \tag{3.8}$$

The equation in question represents a renewal equation with a deficiency. Consequently, it is amenable to the application of conventional renewal theory methodologies to derive the outcomes for Eq. (3.8).

An alternative approach is to take Laplace transforms in (3.8). Here we write:

$$\hat{\xi}(r) = \int_0^\infty e^{-ru} d\xi(u), \hat{\xi}_1(r) = \int_0^\infty e^{-ru} h_1(u) du = \frac{\kappa}{\kappa + r},$$

$$\hat{\xi}_2(r) = \int_0^\infty e^{-ru} h_2(u) du = \frac{1}{\mu} \int_0^\infty e^{-ru} [1 - G(u)] du = \frac{1}{\mu r} \left\{ 1 - \int_0^\infty e^{-ru} dG(u) \right\}.$$

Therefore, multiplying (3.8) by e^{-ru} and integrating over $u \in (0, \infty]$ follows that

$$\hat{\xi}(r) = \frac{\rho \hat{\xi}_1(r)}{1 - (1-\rho)\hat{\xi}_1(r)\hat{\xi}_2(r)}. \tag{3.9}$$

This statement extends the classical outcome to $D = 0$, where $\hat{\xi}_1(r) = 1$. The function $\xi(u)$, and by extension $\zeta(u)$, can be determined through the process of Laplace transform inversion, for example Cohen [3].

4 Numerical Simulation and Application of the EGS Function

Next, Firstly numerical simulations will be conducted on two conclusions.

Formula (3.8) is shown in the figure (Fig. 1). The initial surplus u value is [0,100], with 10 intervals, and the claims achievement rate is between 0.1, 0.2, 0.3, 0.4 and 0.5. For functions $h_1(x)$ and $h_2(x)$, a convolution operation is mainly performed. If the arrival rate of claims λ is set to 0.1, the mean distribution of claims size is 1, and the diffusion coefficient D is 0.1, the initial surplus u can be reversely obtained. It can be seen that when u is constant, the larger the value λ, the smaller $\zeta(u)$ becomes. Conversely, when u is constant, the larger and smaller $\zeta(u)$ becomes. When u is large enough, $\zeta(u)$ will approach 0 and be equal to 0.

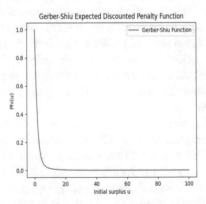

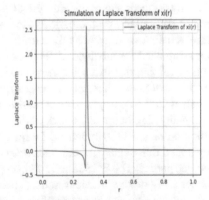

Fig. 1. Formula (3.8)   **Fig. 2.** Formula (3.9)

For formula (3.9), the key to simulation is to simulate the Laplace transform, first $h_1(x)$ and $h_2(x)$ perform the Laplace transform, and then perform the operation. During the simulation process, set $\kappa = 0.1, = 1, = 0.1$, and the output curves of r and $\widehat{\xi}(r)$ can be obtained as shown in the figure (Fig. 2). From the graph, $\widehat{\xi}(r)$ can be seen that as r increases, it first increases, then decreases, and finally remains constant. The values for ρ and r are 0.1, 0.2, 0.3, and 0, 0.2, 0.4, 0.6, and 1 respectively. The scale can be obtained as shown in the table below (Table 1). From the table, it can be seen that as r increases, the final ρ will decrease and stabilize around a constant value. When r is small, if ρ is relatively small, there will be a certain negative growth. When ρ increases, it will become a positive growth. However, when ρ increases to a certain extent, the final $\widehat{\xi}(r)$ will also decrease.

Table 1. Formula (3.9)

value	$r = 0$	$r = 0.25$	$r = 0.5$	$r = 0.75$	$r = 1$
$\rho = 0.1$	0.0	−0.075	0.02	0.013	0.0010
$\rho = 0.2$	0	−0.25	0.04	0.03	0.02
$\rho = 0.3$	0	−1.18	0.06	0.04	0.03
$\rho = 0.4$	0	1.42	0.08	0.05	0.04
$\rho = 0.5$	0	0.61	0.1	0.06	0.05

In order to further apply the EGS function in practice, this paper aims to strengthen the theoretical research activities on the under−explored implementation aspects, and to provide theoretical guidance for further numerical simulations. The EGS function does not involve too many technical details to avoid deviating from the main goal as a theoretical study.

While the EGS function boasts a wealth of features, its application in real-world scenarios remains limited. Our investigation highlights that the primary challenge lies in the intricate processes of numerical approximation and statistical inference, which become particularly problematic as the complexity of the models and their encompassed features increase. To effectively harness the EGS function in practical settings, there is a pressing need for advancements in both numerical and statistical methodologies. We are hopeful that this extensive review will serve as a catalyst for research that not only identifies but also enhances the most fitting numerical and statistical strategies for risk models with the appropriate structural nuances.

Equally significant is the lack of dedicated research focusing on the practical implementation of the EGS function. Despite its recognized potential, the process of mapping actual risk management issues onto the EGS function is not straightforward for all involved parties. The EGS function's comprehensive framework is well-suited to address a wider array of key practical issues, such as assessing solvency risks, managing credit risks, and serving as a measure of risk (expanding beyond the traditional ruin probability paradigm), among others. These practical studies could undoubtedly invigorate and diversify the research landscape surrounding the EGS function.

In conclusion, the development and application of numerical and statistical methods for the EGS function, particularly for the complex, large-scale problems encountered in industry, is far from a straightforward task. To foster the practical use and dissemination of the EGS function, it is not only effective but essential to facilitate the sharing of packaged code (in languages like R, Matlab or Python) through reputable source code repositories such as GitHub, making it freely accessible for public use. This approach could significantly ease the integration of the EGS function into various practical applications and contribute to its broader adoption.

References

1. Cai, J.: Ruin Probabilities and penalty functions with stochastic rates of interest. Stochast. Process. Appl. **112**, 53–78 (2004)
2. Cai, J., Feng, R., Willmot, G.E.: The compound Poisson surplus model with interest and liquid reserves: analysis of the Gerber-Shiu discounted penalty function. Methodol. Comput. Appl. Probab. **11**, 401–423 (2009)
3. Cohen, A.M.: Numerical Methods for Laplace transform inversion, pp. 227–235. Springer, New York (2007)
4. Davis, M.H.A.: Markov Models and Optimization. Chapman and Hall, London (1993)
5. Dynkin, E.B.: Markov Process (I). Springer-Verlag, Berlin (1965)
6. Dufresne, F., Gerber, H.U.: The surpluses immediately before and at ruin, and the amount of the claim causing ruin. Insurance: Math. Econ. **7**(3), 193–199 (1988). https://doi.org/10.1016/0167-6687(88)90076-5
7. Feller, W.: An Introduction to Probability Theory and its Applications, vol. II. Wiley, New York (1971)
8. Gerber, H.U.: An extension of the renewal equation and its application in the collective theory of risk. Scand. Actuar. J. **3–4**, 205–210 (1970)
9. Gerber, H.U.: Martingales in risk theory. Mitt. Ver. Schweiz. Vers. Math **73**, 205–216 (1973)
10. Gerber, H.U., Goovaerts, M.J., Kaas, R.: On the probability and severity of ruin. Astin Bull. **17**, 151–163 (1987)
11. Gerber, H.U., Shiu, E.S.W.: The joint distribution of the time of ruin, the surplus immediately before ruin, and the deficit at ruin. Insurance: Math. Econ. **21**(2), 129–137 (1997)
12. Gerber, H.U., Lin, X.S., Yang, H.: A note on the dividends-penalty identity and the optimal dividend barrier. Astin Bull. **36**, 489–503 (2006)
13. Kuznetsov, A., Morales, M.: Computing the finite-time expected discounted penalty function for a family of Lévy risk processes. Scand. Actuar. J. **1**, 1–31 (2014)
14. Rolski, T., Schmidli, H., Schmidt, V., Teugels, J.: Stochastic Processes for Insurance and Finance. Wiley, Chichester (1999)
15. Schmidli, H.: Extended Gerber-Shiu functions in a risk model with interest. Insurance Math. Econom. **61**, 271–275 (2015)

A Fuzzy-Based System for Assessment of Performance Error in VANETs Considering Environmental Stressors

Ermioni Qafzezi[1]($\boxtimes$), Kevin Bylykbashi[1], Shunya Higashi[2], Phudit Ampririt[2], Keita Matsuo[1], and Leonard Barolli[1]

[1] Department of Information and Communication Engineering, Fukuoka Institute of Technology (FIT), 3-30-1 Wajiro-Higashi, Higashi-Ku, Fukuoka 811-0295, Japan
qafzezi@bene.fit.ac.jp, {kt-matsuo,barolli}@fit.ac.jp
[2] Graduate School of Engineering, Fukuoka Institute of Technology (FIT), 3-30-1 Wajiro-Higashi, Higashi-Ku, Fukuoka 811-0295, Japan
bd21201@bene.fit.ac.jp

Abstract. This research paper presents a fuzzy system tailored for application in Vehicular Ad Hoc Networks (VANETs) to evaluate driver performance error. VANETs rely on real-time communication among vehicles and infrastructure to enhance road safety and traffic efficiency. The proposed system integrates four key input parameters: driver experience, skill level, mental and physical condition, and environmental stressors, with a focus on their relevance to VANET environments. Notably, the inclusion of environmental stressors, such as traffic congestion and adverse weather conditions, acknowledges the dynamic and unpredictable nature of VANET scenarios. Through fuzzy logic-based analysis, the system provides a nuanced evaluation of driver performance error, facilitating proactive measures to mitigate risks and improve overall traffic management in VANETs. This research contributes to the advancement of intelligent transportation systems by offering a comprehensive framework for assessing driver behavior and enhancing safety outcomes within VANET environments.

1 Introduction

In the realm of vehicular safety, human factors play a pivotal role in determining the outcomes of traffic incidents and accidents. While advancements in vehicle technology and infrastructure have undoubtedly contributed to improved safety standards, the behavior and decisions of drivers remain significant determinants of road safety. Understanding the intricate interplay between human factors and driving performance is essential for developing effective strategies to mitigate risks and enhance overall safety on the roads.

Human factors encompass a broad range of elements, including driver experience, skill level, mental and physical condition, and the ability to adapt to varying driving environments. These factors influence drivers' perception, decision-making processes, and reaction times, ultimately shaping their behavior on the road. Neglecting to account for these human-centric variables in the design of safety systems can lead to incomplete or ineffective solutions that fail to address the root causes of traffic incidents.

© The Author(s), under exclusive license to Springer Nature Switzerland AG 2024
L. Barolli (Ed.): IMIS 2024, LNDECT 214, pp. 148–156, 2024.
https://doi.org/10.1007/978-3-031-64766-6_15

In this context, the application of Fuzzy Logic (FL) presents a promising approach to modeling and analyzing human factors in vehicular safety systems. The FL provides a framework for reasoning and decision-making in the presence of uncertainty and imprecision, characteristics inherent in human behavior and environmental conditions. Unlike traditional binary logic systems, which rely on crisp distinctions between true and false states, FL allows for the representation of vague or ambiguous concepts, mirroring the complexities of real-world driving scenarios.

In a prior study [10], we introduced a FL-based framework focused on internal determinants, notably recognition errors, decision errors, and performance errors, as principal contributors to driver errors. Our current investigation aims to enrich the discourse surrounding accident causation and mitigation. This work undertakes an in-depth exploration of driver behavioral analysis within the context of Vehicular Ad Hoc Networks (VANETs), by investing the effect of the environmental stressors in the performance of the driver.

The paper's organization is as follows. Section 2 provides an overview of VANETs, tracing their evolution from traditional systems to cloud-fog-edge Software-Defined Networking (SDN) VANETs. In Sect. 3, we explain and emphasize the advantages of emerging technologies within VANETs, such as Network Slicing (NC) and SDN. In Sect. 4 we explore the design and implementation of a fuzzy system tailored for assessing driver performance error in Vehicular Ad Hoc Networks (VANETs), highlighting its relevance and efficacy in addressing the complexities of human-centric vehicular safety. The simulation results of the proposed fuzzy system for evaluating performance error of the driver are shown in Sect. 5. Lastly, Sect. 6 concludes the paper and offers insights into potential future work.

2 Overview of VANETs

VANETs represent a specialized form of Mobile Ad Hoc Networks (MANETs) where vehicles act as both communication nodes and data relays. In VANETs, vehicles communicate with each other and with roadside infrastructure to exchange real-time information, enhancing road safety, traffic efficiency, and driver convenience. These networks utilize Dedicated Short-Range Communication (DSRC) or Cellular Vehicle-to-Everything (C-V2X) technologies to facilitate communication among vehicles and infrastructure [2,3,9,11].

VANETs have a wide range of applications aimed at improving road safety and traffic management. These include cooperative collision warning systems, intersection collision avoidance, traffic signal optimization, congestion detection and mitigation, and Intelligent Transportation Systems (ITS). Additionally, VANETs enable Vehicle-to-Vehicle (V2V) and Vehicle-to-Infrastructure (V2I) communication, allowing for the dissemination of traffic information, emergency alerts, and navigation assistance.

Despite their potential benefits, VANETs face several challenges that hinder their widespread adoption and effectiveness. These challenges include issues related to network connectivity and coverage, communication latency, security and privacy concerns, scalability, and reliability in dynamic and harsh environments. Additionally, the high mobility of vehicles and the unpredictable nature of traffic patterns pose significant challenges for maintaining stable and robust communication links.

To overcome the challenges faced by VANETs, researchers and engineers are exploring various strategies and solutions. Integration of SDN and other emerging technologies such as Edge Computing, Blockchain, and Artificial Intelligence (AI) can enhance VANET performance and reliability. SDN enables centralized network management and dynamic resource allocation, improving network scalability, flexibility, and efficiency. Additionally, Edge Computing facilitates real-time data processing and decision-making at the network edge, reducing communication latency and improving responsiveness. Blockchain technology ensures data integrity, security, and privacy in VANET transactions, while AI algorithms optimize traffic flow, predict congestion, and enhance driver safety. Integrating these technologies into VANETs can address existing challenges and unlock the full potential of connected and autonomous vehicles in shaping the future of transportation.

3   Technological Advancement in VANETs

The concept of VANETs emerged as researchers explored ways to leverage inter-vehicle communication and roadside infrastructure connectivity for improving road safety and traffic efficiency. Initial efforts focused on developing communication protocols and standards, such as IEEE 802.11p, to enable V2V and V2I communication. These early VANET systems laid the foundation for subsequent advancements in the field.

As VANETs evolved, one notable technological advancement was the integration of Edge, Fog, and Cloud computing paradigms. This integration enables distributed computing and data processing capabilities at multiple network levels, from the network edge to intermediate Fog nodes and centralized Cloud servers. Edge computing facilitates real-time data processing and decision-making at the network periphery, reducing latency and improving responsiveness. Fog computing extends this capability to intermediate nodes, while Cloud computing provides scalable and resource-rich storage and processing capabilities.

The SDN emerged as a transformative technology in VANETs, offering centralized network management and dynamic resource allocation. SDN decouples the control plane from the data plane, enabling centralized control and programmability of network functions. In VANETs, SDN facilitates efficient traffic routing, congestion management, and Quality of Service (QoS) provisioning. It enables network operators to dynamically adapt to changing traffic conditions and optimize network performance in real-time.

AI technologies, including machine learning and deep learning, have significant implications for VANETs. AI algorithms can analyze large volumes of traffic data to extract valuable insights, predict traffic patterns, and optimize traffic flow. In VANETs, AI-powered systems can enhance driver safety by identifying potential hazards, predicting vehicle trajectories, and providing proactive collision avoidance recommendations. AI also plays a crucial role in anomaly detection, security threat analysis, and intrusion detection in VANET environments.

Looking ahead, the integration of Edge-Fog-Cloud computing, SDN, and AI technologies will continue to shape the evolution of VANETs. Future research efforts will focus on optimizing resource allocation, enhancing network scalability, and improving security and privacy mechanisms. Additionally, addressing interoperability issues and standardization challenges will be critical to fostering widespread adoption of VANET technologies. Ultimately, by leveraging technological advancements and addressing emerging challenges, VANETs have the potential to revolutionize transportation systems, making roads safer, more efficient, and environmentally sustainable.

4 Proposed Fuzzy-Based System

The proposed system leverages FL as its foundation for assessing driver performance within the context of VANETs. FL proves to be a robust and suitable approach due to its ability to handle imprecise and uncertain data, a common characteristic of VANET environments [1,4–8,12–14]. Unlike traditional binary logic systems, FL accommodates the inherently vague and fluctuating nature of driver behavior and environmental conditions encountered on the road. This flexibility makes FL well-suited for modeling the complex and dynamic interactions between drivers, vehicles, and infrastructure in VANETs.

In the present study, we present an innovative driving support system named the Fuzzy System for Assessment of Performance Errors (FS-APE). With a primary aim of enhancing driving safety, FS-APE emerges as a cutting-edge solution for real-time assessment of performance errors. In the context of VANETs, FL excels in capturing the nuances of driver performance by considering multiple input parameters, such as driver experience, skill level, mental and physical condition, and environmental stressors. FL's linguistic variables and fuzzy sets allow for the representation of these parameters in a more natural and intuitive manner, mirroring human cognition and decision-making processes. By defining linguistic terms such as "low," "moderate," and "high" for each input parameter, FL effectively handles the uncertainty and variability inherent in driver behavior and environmental conditions. Furthermore, FL's rule-based inference engine enables the system to derive meaningful conclusions about driver performance based on the FL rules and membership functions defined for each parameter. Through this approach, the proposed system provides a comprehensive and nuanced assessment of driver performance in VANETs, contributing to improved road safety and traffic management.

Table 1. FS-APE parameters and their term sets.

Parameters	Term Sets
Experience (EX)	Inexperienced (InE), Moderate (MoE), Experienced (Exp)
Skill (SK)	Low (Lo), Moderate (Mo), High (Hi)
Mental and Physical Condition (MPC)	Excellent (Ex), Good (Go), Poor (Po))
Environmental Stressors (ES)	Mild (Mi), Moderate (Mod), Severe (Se)
Performance Error (PE)	PE1, PE2, PE3, PE4, PE5, PE6, PE7

We explain in detail the input and output parameters in following.

Experience (EX): It captures the driver's familiarity with various driving scenarios and road conditions. It encompasses factors such as years of driving experience, exposure to

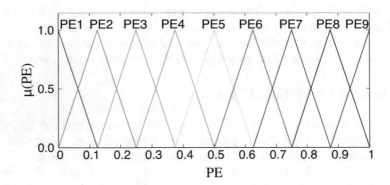

Fig. 1. Proposed system structure.

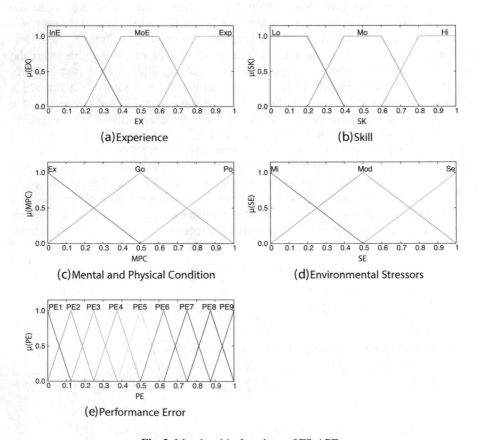

Fig. 2. Membership functions of FS-APE.

different environments (urban, rural, highway), and previous encounters with challenging situations (e.g., inclement weather, heavy traffic). Drivers with extensive experience are often better equipped to anticipate and react to potential hazards, while those with limited experience may exhibit uncertainty or hesitancy in unfamiliar driving situations.

Skill (SK): It reflects the driver's proficiency in executing driving maneuvers and handling diverse road conditions. It considers factors such as vehicle control, spatial awareness, reaction times, and adherence to traffic regulations. Skilled drivers demonstrate precise control over their vehicles, make informed decisions quickly, and exhibit smooth and confident driving behaviors. In contrast, drivers with lower skill levels may struggle with vehicle handling, exhibit erratic driving patterns, or make errors in judgment.

Mental and Physical Condition (MPC): This parameter encompasses the driver's cognitive and physical well-being during the driving task. It considers factors such as alertness, fatigue, stress levels, emotional state, and physical fitness. Drivers in optimal mental and physical condition are more attentive, focused, and capable of responding to changing road conditions effectively. Conversely, drivers experiencing fatigue, stress, or distraction may exhibit impaired reaction times, reduced situational awareness, and increased risk of errors or accidents.

Environmental Stressors (ES): It represents external factors that may impact driver performance and safety on the road. It includes variables such as traffic congestion, adverse weather conditions, road construction, visibility limitations, and the presence of obstacles or hazards. These stressors introduce additional challenges and complexities to driving environments, potentially affecting driver behavior and decision-making processes. Understanding and accounting for these environmental stressors are crucial for assessing driver performance accurately and implementing appropriate interventions to mitigate risks.

Performance Error (PE): It quantifies deviations from expected or optimal driving behavior exhibited by the driver. It serves as a measure of driver errors, lapses in judgment, or deficiencies in performance observed during the driving task. Performance errors can manifest in various forms, including improper lane changes, failure to yield, speeding, erratic braking, or other violations of traffic laws. By analyzing performance errors, the system can identify patterns, trends, and areas for improvement in driver behavior, leading to targeted interventions and strategies aimed at enhancing road safety within the VANET environment.

The architecture of the proposed FS-APE system is illustrated in Fig. 1. The term sets for both input and output parameters are provided in Table 1. Visual representations of the membership functions governing the input and output parameters are shown in Fig. 2, alongside the Fuzzy Rule Base (FRB) shown in Table 2.

5   Simulation Results

The proposed FS-APE system simulations were conducted utilizing FuzzyC, with results across four distinctive scenarios marked by divergent levels of driver experience and skill sets, illustrated in Figs. 3(a)–(d). The findings elucidate the correlation

Table 2. FRB of FS-ARE.

No	EX	SK	MPC	ES	PE	No	EX	SK	MPC	ES	PE	No	EX	SK	MPC	ES	PE
1	InE	Lo	Ex	Mi	PE6	28	MoE	Lo	Ex	Mi	PE3	55	Exp	Lo	Ex	Mi	PE2
2	InE	Lo	Ex	Mod	PE7	29	MoE	Lo	Ex	Mod	PE5	56	Exp	Lo	Ex	Mod	PE3
3	InE	Lo	Ex	Se	PE8	30	MoE	Lo	Ex	Se	PE6	57	Exp	Lo	Ex	Se	PE4
4	InE	Lo	Go	Mi	PE7	31	MoE	Lo	Go	Mi	PE4	58	Exp	Lo	Go	Mi	PE3
5	InE	Lo	Go	Mod	PE8	32	MoE	Lo	Go	Mod	PE6	59	Exp	Lo	Go	Mod	PE4
6	InE	Lo	Go	Se	PE9	33	MoE	Lo	Go	Se	PE7	60	Exp	Lo	Go	Se	PE5
7	InE	Lo	Po	Mi	PE9	34	MoE	Lo	Po	Mi	PE5	61	Exp	Lo	Po	Mi	PE4
8	InE	Lo	Po	Mod	PE9	35	MoE	Lo	Po	Mod	PE7	62	Exp	Lo	Po	Mod	PE5
9	InE	Lo	Po	Se	PE9	36	MoE	Lo	Po	Se	PE8	63	Exp	Lo	Po	Se	PE7
10	InE	Mo	Ex	Mi	PE5	37	MoE	Mo	Ex	Mi	PE3	64	Exp	Mo	Ex	Mi	PE1
11	InE	Mo	Ex	Mod	PE6	38	MoE	Mo	Ex	Mod	PE4	65	Exp	Mo	Ex	Mod	PE2
12	InE	Mo	Ex	Se	PE7	39	MoE	Mo	Ex	Se	PE5	66	Exp	Mo	Ex	Se	PE3
13	InE	Mo	Go	Mi	PE6	40	MoE	Mo	Go	Mi	PE4	67	Exp	Mo	Go	Mi	PE2
14	InE	Mo	Go	Mod	PE7	41	MoE	Mo	Go	Mod	PE5	68	Exp	Mo	Go	Mod	PE3
15	InE	Mo	Go	Se	PE8	42	MoE	Mo	Go	Se	PE6	69	Exp	Mo	Go	Se	PE4
16	InE	Mo	Po	Mi	PE7	43	MoE	Mo	Po	Mi	PE5	70	Exp	Mo	Po	Mi	PE3
17	InE	Mo	Po	Mod	PE8	44	MoE	Mo	Po	Mod	PE6	71	Exp	Mo	Po	Mod	PE4
18	InE	Mo	Po	Se	PE9	45	MoE	Mo	Po	Se	PE7	72	Exp	Mo	Po	Se	PE6
19	InE	Hi	Ex	Mi	PE4	46	MoE	Hi	Ex	Mi	PE2	73	Exp	Hi	Ex	Mi	PE1
20	InE	Hi	Ex	Mod	PE5	47	MoE	Hi	Ex	Mod	PE3	74	Exp	Hi	Ex	Mod	PE1
21	InE	Hi	Ex	Se	PE6	48	MoE	Hi	Ex	Se	PE4	75	Exp	Hi	Ex	Se	PE3
22	InE	Hi	Go	Mi	PE5	49	MoE	Hi	Go	Mi	PE3	76	Exp	Hi	Go	Mi	PE2
23	InE	Hi	Go	Mod	PE6	50	MoE	Hi	Go	Mod	PE5	77	Exp	Hi	Go	Mod	PE2
24	InE	Hi	Go	Se	PE7	51	MoE	Hi	Go	Se	PE6	78	Exp	Hi	Go	Se	PE4
25	InE	Hi	Po	Mi	PE6	52	MoE	Hi	Po	Mi	PE4	79	Exp	Hi	Po	Mi	PE2
26	InE	Hi	Po	Mod	PE7	53	MoE	Hi	Po	Mod	PE6	80	Exp	Hi	Po	Mod	PE3
27	InE	Hi	Po	Se	PE8	54	MoE	Hi	Po	Se	PE7	81	Exp	Hi	Po	Se	PE5

between PE and SE across various MPC values, considering EX and SK as constant parameters.

When dealing with an inexperienced and low-skilled driver (EX = 0.1, SK = 0.1), as shown in Fig. 3(a), the results show the highest values of Performance Error (PE). This suggests that these drivers demonstrate a high likelihood of performance errors, This outcome can be attributed to the lack of driving proficiency and familiarity with diverse road conditions, coupled with limited skills in handling challenging situations. Inexperienced and low-skilled drivers may struggle to adapt to environmental stressors, leading to increased errors and lapses in performance.

Conversely, when drivers possess high skill levels despite their inexperience, lower performance error values are observed. However, when the mental and physical condition is exceptionally high (0.9), shown in Fig. 3(b), output performance error values exceeding 0.5 are observed. This suggests that despite their skillfulness, drivers may still face challenges in maintaining optimal performance under extreme mental or physical stress, potentially placing them in dangerous situations.

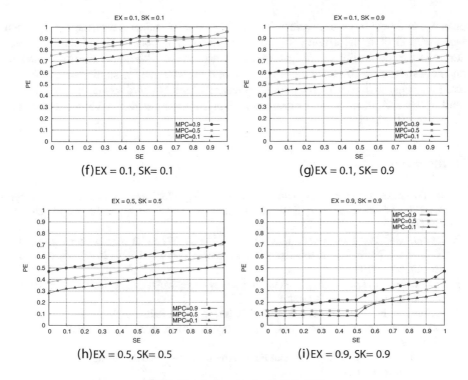

Fig. 3. Simulation results for FS-APE.

For drivers with moderate experience and skill levels (EX = 0.5, SK = 0.5), shown in Fig. 3(c), higher performance error values are observed when both mental and physical condition and environmental stressors are high. This indicates that even drivers with moderate proficiency may struggle to cope with demanding driving conditions, particularly when faced with elevated levels of stress and environmental challenges.

In contrast, for drivers with high experience and skill levels (EX = 0.9, SK = 0.9), shown in Fig. 3(d), the lowest values of performance error are observed. This outcome is consistent with expectations, as drivers with great experience and skill are less likely to encounter challenging situations that could result in performance errors. Their proficiency allows them to effectively manage environmental stressors and minimize errors in performance, resulting in the lowest observed values of performance error in the FL-based system.

6 Conclusions

In conclusion, the results of the analysis highlight the intricate relationship between driver experience, skill level, mental and physical condition, environmental stressors, and performance error within the context of the FL-based system. The findings underscore the critical importance of considering these factors in assessing driver behavior and predicting potential errors on the road. Notably, inexperienced and low-skilled drivers are shown to be particularly vulnerable to performance errors, especially when faced with challenging driving conditions. Conversely, experienced and skilled drivers demonstrate the ability to mitigate performance errors, although extreme mental or physical stress may still pose risks. These insights emphasize the significance of proactive measures, such as driver training and environmental hazard mitigation, to enhance road safety and minimize the occurrence of performance errors in Vehicular Ad Hoc Networks (VANETs).

References

1. Bojadziev, G., Bojadziev, M., Zadeh, L.A.: Fuzzy logic for business, finance, and management. In: Advances in Fuzzy Systems - Applications and Theory, vol. 12. World Scientific (1997)
2. Hartenstein, H., Laberteaux, K.P. (eds.): VANET: Vehicular Applications and Inter-Networking Technologies. Intelligent Transportation Systems. Wiley (2010). https://doi.org/10.1002/9780470740637
3. Hartenstein, H., Laberteaux, L.: A tutorial survey on vehicular ad hoc networks. IEEE Commun. Mag. **46**(6), 164–171 (2008)
4. Kandel, A.: Fuzzy Expert Systems. CRC Press Inc., Boca Raton (1992)
5. Klir, G.J., Folger, T.A.: Fuzzy Sets, Uncertainty, and Information. Prentice Hall, Upper Saddle River (1988)
6. Klir, G.J., Yuan, B.: Fuzzy Sets and Fuzzy Logic - Theory and Applications. Prentice Hall, Upper Saddle River (1995)
7. McNeill, F.M., Thro, E.: Fuzzy Logic: A Practical Approach. Academic Press Professional Inc., San Diego (1994)
8. Munakata, T., Jani, Y.: Fuzzy systems: an overview. Commun. ACM **37**(3), 69–77 (1994)
9. Peixoto, M.L.M., et al.: FogJam: a fog service for detecting traffic congestion in a continuous data stream VANET. Ad Hoc Netw. **140**, 103046 (2023). https://doi.org/10.1016/j.adhoc.2022.103046
10. Qafzezi, E., Bylykbashi, K., Higashi, S., Ampririt, P., Matsuo, K., Barolli, L.: A fuzzy-based error driving system for improving driving performance in VANETs. In: Barolli, L. (ed.) CISIS 2023. LNDECT, vol. 176, pp. 161–169. Springer, Cham (2023). https://doi.org/10.1007/978-3-031-35734-3_16
11. Schünemann, B., Massow, K., Radusch, I.: Realistic simulation of vehicular communication and vehicle-2-x applications. In: Proceedings of the 1st International Conference on Simulation Tools and Techniques for Communications, Networks and Systems & Workshops, SimuTools 2008, Marseille, France, 3–7 March 2008, p. 62. ICST/ACM (2008). https://doi.org/10.4108/ICST.SIMUTOOLS2008.2949
12. Zadeh, L.A., Kacprzyk, J.: Fuzzy Logic for the Management of Uncertainty. Wiley, New York (1992)
13. Zadeh, L.A., Klir, G.J., Yuan, B.: Fuzzy sets, fuzzy logic, and fuzzy systems - selected papers by Lotfi A Zadeh. In: Advances in Fuzzy Systems - Applications and Theory, vol. 6. World Scientific (1996). https://doi.org/10.1142/2895
14. Zimmermann, H.J.: Fuzzy control. In: Fuzzy Set Theory and Its Applications, pp. 203–240. Springer, Dordrecht (1996). https://doi.org/10.1007/978-94-015-8702-0_11

A Comparison Study for Different Number of Mesh Routers and Small Scale WMNs Considering Subway Distribution of Mesh Clients and Three Router Replacement Methods

Admir Barolli[1], Evjola Spaho[2], Shinji Sakamoto[3], Leonard Barolli[4(✉)], and Makoto Takizawa[5]

[1] Department of Information Technology, Aleksander Moisiu University of Durres, L.1, Rruga e Currilave, Durres, Albania
admirbarolli@uamd.edu.al

[2] Department of Electronics and Telecommunication, Faculty of Information Technology, Polytechnic University of Tirana, Mother Teresa Square, No. 4, Tirana, Albania
espaho@fti.edu.al

[3] Department of Information and Computer Science, Kanazawa Institute of Technology, 7-1 Ohgigaoka Nonoichi, Ishikawa 921-8501, Japan
shinji.sakamoto@ieee.org

[4] Department of Information and Communication Engineering, Fukuoka Institute of Technology, 3-30-1 Wajiro-Higashi, Higashi-Ku, Fukuoka 811-0295, Japan
barolli@fit.ac.jp

[5] Department of Advanced Sciences, Faculty of Science and Engineering, Hosei University, 3-7-2, Kajino-machi, Koganei-shi, Tokyo 184-8584, Japan
makoto.takizawa@computer.org

Abstract. In our previous work, we implemented a hybrid intelligent simulation system for Wireless Mesh Networks (WMNs) called WMN-PSOHCDGA by integrating Particle Swarm Optimization (PSO), Hill Climbing (HC) and Distributed Genetic Algorithm (DGA). We carried out many simimulation scenarios, but we used the same number of mesh routers in order to make a fair comparison. However, we found that in many cases some routers were concentrated in some areas. So, it is possible to use less number of mesh routers to cover all mesh clients. In this paper, we consider Subway distribution of mesh clients and samll scale WMNs. We carry out simulations for three router replacement methods: Linearly Decreasing Inertia Weight Method (LDIWM), Linearly Decreasing Vmax Method (LDVM) and Fast Convergence Rational Decrement of Vmax Method (FC-RDVM) and two scenarios where the number of mesh routers is 10 and 16. We evaluate the performance for these two scenarios considering three metrics: Size of Giant Component (SGC), Number of Covered Mesh Clients (NCMC) and Number of Covered Mesh Clients per Router (NCMCpR).

1 Introduction

Wireless Mesh Networks (WMNs) are a promising solution for realizing anywere and anytime connectivity. They have a variety of applications and because of low implementation time they can be used in disaster recover situations. The WMNs are cost-effective

© The Author(s), under exclusive license to Springer Nature Switzerland AG 2024
L. Barolli (Ed.): IMIS 2024, LNDECT 214, pp. 157–169, 2024.
https://doi.org/10.1007/978-3-031-64766-6_16

networks and can be used in different scenarios for last-mile networks and Metropolitan Area Networks (MANs). They are especially suitable for linking edge devices in IoT networks.

In order to have a good connectivity and better client coverage, in the design process of WMNs, it is very important to optimize the position of mesh routers. However, in the design process should be considered many indpended parameters, which makes the problem NP-Hard.

In the optimization problem in WMNs, we consider a specified number of mesh routers and mesh clients scattered within a considered area. The goal is to optimize network connectivity and to have a better coverage of mesh clients, while balancing the load of mesh routers. For the optimization process, we consider three metrics: the Size of Giant Component (SGC), the Number of Covered Mesh Clients (NCMC) and the Number of Covered Mesh Clients per Router (NCMCpR).

The allocation and placement of mesh nodes in WMNs is discussed in many research works [2,4–6,13,14]. In some previous works, different authors have proposed intelligent algorithms to deal with node placement problem [1,3,7,8]. In [9,10], we considered simple heuristic algorithms for implementing intelligent simulations systems.

In our previous work, we implementated a hybrid intelligent simulation system for WMNs called WMN-PSOHCDGA by integrating Particle Swarm Optimization (PSO), Hill Climbing (HC) and Distributed Genetic Algorithm (DGA). We carried out many simimulation scenarios, but we used the same number of mesh routers in order to make a fair comparison. However, we found that in many cases some routers were concentrated in some areas. So, it is possible to use less number of mesh routers to cover all mesh clients.

In this paper, we consider Subway distribution of mesh clients and small scale WMNs. We carry out simulations for three router replacement methods: Linearly Decreasing Inertia Weight Method (LDIWM), Linearly Decreasing Vmax Method (LDVM) and Fast Convergence Rational Decrement of Vmax Method (FC-RDVM) and two scenarios where the number of mesh routers are 10 and 16.

The paper is organized as follows. In Sect. 2, we present the implemented hybrid simulation system. The simulation results are given in Sect. 3. Finally, we give conclusions and future work in Sect. 4.

2   WMN-PSOHCDGA System

We consider the advantages of PSO, HC and DGA algorithms to implement WMN-PSOHCDGA hybrid simulation system. The flowchart of WMN-PSOHCDGA simulation system is illustrated in Fig. 1. The initial solution are generated randomly by *ad hoc* methods [15]. Then, for each island is carried out the optimization process. In the next step, solutions are migrated between islands.

The DGA algorithm is used for the optimization process in each island (see Fig. 2). When a solution is found in DGA part is migrated to PSO part, which updates the velocities and positions of particles by using router replacement methods. In order to improve the convergance of PSO, we consider HC algorithm.

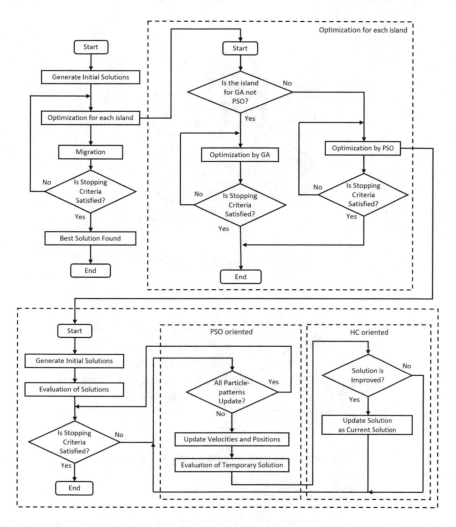

Fig. 1. Flowchart of WMN-PSOHCDGA system.

In our approach, a particle is considered as a mesh router. The position of mesh routers and mesh clients are utilized in order to calculate the fitness value of a particle-pattern shown in Fig. 3. Each individual is a combination of mesh routers that forms a WMN.

The fitness function has the following form.

$$Fitness = \alpha \times SGC(\boldsymbol{x}_{ij}, \boldsymbol{y}_{ij}) + \beta \times NCMC(\boldsymbol{x}_{ij}, \boldsymbol{y}_{ij}) \qquad (1)$$
$$+ \gamma \times NCMCpR(\boldsymbol{x}_{ij}, \boldsymbol{y}_{ij}).$$

In this research work, we consider Subway distribution of mesh clients, which is shown in Fig. 4. In this distribution, the mesh clients are allocated in the same way as the people are distributed in a city subway.

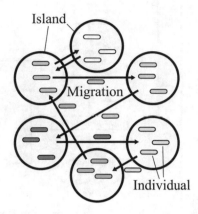

Fig. 2. Model of migration in DGA.

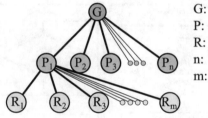

G: Global Solution
P: Particle-pattern
R: Mesh Router
n: Number of Particle-patterns
m: Number of Mesh Routers

Fig. 3. Relation of global solution and particle-patterns.

Fig. 4. Subway distribution of mesh clients.

There are many mesh router replacement methods. We use in this research work: LDIWM, LDVM and FC-RDVM.

In LDIWM, C_1 and C_2 are constant parameters and their value is set to 2.0. While, ω parameter values are changed linearly from 0.9 to 0.4 (from unstable region to stable region).
In LDVM, the PSO parameter values are set to unstable region ($\omega = 0.9$, $C_1 = C_2 = 2.0$). The V_{max} value (the maximum velocity of particles) is decreased linearly with the increase of T value as shown in Eq. (2) [12].

$$V_{max}(k) = \sqrt{W^2 + H^2} \times \frac{T-k}{T} \qquad (2)$$

In Eq. (2), W and H represent the width and the height of the considered area, T is the total number of iterations and k is a variable varying from 1 to T.
In FC-RDVM [11], the V_{max} value decreases with the increase of T, k and δ values as shown in Eq. (3).

$$V_{max}(k) = \sqrt{W^2 + H^2} \times \frac{T-k}{T+\delta k} \qquad (3)$$

In Eq. (3), the variables are the same with other methods, while δ is a curvature parameter.

3   Simulation Results

In this work, we present the simulation results of three router replacement methods implemented in WMN-PSOHCDGA system. The coefficients of the fitness function are set as $\alpha = 0.1$, $\beta = 0.8$, $\gamma = 0.1$. The other parameters considered for simulations are shown in Table 1.

Table 1. Simulation parameters used for simulations.

Parameters	Values	
	Scenario 1	Scenario 2
Coefficients of Fitness Function $\alpha : \beta : \gamma$	1 : 8 : 1	
GA Islands Number	16	
Evolution Steps Number	9	
Migration Number	200	
Mesh Routers Number	16	10
Mesh Clients Number	48	48
Distribution of Mesh Clients	Subway Distribution	
GA Selection Method	Rulette Selection Method	
GA Corssover Method	SPX	
GA Mutation Method	Uniform Mutation	

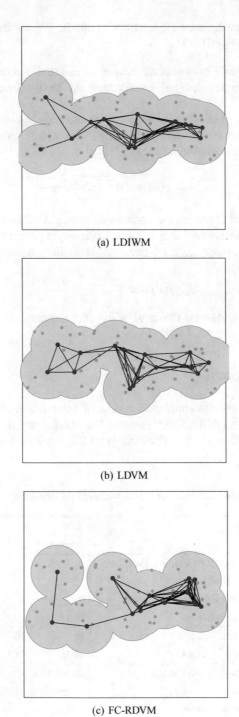

(a) LDIWM

(b) LDVM

(c) FC-RDVM

Fig. 5. Visualization results after optimization (Small Scale WMN: 16 mesh routers).

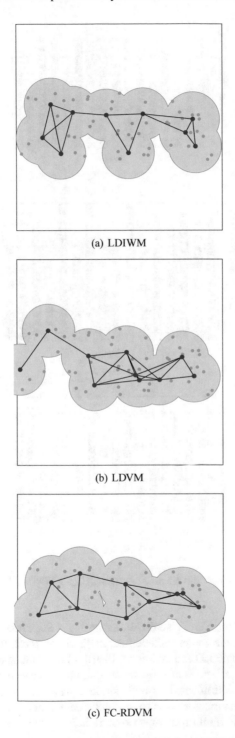

(a) LDIWM

(b) LDVM

(c) FC-RDVM

Fig. 6. Visualization results after optimization (Small Scale WMN: 10 mesh routers).

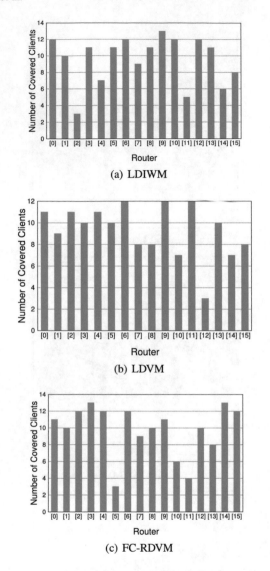

(a) LDIWM

(b) LDVM

(c) FC-RDVM

Fig. 7. Number of covered mesh clients (Small Scale WMN: 16 mesh routers).

In Fig. 5, we show the SGC visualization results for Scenario 1, where we consider 16 mesh routers. For three router replacement methods, all mesh routers are connected, so the SGC is 100%. The LDVM has better distribution of mesh routers than LDIWM and FC-RDVM, but in all cases in some areas, there is a concentration of mesh routers.

We show the simulation results for Scenario 2 where we use 10 mesh routers in Fig. 6. Also, in this scenario for three router replacement methods, all mesh routers are connected. But, the distribution of mesh routers is better than Scenario 1. Thus, with

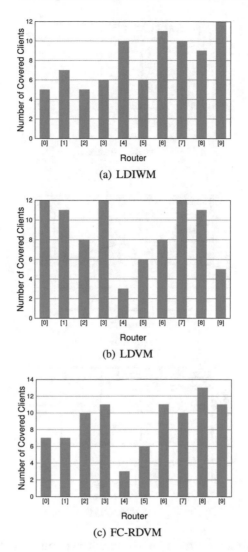

(a) LDIWM

(b) LDVM

(c) FC-RDVM

Fig. 8. Number of covered mesh clients (Small Scale WMN: 10 mesh routers).

less number of mesh routers are covered all mesh clients. This results in lower cost of the WMN compared with case of Scenario 1.

In Fig. 7 and Fig. 8 are shown the simulation results of NCMC for each router in case of Scenario 1 and Scenario 2. In both scenarios, all mesh clients are covered for three methods, but each mesh router covers different number of mesh clients.

The relationship between standard deviation with the number of updates for three methods in case of Scenario 1 and Scenario 2 is shown in Fig. 9 and Fig. 10, respectively. In these figures are shown the data, regression line and r (correlation coefficient).

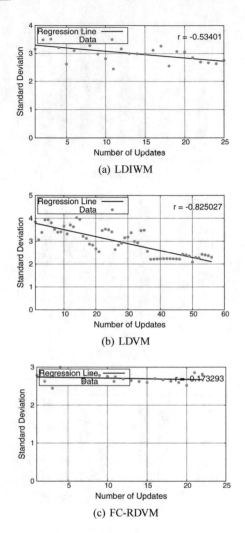

(a) LDIWM

(b) LDVM

(c) FC-RDVM

Fig. 9. Standard deviation, regression line and correlation coefficient (Small Scale WMN).

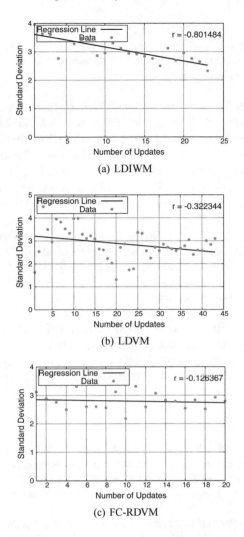

(a) LDIWM

(b) LDVM

(c) FC-RDVM

Fig. 10. Standard deviation, regression line and correlation coefficient (Small Scale WMN).

It should be noted that when the standard deviation is a decreasing line, the load balancing between routers is improving. In Fig. 9, for Scenario 1, the load balancing of LDVM is better than LDIWM and FC-RDVM. While for Scenario 2 in Fig. 10, the load balancing of LDIWM is better than LDVM and FC-RDVM.

4 Conclusions

In this research work, we presented a comparison study of three router replacement methods for optimization of WMNs. We considered Subway distribution of mesh clients and two scenarios for samll scale WMNs. We carried out simulations for three router replacement methods: LDIWM, LDVM and FC-RDVM. In Scenario 1, the number of mesh routers was 16, while in Scenario 2 was 10. We evaluated the performance for these two scenarios considering SGC, NCMC and NCMCpR metrics. From the simulation results, we conclude as follows.

- For SGC in Scenario 1, where we considered 16 mesh routers, for three router replacement methods, all mesh routers were connected. The LDVM had better distribution of mesh routers than LDIWM and FC-RDVM, but in all cases in some areas, we found concentration of mesh routers. In case of Scenario 2 where we used 10 mesh routers also all mesh routers were connected. But, the distribution of mesh routers is better than Scenario 1. Thus, with less number of mesh routers are covered all the mesh clients, which results in lower cost of the WMN compared with case of Scenario 1.
- Considering NCMC, for both scenarios, all mesh clients are covered for three methods, but each mesh router covers different number of mesh clients.
- Considering NCMCpR, for Scenario 1, the load balancing of LDVM is better than LDIWM and FC-RDVM. While for Scenario 2, the load balancing of LDIWM is better than LDVM and FC-RDVM.

In future work, we will consider the implementation of different crossover, mutation and other router replacement methods. We will evaluate the performance of the implemented system for different number of phases and migrations. Furthermore, we plan to implement a testbed and compare the simulation results with experimental results.

References

1. Barolli, A., Sakamoto, S., Ozera, K., Barolli, L., Kulla, E., Takizawa, M.: Design and implementation of a hybrid intelligent system based on particle swarm optimization and distributed genetic algorithm. In: Barolli, L., Xhafa, F., Javaid, N., Spaho, E., Kolici, V. (eds.) EIDWT 2018. LNDECT, vol. 17, pp. 79–93. Springer, Cham (2018). https://doi.org/10.1007/978-3-319-75928-9_7
2. Franklin, A.A., Murthy, C.S.R.: Node placement algorithm for deployment of two-tier wireless mesh networks. In: Proceedings of the Global Telecommunications Conference, pp. 4823–4827 (2007)
3. Girgis, M.R., Mahmoud, T.M., Abdullatif, B.A., Rabie, A.M.: Solving the wireless mesh network design problem using genetic algorithm and simulated annealing optimization methods. Int. J. Comput. Appl. **96**(11), 1–10 (2014)
4. Lim, A., Rodrigues, B., Wang, F., Xu, Z.: k-Center problems with minimum coverage. Theor. Comput. Sci. **332**(1–3), 1–17 (2005)
5. Maolin, T., et al.: Gateways placement in backbone wireless mesh networks. Int. J. Commun. Netw. Syst. Sci. **2**(1), 44–50 (2009)
6. Muthaiah, S.N., Rosenberg, C.P.: Single gateway placement in wireless mesh networks. In: Proceedings of the 8th International IEEE Symposium on Computer Networks, pp. 4754–4759 (2008)
7. Naka, S., Genji, T., Yura, T., Fukuyama, Y.: A hybrid particle swarm optimization for distribution state estimation. IEEE Trans. Power Syst. **18**(1), 60–68 (2003)

8. Sakamoto, S., Kulla, E., Oda, T., Ikeda, M., Barolli, L., Xhafa, F.: A comparison study of simulated annealing and genetic algorithm for node placement problem in wireless mesh networks. J. Mob. Multimed. **9**(1–2), 101–110 (2013)
9. Sakamoto, S., Kulla, E., Oda, T., Ikeda, M., Barolli, L., Xhafa, F.: A comparison study of hill climbing, simulated annealing and genetic algorithm for node placement problem in WMNs. J. High Speed Netw. **20**(1), 55–66 (2014)
10. Sakamoto, S., Oda, T., Ikeda, M., Barolli, L., Xhafa, F.: Implementation and evaluation of a simulation system based on particle swarm optimisation for node placement problem in wireless mesh networks. Int. J. Commun. Netw. Distrib. Syst. **17**(1), 1–13 (2016)
11. Sakamoto, S., Barolli, A., Liu, Y., Kulla, E., Barolli, L., Takizawa, M.: A fast convergence RDVM for router placement in WMNs: performance comparison of FC-RDVM with RDVM by WMN-PSOHC hybrid intelligent system. In: Barolli, L. (eds.) Complex, Intelligent and Software Intensive Systems. CISIS 2022. LNCS, vol. 497, pp. 17–25. Springer, Cham (2022). https://doi.org/10.1007/978-3-031-08812-4_3
12. Schutte, J.F., Groenwold, A.A.: A study of global optimization using particle swarms. J. Glob. Optim. **31**(1), 93–108 (2005)
13. Vanhatupa, T., Hannikainen, M., Hamalainen, T.: Genetic algorithm to optimize node placement and configuration for WLAN planning. In: Proceedings of the 4th IEEE International Symposium on Wireless Communication Systems, pp. 612–616 (2007)
14. Wang, J., Xie, B., Cai, K., Agrawal, D.P.: Efficient mesh router placement in wireless mesh networks. In: Proceedings of the IEEE International Conference on Mobile Adhoc and Sensor Systems (MASS-2007), pp. 1–9 (2007)
15. Xhafa, F., Sanchez, C., Barolli, L.: Ad hoc and neighborhood search methods for placement of mesh routers in wireless mesh networks. In: Proceedings of the 29th IEEE International Conference on Distributed Computing Systems Workshops (ICDCS-2009), pp. 400–405 (2009)

Safety Assurance of Omnidirectional Wheelchair Robot for Playing Badminton Game

Keita Matsuo[✉][iD] and Leonard Barolli[iD]

Department of Information and Communication Engineering, Fukuoka Institute of Technology
(FIT), 3-30-1 Wajiro-Higashi, Higashi-Ku, Fukuoka 811-0295, Japan
{kt-matsuo,barolli}@fit.ac.jp

Abstract. Globally, there are many people with disabilities. Recently, a variety of facilities and equipment have been innovated, catering not only the disabled individuals but also the aging population. Among these, the wheelchair is one of the most commonly utilized tools in diverse scenarios for disabled and elderly individuals. In this paper, we deal with the safety assurance of omniditectional wheelchair robot for playing badminton. We conducted experiments to control multiple omnidirectional robots and prevent collisions. We conducted experiments by placing reference points, where the robot should pass through with an error margin of a few centimeters at 1 [m] intervals. Additionally, we investigated the trajectory when the reference points were set at 50 [cm] intervals and when there are no reference points. The evaluation results in case when we did not use intervals are better than other results (when using point intervals of 0.5 [m] and 1 [m]).

1 Introduction

As per the World Health Organization (WHO) [1], there are more than 1.3 billion people with disabilities globally, constituting around 15% of the world population. Recently, a variety of facilities and equipment have been innovated, catering not only to disabled individuals but also to the aging population. Among these, the wheelchair is one of the most commonly utilized tools in diverse scenarios for disabled and elderly individuals.

Approximately 132 million people with disabilities, constituting around 1.86% of the global population, are in need of wheelchair across 34 developed and 156 developing countries. Additionally, numerous individuals face health challenges associated with motor disabilities, affecting their control over limbs or even head movements [6]. Maintaining a healthy and active lifestyle is crucial for everyone. Engaging in sports can significantly contribute to a better quality of life, and there are various activities accessible for individuals who use wheelchairs.

ⓒ The Author(s), under exclusive license to Springer Nature Switzerland AG 2024
L. Barolli (Ed.): IMIS 2024, LNDECT 214, pp. 170–178, 2024.
https://doi.org/10.1007/978-3-031-64766-6_17

Until now, various wheelchairs with diverse functionalities have been developed and they can support humans for the daily life. For example, the wheelchair with navigation system [4,16,17] can help users by taking them to a place they want. Head moving interface may drive wheelchair based on inertial sensors [7]. The Brain Computer Interface (BCI) is particularly beneficial for disabled individuals who are unable to move their hands and legs or are experiencing cerebromedullospinal disconnection [2,21]. The Human Machine Interface (HMI) is using piezoelectric sensors to sense the face and tongue movements [3]. Also, there is an electrooculography based interface for eyes control [5,8].

On the other hand, considering the world sports, the wheelchair for sports has gained global recognition as a Paralympic sport. The number of wheelchair sports players has been steadily increasing each year, with numerous competitions held worldwide to enhance the player skills. Nevertheless, wheelchair sports poses significant challenges because players need advanced techniques and strength to effectively control the wheelchair during the game such as wheelchair tennis and badminton. Furthermore, we had previously introduced the concept of Machine Tennis [9,11], which uses an electric wheelchair with omnidirectional movement capabilities.

So far, we have explored the capabilities of omnidirectional robots for various applications such as the utilization of the omnidirectional moving robot as a mobile router to offer network support [18–20] and for supporting individuals who use wheelchairs [10,14,15]. Additionally, we implemented the omnidirectional moving robot as a wheelchair (which we called the Omnidirectional Wheelchair Robot) for sports activities like wheelchair tennis or badminton [12,13].

Recently, the wheelchairs for badminton have been one of the attractive sports. Therefore, we proposed Machine Badminton which utilizes Omnidirectional Wheelchair Robots. However, the performance of omnidirectional wheelchair and user skills need to be enhanced. Also, to ensure safety during game, it is crucial to prevent collisions between wheelchairs and objects like nets, pole, other players and so on.

In this paper, we deal with safety assurance of omniditectional wheelchair robot for playing badminton. We conducted experiments to control multiple omnidirectional robots to prevent collisions.

The rest of this paper is structured as follows. In Sect. 2, we introduce the implemented omnidirectional wheelchair robot. In Sect. 3, we describe safety assurance system to avoid collision for omnidirectional wheelchair robot. In Sect. 4, we explain experimental results. Finally, conclusions and future work are given in Sect. 5.

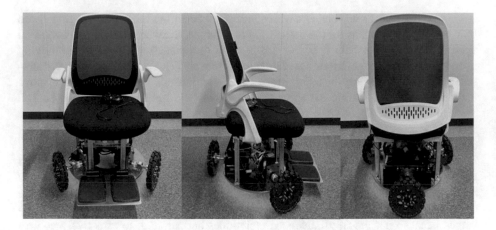

Fig. 1. Image of an omnidirectional wheelchair robot (Front, Side, Rear view).

Table 1. Component of omnidirectional wheelchair robot.

Items	Components
Brushless Motor	ORIENTAL MOTOR CO., BLHM450K-30
Motor Driver	ORIENTAL MOTOR, BLH2D50-KD
CPU	Raspberry Pi 4 Model B
Signal Generator	pigpio (Pulse Width Modulation)
Controller	GameSir T4 Mini pc controller
Display	ELECROW, 7 in. touch panel 1024 × 600 60 Hz
Battery 12V for Motor	Ampere Time, 12V 12Ah (LiFePO4)
Battery 5V for CPU	Z & Y, 5V 2Ah (Li-ion)
Omniwheel	8 in. roller bearing

2   Implemented Omnidirectional Wheelchair Robot

In this section, we explain the implemented omnidirectional wheelchair robot for badminton. In Fig. 1 is shown the image of omnidirectional wheelchair robot (Front, Side and Rear view). In Fig. 2 is shown the image of movement unit of the Omnidirectional wheelchair robot.

In Table 1, we show the component and the dimensions of the omnidirectional wheelchair robot: Length: 75[cm], Width: 65[cm], Height: 87[cm]. The omnidirectional wheelchair robot is able to move in any direction while keeping the same direction because it is equipped with 3 omniwheels. The omnidirectional wheelchair robot can be controlled by any user through a user-side interface using a TCP/IP connection or controller. The diagram of control system is shown in Fig. 3.

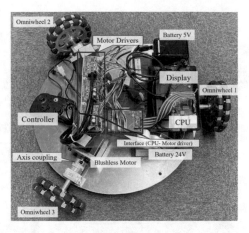

Fig. 2. Image of movement unit of omnidirectional wheelchair robot.

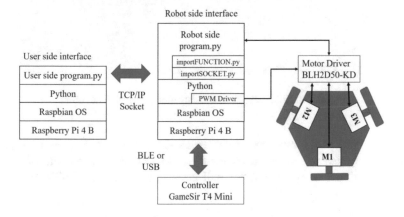

Fig. 3. Diagram of control system for omnidirectional wheel chair robot.

3   Safety Assurance System for Collision Avoidance of Omnidirectional Wheelchair Robot

In this section, we introduce the safety assurance system to avoid collision of omnidirectional wheelchair robots. The image of safety assurance syste is shown in Fig. 4. The robot has two sensors (right and left) for deciding the directions. The circles in Fig. 4 indicate other robots, and the small square shows the robot position.

This safety assurance system can avoid collision between wheelchairs. In a badminton game, for doubles games might occur collisions between teammates. This may have serious injuries to players.

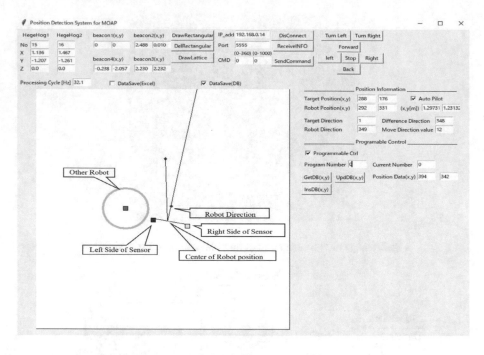

Fig. 4. Control screen for omnidirectional wheelchair robot.

In the normal play mode, the omnidirectional wheelchair robot can be easly controlled by using joystick. However, when there is a risk of collision with other robots, the omnidirectional wheelchair robot will automatically activate collision avoidance system.

4   Experimental Results

In order to measure the precision of robot control, we investigated the trajectory of the omnidirectional robot movement in a 4 [m] × 4 [m] space, as shown in Fig. 5. The beacons B1, B2, B3, B4 are used to measure the robot position and dots between beacons are reference points. We conducted experiments by placing reference points were the robot must pass through with an error margin of a few centimeters at 1[m] intervals. We investigated the trajectory for two cases when the reference points were set at 50 [cm] intervals and when there are no reference points.

The experimental results are shown in Fig. 6. In Fig. 6(a) the robot moves based on the reference line. In Fig. 6(b) is shown the trajectory when the robot is moving without placing reference points, Fig. 6(c) is the case when we set the reference points at 1 [m], while in Fig. 6(d) the reference points are set at 0.5 [m]. The experimental results show that in the case of no reference points the results are better than other results (0.5 [m] and 1 [m]).

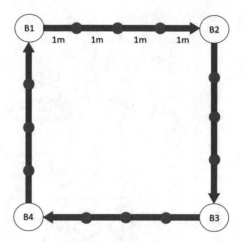

Fig. 5. Experimental environment.

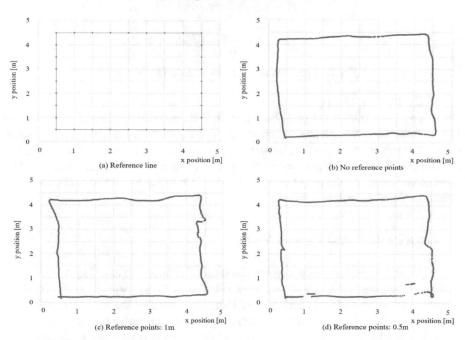

Fig. 6. Experimental results for robot trajectory.

We also conducted experiments on collision avoidance between wheelchairs. Figure 7 illustrates the results of the collision avoidance experiments. The central coordinate of the wheelchair are represented by the center of the circle, while the random trajectories show situations where other wheelchairs attempted to collide towards the center of the circle.

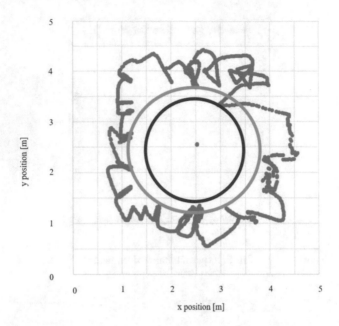

Fig. 7. Experimental results for collision avoidance.

There are two circles in Fig. 7, the inner circle shows collision line. If there is something over this line the robot will collide. The outside circle is collision avoidance line, when other robots cross this line, the system controls the other robots to avoid collision.

The results in Fig. 7 show that we were able to control the omnidirectional robots to avoid collision. So, the proposed system provide a safety assurance environment for Omniditectional Wheelchair Robot.

5   Conclusions and Future Work

In this paper, we explained the current situation of elderly and disabled people in the world and discussed the necessity of wheelchairs. Also, we introduced the proposed and implemented wheelchair robot for sports. Then, we presented the wheelchairs for badminton, which have been one of the attractive sports. In order to ensure the safety during game, it is crucial to prevent collisions between wheelchairs and objects like nets, pole, other players and so on. Therefore, we proposed a system that can effectively avoid collision. Experimental results show that the proposed system accurately track and control the omnidirectional robot position. Also, we investigated the trajectory when the reference points were set at 50 [cm] intervals and when there are no reference points. We found that the results in case when we did not use intervals are better than other results (when using point intervals of 0.5 [m] and 1 [m]).

In the future work, we would like to improve the proposed system and consider other applications of the implemented omnidirectional wheelchair.

Acknowledgements. This work is supported by JSPS KAKENHI Grant Number JP22K11598.

References

1. World health organization. https://www.who.int/
2. Ansari, M.F., Edla, D.R., Dodia, S., Kuppili, V.: Brain-computer interface for wheelchair control operations: an approach based on fast fourier transform and on-line sequential extreme learning machine. Clin. Epidemiol. Glob. Health **7**(3), 274–278 (2019)
3. Bouyam, C., Punsawad, Y.: Human–machine interface-based wheelchair control using piezo-electric sensors based on face and tongue movements. Heliyon **8**(11) (2022)
4. Chatterjee, S., Roy, S.: Multiple control assistive wheelchair for lower limb disabilities & elderly people (2021)
5. Choudhari, A.M., Porwal, P., Jonnalagedda, V., Mériaudeau, F.: An electrooculography based human machine interface for wheelchair control. Biocybern. Biomed. Eng. **39**(3), 673–685 (2019)
6. Dahmani, M., et al.: An intelligent and low-cost eye-tracking system for motorized wheelchair control. Sensors **20**(14), 3936 (2020)
7. Gomes, D., Fernandes, F., Castro, E., Pires, G.: Head-movement interface for wheelchair driving based on inertial sensors. In: 2019 IEEE 6th Portuguese Meeting on Bioengineering (ENBENG), pp. 1–4 (2019)
8. Kaur, A.: Wheelchair control for disabled patients using EMG/EOG based human machine interface: a review. J. Med. Eng. Technol. **45**(1), 61–74 (2021)
9. Matsuo, K., Barolli, L.: Design and implementation of an omnidirectional wheelchair: control system and its applications. In: 2014 Ninth International Conference on Broadband and Wireless Computing, Communication and Applications, pp. 532–535 (2014)
10. Matsuo, K., Barolli, L.: Design and implementation of an omnidirectional wheelchair: control system and its applications. In: 2014 Ninth International Conference on Broadband and Wireless Computing, Communication and Applications, pp. 532–535. IEEE (2014)
11. Matsuo, K., Barolli, L.: Design of an omnidirectional wheelchair for playing tennis. In: 2016 10th International Conference on Complex, Intelligent, and Software Intensive Systems (CISIS), pp. 377–381. IEEE (2016)
12. Matsuo, K., Barolli, L.: Prediction of rssi by scikit-learn for improving position detecting system of omnidirectional wheelchair tennis. In: Advances on Broad-Band Wireless Computing, Communication and Applications: Proceedings of the 14th International Conference on Broad-Band Wireless Computing, Communication and Applications (BWCCA-2019) 14. pp. 721–732. Springer (2020)
13. Izrailov, K., Levshun, D., Kotenko, I., Chechulin, A.: Classification and analysis of vulnerabilities in mobile device infrastructure interfaces. In: You, I., Kim, H., Youn, T.-Y., Palmieri, F., Kotenko, I. (eds.) MobiSec 2021. CCIS, vol. 1544, pp. 301–319. Springer, Singapore (2022). https://doi.org/10.1007/978-981-16-9576-6_21
14. Mitsugi, K., Matsuo, K., Barolli, L.: A comparison study of control devices for an omnidirectional wheelchair. In: Barolli, L., Amato, F., Moscato, F., Enokido, T., Takizawa, M. (eds.) WAINA 2020. AISC, vol. 1150, pp. 651–661. Springer, Cham (2020). https://doi.org/10.1007/978-3-030-44038-1_60
15. Mitsugi, K., Matsuo, K., Barolli, L.: Evaluation of a user finger movement capturing device for control of self-standing omnidirectional robot. In: Barolli, L., Woungang, I., Enokido, T. (eds.) AINA 2021. LNNS, vol. 227, pp. 30–40. Springer, Cham (2021). https://doi.org/10.1007/978-3-030-75078-7_4

16. Ngo, B.V., Nguyen, T.H., Ngo, V.T., Tran, D.K., Nguyen, T.D.: Wheelchair navigation system using EEG signal and 2d map for disabled and elderly people. In: 2020 5th International Conference on Green Technology and Sustainable Development (GTSD), pp. 219–223. IEEE (2020)
17. Nudra Bajantika Pradivta, I.W., Arifin, A., Arrofiqi, F., Watanabe, T.: Design of myoelectric control command of electric wheelchair as personal mobility for disabled person. In: 2019 International Biomedical Instrumentation and Technology Conference (IBITeC), vol. 1, pp. 112–117 (2019)
18. Toyama, A., Mitsugi, K., Matsuo, K., Barolli, L.: Implementation of a moving omnidirectional access point robot and a position detecting system. In: Barolli, L., Poniszewska-Maranda, A., Park, H. (eds.) IMIS 2020. AISC, vol. 1195, pp. 203–212. Springer, Cham (2021). https://doi.org/10.1007/978-3-030-50399-4_20
19. Toyama, A., Mitsugi, K., Matsuo, K., Barolli, L.: Implementation of control interfaces for moving omnidirectional access point robot. In: Barolli, L., Takizawa, M., Enokido, T., Chen, H.-C., Matsuo, K. (eds.) BWCCA 2020. LNNS, vol. 159, pp. 436–443. Springer, Cham (2021). https://doi.org/10.1007/978-3-030-61108-8_43
20. Toyama, A., Mitsugi, K., Matsuo, K., Kulla, E., Barolli, L.: Implementation of an indoor position detecting system using mean BLE RSSI for moving omnidirectional access point robot. In: Barolli, L., Yim, K., Enokido, T. (eds.) CISIS 2021. LNNS, vol. 278, pp. 225–234. Springer, Cham (2021). https://doi.org/10.1007/978-3-030-79725-6_22
21. Zubair, Z.R.S.: A deep learning based optimization model for based computer interface of wheelchair directional control. Tikrit J. Pure Sci. 26(1), 108–112 (2021)

A Fuzzy-Based System for Assessment of Relational Trust Considering Reputation as a New Parameter

Shunya Higashi[1]($\boxtimes$), Phudit Ampririt[1], Ermioni Qafzezi[2], Makoto Ikeda[2], Keita Matsuo[2], and Leonard Barolli[2]

[1] Graduate School of Engineering, Fukuoka Institute of Technology, 3-30-1 Wajiro-Higashi, Higashi-Ku, Fukuoka 811-0295, Japan
{mgm23108,bd21201}@bene.fit.ac.jp
[2] Department of Information and Communication Engineering, Fukuoka Institute of Technology, 3-30-1 Wajiro-Higashi, Higashi-Ku, Fukuoka 811-0295, Japan
qafzezi@bene.fit.ac.jp, makoto.ikd@acm.org, {kt-matsuo,barolli}@fit.ac.jp

Abstract. As digital technologies pervade every facet of our lives, establishing trust for digital interactions has become paramount. This paper explores the dynamics of trust computing within the digital realm, leveraging Fuzzy Logic (FL) to address the inherent ambiguities and uncertainties. We introduce a Fuzzy-based System for Assessment of Relational Trust (FSART), which evaluates the relational trust by considering four input parameters: Influence (If), Importance (Ip), Similarity (Sm), and Reputation (Rp), which is a new parameter. The Relational Trust (RT) is the output parameter. We assessed the performance of the proposed system by computer simulations. The evaluation results show that the values of RT increase with increasing of Sm, Rp, Ip, and If, confirming their positive impact on RT. In the case when If $= 0.5$, Ip $= 0.9$, Rp $= 0.9$, and when If $= 0.9$, Ip $= 0.9$, and Rp is 0.5 and 0.9, for all values of Sm, all RT values exceed 0.5, indicating that entities are considered trustworthy.

1 Introduction

The digital technologies significantly influence our interactions, provide access to extensive information and even shape our identities. However, this integration brings complex challenges. The convenience and interconnectedness offered by technology coexist with persistent concerns about trust.

Recent studies, such as the Edelman Trust Barometer [1], indicate a notable decrease in public trust towards tech organizations, driven by worries related to misinformation, breaches of data privacy and the lack of transparency in algorithms that guide our digital experiences [2]. Despite these concerns, the continuous adoption of digital services [3], like online healthcare, highlights the critical need to reconcile technological advances with trust enhancement.

Recently new technologies such as Blockchain technology is being explored to decentralize data control, aiming to improve data security and user autonomy [4].

L. Barolli (Ed.): IMIS 2024, LNDECT 214, pp. 179–188, 2024.
https://doi.org/10.1007/978-3-031-64766-6_18

In Artificial Intelligence (AI) field, there is a trend towards explainable AI, which is trying to clarify algorithmic decisions, intending to rebuild trust [5]. Nonetheless, aligning these technological advances with ethical standards and fostering responsible AI require continuous cooperation among researchers, policymakers, and the public [6].

For trust in digital interactions not only the reliability and security but also the perception of entities play a pivotal role. Another important concept is the reputation, which significantly influences the trust dynamics in digital and decentralized environments. In [7], the metrics are quantified and embedded into digital frameworks to alleviate uncertainties and bolster the trust. This approach demonstrates how digital systems can dynamically adjust to changes in behavior and maintain robust configurations in the case of unreliable components.

This paper proposes a new approach using Fuzzy Logic (FL) to assess relational trust, considering Influence, Importance, Similarity, and Reputation in order to build more trustworthy digital systems. Computer simulations have confirmed a positive relationship between these variables and the relational trust, suggesting that when these metrics are increased the trust is improved.

The structure of this paper is organized as follows. Section 2 provides an overview of trust computing. In Sect. 3 is give the outline of FL. In Sect. 4 is presented the proposed fuzzy-based system. The simulation results are discussed in Sect. 5. The concluding remarks and future work are given in Sect. 6.

2   Trust Computing

The concept of trust computing is critical in shaping user interactions, data exchanges, and the overall digital experience. Trust, in this context, extends beyond just expectations of reliability and security to include a complex structure of transparency, accountability, and user confidence [8]. Also, the management of trust and reputation plays a pivotal role in Internet and social computing systems, enhancing the quality of user interactions and contributing to the construction of a trustworthy digital environment [9].

A key element of the trust is the accountability of digital entities for their actions. Systems considered trustworthy are founded on accountability principles, guaranteeing timely and proper responses to concerns about security, privacy, or ethics. These systems ensure that the responsible parties are held accountable. The main aim of trust computing is to build confidence among users, offering them a sense of security and reliability. Achieving this goal requires a careful balance between technological integrity and a user-centered design approach.

As reliance on digital platforms increases, the significance of trust computing grows, influencing user behavior, business transactions, and societal trust in digital infrastructure. A thorough understanding and implementation of trust computing principles not only enhance the user experience but also contribute to creating a resilient and adaptable digital ecosystem [10].

In Fig. 1 is presented the model of the trust framework [11], clarifying the interactions between trustors and trustees. In this model, trustors begin the process by generating a 'trust object' based on their past experiences for trustees. Trustees then assess the trustworthiness of this object based on their experiences and the current trust environment. This decision-making process carries inherent risks, particularly due to possible miscommunications or misunderstandings of actions.

Trust computing involves a comprehensive evaluation of trust, incorporating a wide range of parameters. As shown in Fig. 2, this includes individual trust, determined by analyzing specific trustee characteristics through logical analysis and evidence (including emotional responses and predispositions), and relational trust, which assesses the nature of the relationship between the trustor and trustee [12].

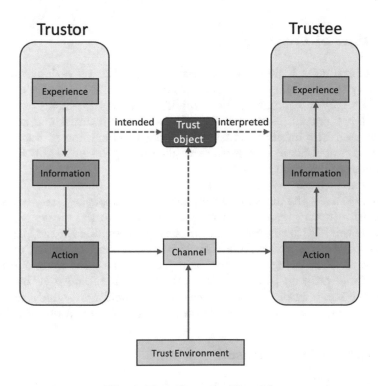

Fig. 1. Trust framework model.

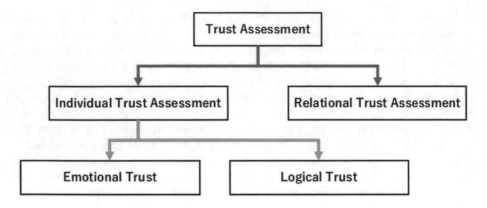

Fig. 2. Trust assessment process.

3   Outline of Fuzzy Logic

The FL represents a logic system for handling continuous variables, providing adaptability and flexibility in scenarios where traditional binary logic might not suffice, especially in computing the reliability. For the assessment of trust computing, FL is a crucial instrument for addressing uncertainties and ambiguities. It can be used in reliability modeling allowing the inclusion of imprecise conditions and probabilistic elements, thus integrating the uncertain factors and improving overall system reliability.

In the architecture of the proposed fuzzy-based system, the Fuzzy Logic Controller (FLC) is the most important element. By applying fuzzy control rules together with input parameters, the controller derives the output values. The structure of the FLC is composed of four principal components: the Fuzzifier, the Inference Engine, the Fuzzy Rule Base (FRB), and the Defuzzifier, as depicted in the Fig. 3. Employing fuzzy sets and control rules, the system progresses from inputs to outputs [13,14]. This methodology is intended to be versatile, enabling the handling of diverse parameters and conditions, which increase the adaptability and efficacy of trust evaluations in the proposed fuzzy-based system.

4   Proposed Fuzzy-Based System

This section introduces our system designed for the assessment of relational trust, named the Fuzzy-based System for Assessment of Relational Trust (FSART). For the implementation of FSART, we consider four input parameters: Influence (If), Importance (Ip), Similarity (Sm), and Reputation (Rp), while the Relational Trust (RT) is the output parameter. The structure of the proposed system is shown in Fig. 4.

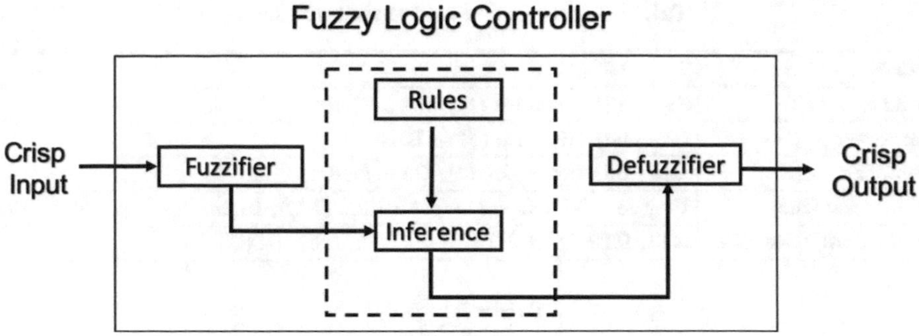

Fuzzy Logic Controller

Fig. 3. FLC structure.

In Fig. 5 are shown the membership functions. The linguistic values and their specific ranges for each parameter are presented in Table 1. The decision-making process in FSART operates based on the FRB shown in Table 2. This FRB includes 135 rules, in the format: "IF condition THEN control action" for evaluation of the relational trust.

5   Simulation Results

This section presents the simulation results. We show the influence of four input parameters on Relational Trust (RT). The simulation results are illustrated in Fig. 6, Fig. 7, and Fig. 8. These figures show the relationship between the RT and Sm, while If, Ip and Rp are kept constant.

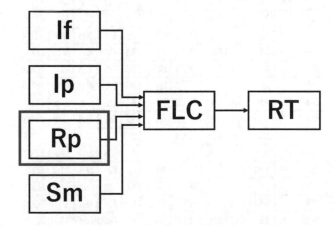

Fig. 4. Proposed system structure.

Table 1. Parameter and their term sets.

Parameters	Term set
Influence (If)	Small (S), Medium (M), Large (L)
Importance (Ip)	Low (Lo), Medium (Me), High (Hi)
Reputation (Rp)	Bad (Bd), Normal (No), Good (Gd)
Similarity (Sm)	Very low (Vl), Low (Lw), Medium (Md), High (Hg), Very high (Vh)
Relational Trust (RT)	RT1, RT2, RT3, RT4, RT5, RT6, RT7, RT8, RT9

Table 2. FRB.

Rule	If	Ip	Rp	Rp	RT	Rule	If	Ip	Rp	Rp	RT
1	S	Lo	Bd	Vl	RT1	68	M	Me	No	Md	RT5
2	S	Lo	Bd	Lw	RT1	69	M	Me	No	Hg	RT7
3	S	Lo	Bd	Md	RT1	70	M	Me	No	Vh	RT8
4	S	Lo	Bd	Hg	RT1	71	M	Me	Gd	Vl	RT5
5	S	Lo	Bd	Vh	RT2	72	M	Me	Gd	Lw	RT6
6	S	Lo	No	Vl	RT1	73	M	Me	Gd	Md	RT7
7	S	Lo	No	Lw	RT1	74	M	Me	Gd	Hg	RT8
8	S	Lo	No	Md	RT1	75	M	Me	Gd	Vh	RT9
9	S	Lo	No	Hg	RT2	76	M	Hi	Bd	Vl	RT3
10	S	Lo	No	Vh	RT3	77	M	Hi	Bd	Lw	RT5
11	S	Lo	Gd	Vl	RT1	78	M	Hi	Bd	Md	RT6
12	S	Lo	Gd	Lw	RT2	79	M	Hi	Bd	Hg	RT7
13	S	Lo	Gd	Md	RT3	80	M	Hi	Bd	Vh	RT8
14	S	Lo	Gd	Hg	RT4	81	M	Hi	No	Vl	RT5
15	S	Lo	Gd	Vh	RT5	82	M	Hi	No	Lw	RT6
16	S	Me	Bd	Vl	RT1	83	M	Hi	No	Md	RT7
17	S	Me	Bd	Lw	RT1	84	M	Hi	No	Hg	RT8
18	S	Me	Bd	Md	RT2	85	M	Hi	No	Vh	RT9
19	S	Me	Bd	Hg	RT2	86	M	Hi	Gd	Vl	RT6
20	S	Me	Bd	Vh	RT4	87	M	Hi	Gd	Lw	RT8
21	S	Me	No	Vl	RT1	88	M	Hi	Gd	Md	RT8
22	S	Me	No	Lw	RT2	89	M	Hi	Gd	Hg	RT9
23	S	Me	No	Md	RT3	90	M	Hi	Gd	Vh	RT9
24	S	Me	No	Hg	RT4	91	L	Lo	Bd	Vl	RT1
25	S	Me	No	Vh	RT5	92	L	Lo	Bd	Lw	RT2
26	S	Me	Gd	Vl	RT2	93	L	Lo	Bd	Md	RT3
27	S	Me	Gd	Lw	RT3	94	L	Lo	Bd	Hg	RT5
28	S	Me	Gd	Md	RT4	95	L	Lo	Bd	Vh	RT6
29	S	Me	Gd	Hg	RT6	96	L	Lo	No	Vl	RT2
30	S	Me	Gd	Vh	RT7	97	L	Lo	No	Lw	RT4

(*continued*)

Table 2. (*continued*)

Rule	If	Ip	Rp	Rp	RT	Rule	If	Ip	Rp	Rp	RT
31	S	Hi	Bd	Vl	RT1	98	L	Lo	No	Md	RT5
32	S	Hi	Bd	Lw	RT2	99	L	Lo	No	Hg	RT6
33	S	Hi	Bd	Md	RT3	100	L	Lo	No	Vh	RT8
34	S	Hi	Bd	Hg	RT4	101	L	Lo	Gd	Vl	RT4
35	S	Hi	Bd	Vh	RT5	102	L	Lo	Gd	Lw	RT6
36	S	Hi	No	Vl	RT2	103	L	Lo	Gd	Md	RT7
37	S	Hi	No	Lw	RT3	104	L	Lo	Gd	Hg	RT8
38	S	Hi	No	Md	RT4	105	L	Lo	Gd	Vh	RT9
39	S	Hi	No	Hg	RT6	106	L	Me	Bd	Vl	RT3
40	S	Hi	No	Vh	RT7	107	L	Me	Bd	Lw	RT4
41	S	Hi	Gd	Vl	RT3	108	L	Me	Bd	Md	RT5
42	S	Hi	Gd	Lw	RT5	109	L	Me	Bd	Hg	RT6
43	S	Hi	Gd	Md	RT6	110	L	Me	Bd	Vh	RT8
44	S	Hi	Gd	Hg	RT7	111	L	Me	No	Vl	RT4
45	S	Hi	Gd	Vh	RT8	112	L	Me	No	Lw	RT5
46	M	Lo	Bd	Vl	RT1	113	L	Me	No	Md	RT7
47	M	Lo	Bd	Lw	RT2	114	L	Me	No	Hg	RT8
48	M	Lo	Bd	Md	RT2	115	L	Me	No	Vh	RT9
49	M	Lo	Bd	Hg	RT3	116	L	Me	Gd	Vl	RT6
50	M	Lo	Bd	Vh	RT5	117	L	Me	Gd	Lw	RT7
51	M	Lo	No	Vl	RT2	118	L	Me	Gd	Md	RT8
52	M	Lo	No	Lw	RT3	119	L	Me	Gd	Hg	RT9
53	M	Lo	No	Md	RT4	120	L	Me	Gd	Vh	RT9
54	M	Lo	No	Hg	RT5	121	L	Hi	Bd	Vl	RT4
55	M	Lo	No	Vh	RT6	122	L	Hi	Bd	Lw	RT6
56	M	Lo	Gd	Vl	RT3	123	L	Hi	Bd	Md	RT7
57	M	Lo	Gd	Lw	RT4	124	L	Hi	Bd	Hg	RT8
58	M	Lo	Gd	Md	RT5	125	L	Hi	Bd	Vh	RT9
59	M	Lo	Gd	Hg	RT7	126	L	Hi	No	Vl	RT6
60	M	Lo	Gd	Vh	RT8	127	L	Hi	No	Lw	RT7
61	M	Me	Bd	Vl	RT2	128	L	Hi	No	Md	RT8
62	M	Me	Bd	Lw	RT3	129	L	Hi	No	Hg	RT9
63	M	Me	Bd	Md	RT4	130	L	Hi	No	Vh	RT9
64	M	Me	Bd	Hg	RT5	131	L	Hi	Gd	Vl	RT7
65	M	Me	Bd	Vh	RT7	132	L	Hi	Gd	Lw	RT9
66	M	Me	No	Vl	RT3	133	L	Hi	Gd	Md	RT9
67	M	Me	No	Lw	RT4	134	L	Hi	Gd	Hg	RT9
						135	L	Hi	Gd	Vh	RT9

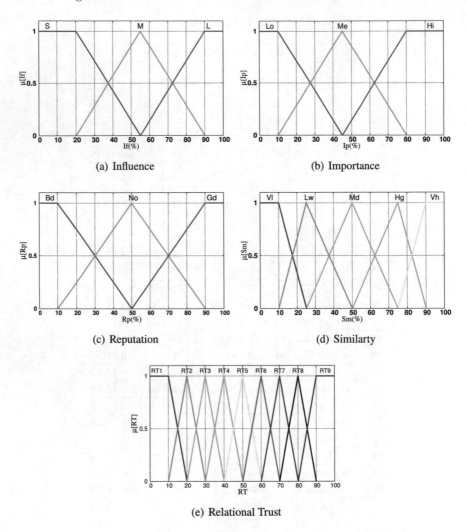

(a) Influence

(b) Importance

(c) Reputation

(d) Similarty

(e) Relational Trust

Fig. 5. Membership functions.

In Fig. 6(a), when If and Ip are set to a low value (0.1), and Rp is increased from 0.1 to 0.5 and then to 0.9, the RT is increased by 12% and 20% when Sm value is 0.8, respectively. This shows that the RT is increased when the Rp is increased. To explore the impact of Ip on RT, the Ip value is changed from 0.1 to 0.9, as illustrated in Fig. 6(a) and Fig. 6(b). The RT is improved by 40% when Rp and Sm are set to 0.5 and 0.8, respectively.

In Fig. 6(a) and Fig. 7(a), we see that by changing If from 0.1 to 0.5 and Ip is 0.1, the RT is increased by 23% when Rp and Sm are 0.5 and 0.8, respectively. This indicates a positive correlation between If and RT.

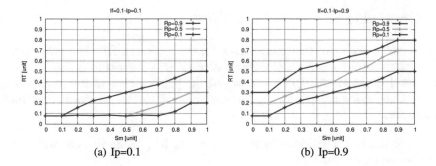

Fig. 6. Simulation results for If = 0.1.

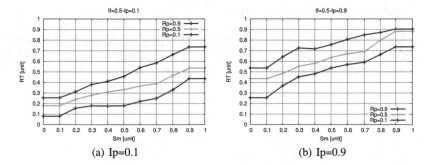

Fig. 7. Simulation results for If = 0.5.

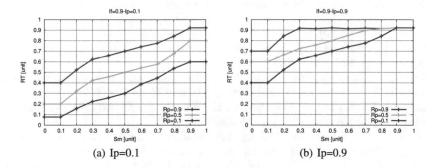

Fig. 8. Simulation results for If = 0.9.

In Fig. 8, the If value is increased to 0.9. The comparison with Fig. 6 and Fig. 7 shows a substantial increase in RT values. Specifically, in Fig. 7(b), when If is 0.5, Ip = 0.9, Rp = 0.9, and in Fig. 8(b), when If is 0.9, Ip = 0.9, and Rp is 0.5 and 0.9, for all values of Sm, all RT values exceed 0.5, indicating that entities are considered trustworthy.

6   Conclusions and Future Work

In this paper, we presented the FSART for evaluation of RT. The proposed system considers four input parameters: Influence (If), Importance (Ip), Similarity (Sm), and Reputation (Rp). We assessed the performance of the proposed system by computer simulations. From the simulation results, we draw the following conclusions.

- The values of RT increase by increasing Sm, Rp, Ip, and If, confirming their positive impact on RT.
- In the case when If $= 0.5$, Ip $= 0.9$, Rp $= 0.9$, and when If $= 0.9$, Ip $= 0.9$, and Rp is 0.5 and 0.9, for all values of Sm, all RT values exceed 0.5, indicating that entities are considered trustworthy.

In the future work, we plan to introduce new parameters and conduct extensive simulations to evaluate the performance and effectiveness of the FSART system and expand its applicability in trust computation.

References

1. Edelman. Edelman trust barometer (2023). https://www.edelman.com/trust/2023/trust-barometer. Accessed Jan 2024
2. Ziewitz, M.: Governing algorithms. Sci. Technol. Hum. Values **41**(1), 16–30 (2016)
3. Heponiemi, T., Jormanainen, V., Leemann, L., Manderbacka, K., Aalto, A., Hyppönen, H.: Digital divide in perceived benefits of online health care and social welfare services: national cross-sectional survey study. J. Med. Internet Res. **22**(7), 1–12, e17616 (2020)
4. Chen, N., Cho, D.S.-Y.: A blockchain based autonomous decentralized online social network, pp. 186–190 (2021)
5. Adadi, A., Berrada, M.: Peeking inside the black-box: a survey on explainable artificial intelligence (XAI). IEEE Access **6**, 52 138–52 160 (2018)
6. Zhu, L., Xu, X., Lu, Q., Governatori, G., Whittle, J.: AI and ethics - operationalising responsible AI. arXiv, vol. abs/2105.08867 (2021)
7. Kiefhaber, R.: Calculating and aggregating direct trust and reputation in organic computing systems (2014). https://api.semanticscholar.org/CorpusID:11155524
8. Frank, R.D., Chen, Z., Crawford, E., Suzuka, K., Yakel, E.: Trust in qualitative data repositories. Proc. Assoc. Inf. Sci. Technol. **54**(1), 102–111 (2017)
9. Liu, L., Shi, W.: Trust and reputation management. IEEE Internet Comput. **14**, 10–13 (2010)
10. Jayasinghe, U., Lee, G., Um, T.-W., Shi, Q.: Machine learning based trust computational model for IoT services. IEEE Trans. Sustain. Comput. **4**(1), 39–52 (2019)
11. Schultz, C.D.: A trust framework model for situational contexts, pp. 1–7 (2006)
12. Cho, J.-H., Chan, K., Adali, S.: A survey on trust modeling. ACM Comput. Surv. (CSUR) **48**(2), 1–40 (2015)
13. Lee, C.-C.: Fuzzy logic control systems: fuzzy logic controller - Part I. IEEE Trans. Syst. Man Cybern. **20**(2), 404–418 (1990)
14. Mendel, J.: Fuzzy logic systems for engineering: a tutorial. Proc. IEEE **83**(3), 345–377 (1995)

Novel Dynamic Difficulty Adjustment Methods for Niche Games

Qingwei Mi and Tianhan Gao[✉]

Software College, Northeastern University, Shenyang, China
2110491@stu.neu.edu.cn, gaoth@mail.neu.edu.cn

Abstract. As a research hotspot in Game Artificial Intelligence (Game AI), Dynamic difficulty adjustment (DDA) is a technology that adapts the game's challenge to match the player's skill. DDA is the core mechanic to provide continuous motivation and immersion to players during game design and development. This paper proposes general methods for niche games to improve the current research system of DDA. Advantages on adaptive parameters, gameplay time, and Immersive Experience Questionnaire (IEQ) results demonstrate the effectiveness in enhancing player experience and retention of the proposed methods with a high adaptability. The robust DDA system can help game designers to ensure experience and avoid churn of players.

1 Introduction

Artificial Intelligence (AI) refers to the intelligence displayed by machines made by humans, which corresponds to Natural Intelligence (NI). It represents the ability of machines to perform cognitive function associated with the human mind, such as perception, reasoning, learning, interacting, problem-solving, and even creating visions of the future. The focus field in AI is how to optimize the decision-making by intelligent agents that can process information according to the environment [1–5]. At present, AI is dominating board games gradually. Meanwhile, it is also changing video games and defining a new mode of human-computer interaction [6–9].

The video game is an electronic game that involves interaction with input devices to generate visual feedback from a display device. It's been more than 70 years since the first video game appeared. Nowadays, video games are typically categorized according to their gameplay interaction. The Role-playing Game (RPG), Racing Game (RAC), Fighting Game (FTG), Shooting Game (STG), Adventure Game (AVG), and so on are major genres widely recognized by the game industry [10–13]. With the evolving of game design patterns, more genres led by the idle game and storytelling game for niche players have emerged. Although there are differences between the gamers targeted by each genre, ensuring the best experience for players is the common core goal, which reflects the importance of Game Artificial Intelligence (Game AI).

Game AI is a distinct field and differs from academic AI. It centers on optimizing player experience with great gameplay [14–16]. As the research hotspot in Game AI, Dynamic Difficulty Adjustment (DDA) is a useful method to ensure experience. DDA can modify the parameters in games in real-time according to player skill level, so as to

L. Barolli (Ed.): IMIS 2024, LNDECT 214, pp. 189–200, 2024.
https://doi.org/10.1007/978-3-031-64766-6_19

support retaining players by game designers [17–19]. However, in the scarce research on DDA, most of them only apply to major genres and nearly no solution for niche ones.

To improve the current research system as well as expand the applied range of DDA, this paper proposes a series of novel dynamic difficulty adjustment methods based on the flow model for niche games. Each Scheme is designed centering on the general Game Completion Threshold (GCT). The authors invited players to participate in gameplay and questionnaires to test the adaptability of methods and the immersion of players respectively. The results demonstrate that the adaptability, gameplay time, and immersion provided of the proposed DDA methods is higher than without DDA.

The remainder is organized as follows. Section 2 introduces the definitions of the flow model, player churn, and niche game genres. The details of the proposed DDA methods and schemes are elaborated on in Sect. 3. The tests are shown in Sect. 4, and the adaptability of schemes as well as player experience in different games is analyzed. Section 5 concludes the contributions and future work of the paper finally.

2 Flow, Churn and Genres

Players are the most essential entity in video games. The gaming experience of players determines key metrics such as sales, reviews, and retention directly. A good game must be able to provide players with a good experience so that it can be recognized by players and the market [20, 21]. This can be achieved by keeping players in the flow state constantly. The flow represents the state where a player is neither anxious nor bored. No matter for major or niche game genres, realizing the flow based on DDA is an effective means to reduce player churn rate.

Csikszentmihalyi's Flow Model. The psychologist Mihaly Csikszentmihalyi put forward the flow model in 1975 [22]. So far, the model has been widely referred to across various fields [23–26]. For video games, the primary goal is to create entertainment through intrinsic motivation, which is highly correlated with the flow. In games, the flow tends to be achieved through a balance between player skill level and game challenge level (shown in Fig. 1). Players whose current skill level is certain below the game challenge level will feel anxious. Whereas, boredom will occur when their skill level exceeds the challenge level to some extent. Either of the situations above will lead players to leave the game.

In order to stay in the flow state, it's necessary to keep a dynamic balance between player skill level and game challenge level in real-time. DDA can help game designers to implement it and increase continuous motivation to keep players playing. The enjoyable flow experience for players provided by DDA can prevent churn effectively.

Player Churn. In the game industry, player churn generally refers to players leaving the game permanently due to factors such as game mechanics, charging pressure, personal reasons, etc. [27, 28]. It has always been a state that all practitioners must try to avoid. As players may leave at any stage of the game throughout their game life cycle, designing DDA mechanics based on game completion time is key to making players immerse themselves in the personalized experience.

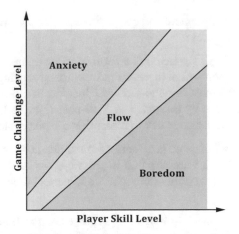

Fig. 1. Csikszentmihalyi's Flow Model. The model shows the map between player skill level and game challenge level in a game.

Niche Game Genres. The gameplay interaction is a typical classification method currently. After surveying, the idle game, storytelling game, survival game, party game, logic game, rhythm game, and sport game are selected for market-oriented. However, it should be noted that a game may belong to several genres at once. The authors take this into account to improve the generality of the schemes.

Idle Game. The idle game is described as being tuned for a never-ending sense of escalation, which is also known as incremental game. Its gameplay consists of players doing simple actions with limited interaction. This mechanic offers a low-pressure experience with constant growth and feedback. The rapid growth of cost, power and rewards makes the game fun and attractive. Various achievements are set to neutralize boredom feeling. As for the win condition, open-ended or closed loops are both feasible. "Clicker Heroes", "Melvor Idle", "Cookie Clicker", and "Dragon Cliff" are well-known idle games.

Storytelling Game. The storytelling game is also called interactive fiction with the story line as the core. It mainly uses text narration supplemented by Computer Graphics (CG) or animations to perform the story. The game generally has some puzzle and reasoning elements with branches and multiple endings. Besides, some specific storytelling games hide the story and run it throughout the entire gaming experience, so as to give players food for thought through recalls after they finish the games. "Inside", "Ace Attorney", "Far Away", and "Turlock Holmes", are representative storytelling games.

Survival Game. In the survival game, players are required to survive as long as possible in hostile and intense environments. Players usually start with few resources. It encourages players to explore unknown areas and use the resources found to make items. In terms of handling character death, players may lose equipment or items or end the game directly.

Party Game. The party game is a multiplayer game and is generally designed as a collection of simple minigames, features intuitive designs and easy to control. It intends to be played socially and is suitable for players of any skill level. That is, the game is friendly for novices. Entertainment and competitiveness coexist among players, so as to provide different atmospheres.

Logic Game. Display logic puzzles in the form of video games, the logic game genre appears. It's commonly non-verbal in nature and requires strong logical reasoning ability to complete. One of the most famous forms is Sudoku. In a Sudoku, players need to place numbers in each grid by inference. A well-posed puzzle has a unique solution. The solving time of each puzzle is key to measuring the player skill level.

Rhythm Game. The rhythm game is a genre of music-themed that challenges the sense of rhythm of players. Players are required to press buttons precisely and continuously to finish levels without interruption. Full combo and perfect time is the final goal for all players.

Sport Game. As simulations of sports, the sport game has a wide coverage. Some games emphasize playing the sport, while others focus on the management. Certain copyrighted game series feature the names and characteristics of real players and teams. The rosters will be updated periodically to reflect changes in reality.

3 Dynamic Difficulty Adjustment Methods

Based on the flow model, the DDA methods are proposed to avoid players getting into anxiety and boredom states. The authors define GCT as the general variable to determine whether to active DDA. Furthermore, detailed schemes are provided for the idle game, storytelling game, survival game, party game, logic game, rhythm game, and sport game to realize DDA effectively.

GCT aims to improve the efficiency of algorithms and reduce the probability of DDA being perceived by players by dividing their game life cycle into early/late stage. For the idle game and storytelling game, the process from start to complete the game exists as a whole, while that of the other five genres can be set by total play times. The methods will adjust the difficulty to change players into flow from feeling anxious. At the same time, DDA will reverse to reduce boredom. The value of GCT can be set adaptively by game designers. On the basis of this, the forward DDA schemes are designed as follows.

Idle Game Scheme. In order to improve the pertinence of each detailed scheme for the seven genres, it's necessary to clarify the core motivation of players. The purpose of players playing a certain game can be highly summarized in one word. For the idle game, it's the "value" undoubtedly.

The idle game offers an attractive gaming experience that requires little effort from players, which makes it suitable for casual players who don't have time to invest in a more complex game. The game is designed to be played over a long time, with players slowly progressing throughout their game life. It allows players to continue increasing various values when they are offline. That is, players can also make progress even they have no time to play. The maximum duration for obtaining resources and rewards offline should be set precisely by game designers.

Since the early game process of the idle game is relatively simple and characters and items are easy to progress, DDA is supposed to be enabled when the game process is over

GCT. The growth of resources and rewards should be improved. The DDA scheme for idle games will increase the standard probability of getting helpful items under current circumstances. Rapid growth of values makes it easier for players to reach predetermined game achievements or personal goals, which can increase their motivation to go on.

Storytelling Game Scheme. The "inspiration" can be concluded as what players would like to get in a storytelling game. The storytelling game relies on its story and atmosphere to make players satisfied. In fact, a good mix of story and gameplay is the ideal balance to strive for. A well-designed story with straight gameplay can turn a good game into an unforgettable experience, and memorable inspiration is what players seek. Overly complex narrative may interrupt the flow state.

If the game process exceeds GCT, DDA in the storytelling game is supposed to be enabled as the story won't go rapidly during the early sections. When players get stuck for a long time in the game, the scheme will give players a bit of spoiler without affecting the attraction of the story which is presented through non-essential characters. Meanwhile, a hint will appear in the User Interface (UI). The spoiler and hint are helpful in continuing experiencing the story for players so that they can gain inspiration earlier.

Survival Game Scheme. The survival game encourages creativity and exploration. It's challenging by providing strained as well as relaxing experience. Players are required to adapt to changing environments and manage limited resources effectively. They need to strategize to find resources, manage hunger, and counter dangers. So, the "plan" is vital for players to survive in the game.

Each survival game differs in the design of resources and scenes. This results in even with certain experience in survival games, it takes a period of time for players to become familiar with the specific mechanics. Initial resources that can provide full security are generally difficult to get quickly. The DDA mechanics will reduce the attributes and occurrence frequency of enemies as well as increase the generate probability of useful resources around the player's spawn point when the player plays first GCT times. And for games where the character may lose equipment or items when dies, DDA will decrease the probability of losing important things. Based on the scheme, players can survive more easily in the early game stage. These offer players space to show their planning and problem-solving skills.

Party Game Scheme. The party game is designed for players to play together with their relatives and friends. It adds fun, laughter, and friendly competition to the participants, making it enjoyable for everyone involved. The game requires skill level below average to attract "socializing" players of all ages.

Although the characteristic of the party game leads to good game experience even if the player does not win lastly, players who are losing consistently will also not stay in the game. To address this issue, the DDA scheme will adjust the generate probability of items if the player play times are below GCT. In each round, the items in the scene can influence the outcome to a certain extent. The spawn points of good items will be closer to disadvantaged individuals and teams. At the same time, it will be easier for them to get good items when they open random boxes. The scheme only increases the probability for participants in the lower rankings and doesn't reduce that for ones in the higher rankings directly. The reason is that reducing the parameters numerically which are personal attributes of leaders directly doesn't meet the core requirements of DDA. It's a punishment mechanism for better-performing players. DDA for party games lets players with low skill level taste the joy of victory.

Logic Game Scheme. Entertaining with a sense of relaxation and mindfulness, that is the logic game. Puzzle solving requires focus, patience, and attention to detail. It allows players to detach from the real world and engage in a hands-on activity that exercises their brains. The game emphasizes the process of solving a puzzle should be meditative, allowing players "thinking" to their heart's content to achieve a sense of accomplishment.

Procedural Content Generation (PCG) is a method of creating data algorithmically as opposed to manually which is widely used in game level design [29]. In the logic game, PCG is usually used to generate puzzles automatically. In this case, when the player plays first GCT times, DDA is allowed to increase the number of related minimum elements to be solved so that players. When solving an element, its related elements are easier to solve than the unrelated ones, or the adjacent related elements can be solved as a whole. The scheme for logic games makes it further for players to show the ability to think within their capabilities.

Rhythm Game Scheme. Music relaxes people, and so should the rhythm game make players "relaxation". Most rhythm games give players interactive notes commonly. The music and the button pressing are the two keys to make the game great. If the designers master the music, it's much easier to attract players. But to make button pressing fun, they required very accurate input timings so that players would literally have to follow the rhythm perfectly. The two points together create real rhythm game experience.

The music in the rhythm game conveys orderly vibrations and information to players so that they are more likely to enter into flow compared to other genres. If the player play times are below GCT, the decision time of full combo and perfect are improved slightly by DDA. Set a variable named Condition Trigger Threshold (CTT). The improvement of combo and perfect will only be triggered independently when either's continuous finish times exceeds CTT in each round. DDA reduces the occurrence of failure to achieve the final goal for players due to a tiny error at the last moment of the round.

Sport Game Scheme. The sports game involves physical and tactical challenges. It attempts to simulate athletic characteristics, including speed, strength, accuracy, and so on. In addition, career and dynasty modes for customizable characters and teams as well as play-by-play are usually added to enhance the "substitution" of players.

Rubber-banding is a common DDA technique used in RAC to control AI opponents to change their game style in real-time [16]. Use the point difference between the player and AI opponent instead of the distance between the two to complete the adaption of rubber-banding for the competition-based sport game. When the player plays first GCT times, AI will reduce its skill level when the player's point is behind that of it. The degree of adjustment is positive correlation with the difference. The DDA scheme for sport games is able to give hope to players chasing points.

4 Results and Discussion

To fully test the actual application effect of the DDA methods proposed, a system is designed and implemented which introduces prototypes of the idle game, storytelling game, survival game, party game, logic game, rhythm game, and sport game. Firstly, the authors create the player and other necessary classes. The related variables, functions, and interfaces are set subsequently. After all algorithms are implemented, the authors do art and sound effects. The levels for tests will be built finally.

The authors invited 9 players to participate in system tests, who were divided into three groups equally to represent the three ranks from low to high based on their skill level to participate in methods tests.

Methods Tests. To confirm the feasibility and availability of the core DDA algorithms, the execution during gameplay is tested systematically and successively. Table 1 shows the adaptability of the methods with/without DDA. The results prove that the proposed methods with a higher adaptability. DDA is quite feasible and effective in modifying relevant parameters in the niche games.

Table 1. Adaptability of the schemes for the seven game genres.

Genres	DDA	Parameters
Idle	Without	The times of getting helpful items are standard
	With	The times of getting helpful items increase
Storytelling	Without	Advance the story process normally
	With	Advance the story process more
Survival	Without	Defeat certain enemies, the times of enemy appearance are standard, normal useful resources are around the spawn point, the times of losing important things are standard
	With	Defeat more enemies, the times of enemy appearance decrease, more useful resources are around the spawn point, the times of losing important things decrease
Party	Without	The times of getting good items from spawning directly and random boxes are standard
	With	The times of getting good items from spawning directly and random boxes increase
Logic	Without	Solve the puzzle regularly
	With	Solve the puzzle faster
Rhythm	Without	The times of full combo and perfect are standard
	With	The times of full combo and perfect increase
Sport	Without	The point difference is certain
	With	The point difference is lower

Experience Tests. To verify the effect of the system on reducing player churn rate as well as improving the player experience, 205 players were invited to participate in the four-week gameplay experience voluntarily. The players need to experience the seven prototypes with DDA disabled/enabled separately for two weeks.

Gameplay time is an applicable measure of player retention. The Immersive Experience Questionnaire (IEQ) is one of the most widely used player experience questionnaires in gaming [30]. The questions in the IEQ are shown in Table 2. All questions have a 5-point Likert scale, anchored at the ends with strongly disagree and strongly agree. Among the 32 questions, items with odd IDs are scored forward, while even ones are scored reversely. Players are required to rate how far they would agree with the statements just before they were interrupted.

Table 2. Questions in the IEQ.

ID	Questions
1	I felt that I really empathized/felt for with the game
2	I did not feel any emotional attachment to the game
3	I was interested in seeing how the game's events would progress
4	It did not interest me to know what would happen next in the game
5	I was in suspense about whether I would win or lose the game
6	I was not concerned about whether I would win or lose the game
7	I sometimes found myself to become so involved with the game that I wanted to speak to the game directly
8	I did not find myself to become so caught up with the game that I wanted to speak to directly to the game
9	I enjoyed the graphics and imagery of the game
10	I did not like the graphics and imagery of the game
11	I enjoyed playing the game
12	Playing the game was not fun
13	The controls were not easy to pick up
14	There were not any particularly frustrating aspects of the controls to get the hang of
15	I became unaware that I was even using any controls
16	The controls were not invisible to me
17	I felt myself to be directly travelling through the game according to my own volition
18	I did not feel as if I was moving through the game according to my own will
19	It was as if I could interact with the world of the game as if I was in the real world
20	Interacting with the world of the game did not feel as real to me as it would be in the real world

(continued)

Table 2. (*continued*)

ID	Questions
21	I was unaware of what was happening around me
22	I was aware of surroundings
23	I felt detached from the outside world
24	I still felt attached to the real world
25	At the time the game was my only concern
26	Everyday thoughts and concerns were still very much on my mind
27	I did not feel the urge at any point to stop playing and see what was going on around me
28	I was interested to know what might be happening around me
29	I did not feel like I was in the real world but the game world
30	I still felt as if I was in the real world whilst playing
31	To me it felt like only a very short amount of time had passed
32	When playing the game time appeared to go by very slowly

Four weeks later, gameplay time of the 205 players was collected. Besides, the authors distributed two IEQs to the players. They should fill in the two questionnaires according to their individual experience with/without DDA. The mean gameplay time and IEQ scores of the players are shown in Fig. 2 and Fig. 3 respectively.

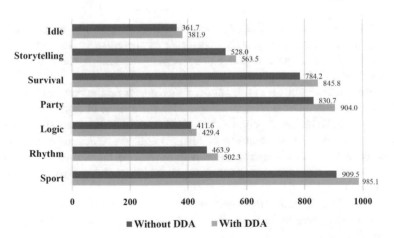

Fig. 2. Mean gameplay time of the players in each game genre (unit: minute).

The results in the figures above demonstrate that the mean gameplay time and IEQ scores of the players with DDA have different degrees of improvement compared with those without DDA in each niche game. In terms of the increase of mean gameplay time, the sport game, party game, and survival game are the top three. While, the party

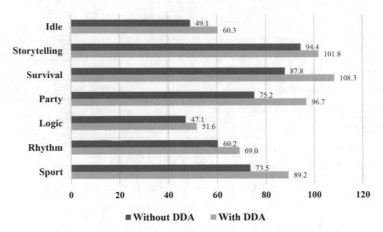

Fig. 3. Mean IEQ scores of the players in each game genre.

game, survival game, and sport game occupy the top three on improvement of IEQ scores. Among the seven genres, the sport game and party game have 75.6 time and 21.5 scores greater respectively, which are the largest. 17.8 and 4.5 of the logic game are the smallest. Overall, our proposed DDA methods enhance player retention as well as experience effectively.

5 Conclusion

This paper proposes a series of DDA methods with high adaptability and customizability for niche games to improve the gaming experience of players as well as reduce the churn rate. The methods are suitable for idle games, storytelling games, survival games, party games, logic games, rhythm games, and sport games.

The novel methods for the seven genres above center on the flow model and define GCT as the core of detailed schemes to improve the efficiency of algorithms and reduce the probability of DDA being perceived by players. All algorithms and mechanics are designed adaptively by analyzing the challenges and interaction of each genre. The results of systematic tests demonstrate that the proposed DDA methods can provide efficient solutions for improving player experience and preventing player churn in various niche games.

In future studies, the authors will continue to focus on the improvement of mechanics of DDA, especially schemes for the logic game, thus promoting the development of DDA and Game AI.

Acknowledgments. This work was supported by National Natural Science Foundation of China under Grant Number: 52130403.

References

1. Kaplan, A., Haenlein, M.: Rulers of the world, unite! The challenges and opportunities of artificial intelligence. Bus. Horiz. **63**(1), 37–50 (2020)
2. Oke, S.A.: A literature review on artificial intelligence. Int. J. Inf. Manag. Sci. **19**(4), 535–570 (2008)
3. Kandlhofer, M., et al.: Artificial intelligence and computer science in education: from kindergarten to university. In: Proceedings of the 2016 IEEE Frontiers in Education Conference, pp. 1–9. IEEE (2016)
4. Wang, Y.: The theoretical framework of cognitive informatics. Int. J. Cogn. Inform. Nat. Intell. **1**(1), 1–27 (2007)
5. Zhang, C., Lu, Y.: Study on artificial intelligence: the state of the art and future prospects. J. Ind. Inf. Integr. **23**, 100224 (2021)
6. Chen, J.X.: The evolution of computing: AlphaGo. Comput. Sci. Eng. **18**(4), 4–7 (2016)
7. Somers, J.: How the artificial-intelligence program AlphaZero mastered its games. New Yorker, vol. 3 (2018)
8. Harper, R.H.: The role of HCI in the age of AI. Int. J. Hum. Comput. Interact. **35**(15), 1331–1344 (2019)
9. Vishwarupe, V., et al.: Bringing humans at the epicenter of artificial intelligence: a confluence of AI, HCI and human centered computing. Proc. Comput. Sci. **204**, 914–921 (2022)
10. Adams, E.: Fundamentals of Game Design. Pearson Education, Upper Saddle River (2014)
11. Gregory, J.: Game Engine Architecture. AK Peters, Natick (2018)
12. Vargas-Iglesias, J.J.: Making sense of genre: the logic of video game genre organization. Games Cult. **15**(2), 158–178 (2020)
13. Clearwater, D.: What defines video game genre? Thinking about genre study after the great divide. Loading **5**(8) (2011)
14. Millington, I., Funge, J.: Artificial Intelligence for Games. CRC Press, Boca Raton (2009)
15. Yannakakis, G.N., Togelius, J.: Artificial Intelligence and Games. Springer, Switzerland (2018). https://doi.org/10.1007/978-3-319-63519-4
16. Rabin, S.: Game AI Pro: Collected Wisdom of Game AI Professionals. CRC Press, Boca Raton (2013)
17. Dziedzic, D., Włodarczyk, W.: Approaches to measuring the difficulty of games in dynamic difficulty adjustment systems. Int. J. Hum. Comput. Interact. **34**(8), 707–715 (2018)
18. Moon, H.S., Seo, J.: Dynamic difficulty adjustment via fast user adaptation. In: Proceedings of the 33rd Annual ACM Symposium on User Interface Software and Technology, pp. 13–15. ACM (2020)
19. Constant, T., Levieux, G.: Dynamic difficulty adjustment impact on players' confidence. In: Proceedings of the 2019 CHI Conference on Human Factors in Computing Systems, pp. 1–12. ACM (2019)
20. Wiemeyer, J., Nacke, L., Moser, C., 'Floyd' Mueller, F.: Player experience. In: Dörner, R., Göbel, S., Effelsberg, W., Wiemeyer, J. (eds.) Serious Games, pp. 243–271. Springer, Cham (2016). https://doi.org/10.1007/978-3-319-40612-1_9
21. Abeele, V.V., et al.: Development and validation of the player experience inventory: a scale to measure player experiences at the level of functional and psychosocial consequences. Int. J. Hum. Comput. Stud. **135**, 102370 (2020)
22. Csikszentmihalyi, M.: Beyond Boredom and Anxiety. Jossey-Bass, London (1975)
23. Klarkowski, M., et al.: Operationalising and measuring flow in video games. In: Proceedings of the Annual Meeting of the Australian Special Interest Group for Computer Human Interaction, pp. 114–118. ACM (2015)

24. Harmat, L., de Manzano, Ö., Ullén, F.: Flow in music and arts. In: Peifer, C., Engeser, S. (eds.) Advances in Flow Research, pp. 377–391. Springer, Cham (2021). https://doi.org/10.1007/978-3-030-53468-4_14
25. Csikszentmihalyi, M.: Applications of Flow in Human Development and Education. Springer, New York (2014). https://doi.org/10.1007/978-94-017-9094-9
26. Keller, J., Bless, H.: Flow and regulatory compatibility: an experimental approach to the flow model of intrinsic motivation. Pers. Soc. Psychol. Bull. 34(2), 196–209 (2008)
27. Fernández del Río, A., Guitart, A., Periáñez, Á.: A time series approach to player churn and conversion in videogames. Intell. Data Anal. 25(1), 177–203 (2021)
28. Xue, S., et al.: Dynamic difficulty adjustment for maximized engagement in digital games. In: Proceedings of the 26th International Conference on World Wide Web Companion, pp. 465–471. ACM (2017)
29. Liu, J., et al.: Deep learning for procedural content generation. Neural Comput. Appl. 33(1), 19–37 (2021)
30. Jennett, C., et al.: Measuring and defining the experience of immersion in games. Int. J. Hum. Comput. Stud. 66(9), 641–661 (2008)

Comparative Analysis of Fine-Tuned MobileNet Versions on Fish Disease Detection

Hien Van Nguyen, Thinh Quoc Huynh, Nhat Minh Nguyen, Anh Kim Su, and Hai Thanh Nguyen$^{(\boxtimes)}$

Can Tho University, Can Tho, Vietnam
{hienb2014833,thinhb2014880,nhatb2014865}@student.ctu.edu.vn,
{sukimanh,nthai.cit}@ctu.edu.vn

Abstract. Smart aquaculture farming is gradually becoming more popular and widely adopted in many fish farms, ranging from small to medium scale. To ensure that the final harvest quantity is not significantly reduced due to diseases, early detection of diseased fish individuals is crucial to remove them from the farming environment, thus preventing rapid spread to other individuals. The main objective of addressing this issue is to utilize identification and classification technologies as alternatives to manual classification methods, which help save time and significantly reduce costs in aquaculture. The various versions of MobileNet are leveraged to classify fish disease. MobileNet is a lightweight network due to its architecture, which can be analyzed depthwise, reducing the model size and computational cost. Training and evaluation were conducted on the SalmonScan dataset; the input image was resized to 224×224, which several research teams had previously used. Experimental results show that MobileNetV1 achieves effective disease classification on fish with (96.30%) and (95.47%) accuracies, respectively, with and without data augmentation, indicating success in identifying disease symptoms in fish. Moreover, compared to previous studies on the dataset where manual processing was required to extract features from images to clarify disease points on fish, MobileNets demonstrate more robust support as it can autonomously learn these features on the network without the need for such complex processing.

Keywords: Fish disease · MobileNets · Data augmentation · Salmon

1 Introduction

As reported in [1], fish is one of the food sources that brings many benefits to human health needs. It has been proven to bring many advantages, such as cardiovascular protection and liver protection, enhancing the ability to fight against viruses and bacteria thanks to immunoglobins in fish proteins. Inside the fish, there is abundant protein content, low-fat content, and accompanied by Omega-3 fatty acids that reduce cholesterol levels in the blood, reducing the risk

L. Barolli (Ed.): IMIS 2024, LNDECT 214, pp. 201–212, 2024.
https://doi.org/10.1007/978-3-031-64766-6_20

of cardiovascular diseases and stroke [2]. Adding fish to the diet menu also helps increase anti-inflammatory capacity, aiding in controlling conditions related to inflammation with active ingredients found in fish.

As of 2019, global fish consumption per capita reached a record high of 20.5 kg compared to only 19.6 kg in the 2010s [3]. In addition, China is the world's largest consumer of freshwater fish, with this country alone having a recorded fish consumption per capita of 40.1 kg in 2019 [3,4]. With the consumption recorded above, the aquaculture industry is a source of supply that meets demand and avoids overfishing in the wild, affecting biodiversity conservation. Aquaculture is the fastest-growing sector in global food production, with particular growth in Asia, contributing 89% of world aquaculture production in 2016 [5]. According to the Food and Agriculture Organization of the United Nations (FAO) records up to the year 2020, the global total fish production has maintained an increasing trend, reaching 87.5 million tonnes of mainly aquatic animals primarily used as food for humans; among these, the production of finfish reached 57.5 million tonnes [3].

The volume of aquatic products supplied has increased to meet market demand, increasing fish stocking density. Along with that, food residues and antibiotics during this period also increased. Most aquaculture operations employ intensive and semi-intensive farming methods, resulting in limited living environments for organisms due to high stocking densities; this increases the risk of disease outbreaks among the animals in such environments [6]. Fish food mixed with high doses of antibiotics will cause residues in the water environment where fish are raised, caused by excretions from fish and some undigested food. These substances persist for a long time and gradually accumulate, forming sediments where bacteria and diseases accumulate that are harmful to organisms living in this environment [7].

In aquaculture, failure to control disease-causing agents such as pathogens or invasive bacteria can lead to mass fish mortality in farms, resulting in significant losses in harvested yield [8]. Some common diseases that significantly affect fish include those caused by fungi, bacteria, hemorrhagic diseases, and exophthalmia [9]. External damage diseases such as necrosis, ulcers, or exophthalmia on fish reduce the acceptability of buyers, making it challenging to market fish products, thereby reducing supply yields. The current solution for ensuring fish harvest yields and ensuring food safety for human health involves rapid detection and classification of diseased fish in aquaculture to quickly eliminate infected individuals, thus preventing rapid spread and impact on other individuals [10]. Classifying fish diseases through manual methods such as dissection incurs high costs and often yields ineffective results. Diagnosing fish diseases underwater requires advanced technical expertise and knowledge, while disease factors spread quickly in the aquatic environment [11]. This necessitates the integration of modern technologies into aquaculture practices to ensure timely detection of diseased individuals.

The Convolutional Neural Network (CNN) has long been widely applied in computer vision. The CNN architecture comprises alternatively stacked convolutional and spatial pooling layers, enabling it to learn features better and improve processing capabilities [12]. For the problem of disease classification in fish, we propose using the MobileNet architecture, a lightweight deep CNN designed based on convolutional layers that can analyze depth except for the first layer, a fully connected layer, thus making the model smaller and reducing computation time [13].

In the remainder of this article, Sect. 2 reviews some related studies on disease fish classification using many different machine learning models. Section 3 outlines how we collect and process data, propose a suitable machine learning model to apply, and present the overall structure of that model. Section 4 describes the experimental process to determine the set of hyper-parameter combinations that help the model achieve optimal results and compare the performance between models to find the most suitable model. Conclusions and future development directions are presented in Sect. 5.

2   Related Work

In the context of the evolving technological era, accompanied by various fields in life, machine learning, and deep learning techniques have been applied to industries in general and the aquaculture industry, in particular, [14]. The development of these technologies has contributed to detecting and classifying diseases in fish based on their manifestations, replacing manual methods and reducing significant costs in aquaculture.

According to Jueal Mia and colleagues in [10], the research team applied the K-means clustering algorithm on images after enhancing the contrast to extract image features and obtain clear images of the affected areas. The classification report was based on self-collected images at fish markets and fisheries captured using digital cameras and mobile phones. The image data included 211 images of fish with Epizootic Ulcerative Syndrome (EUS), 99 images of fish with tail and fin rot, and 175 images of healthy fish. Eight classifiers, including Logistic Regression, Backpropagation Neural Network (BPN), Counterpropagation Neural Network (CPN), Gradient Boosting, Support Vector Machine (SVM), Random Forest, K-Star, and K-nearest neighbors (KNN), were applied to train and evaluate the dataset. The accuracy recorded in the report was quite good, with nearly 88.87% achieved by the Random Forest algorithm in classification. Since fish diseases are pretty diverse, the research team suggests that future work should focus on enhancing the collection of diverse disease images on fish to classify more fish diseases.

Another study in [15] by Md. Rashedul Islam Mamun and their research team utilized image segmentation techniques to identify affected areas, thereby diagnosing diseases in fish. The research team collected 1382 images, including images of fish with red spots, black spots, white spots, and healthy fish. These images were primarily gathered from social media platforms, free image repositories, and publicly available data on the Kaggle platform. An augmented dataset was generated and used to expand the training set by creating modified copies of the collected dataset images. Performance measurement was achieved through the support of performance evaluation matrices, with nine popular classification algorithms applied in the study. The recorded results showed the highest accuracy of 99.64% achieved by the VGG16 and VGG19 ensemble models and 99.28% for ResNet-50. These values demonstrate high accuracy in classifying various diseases in fish, as mentioned above.

The authors in [16] conducted the detection of diseased salmon using an SVM model. The research process was divided into two stages: in the first stage, they performed image segmentation to reduce noise and amplify images based on the Cubic Splines interpolation method. In the next stage, they extracted relevant features for disease classification using the SVM model [17] and kernel function while employing augmented datasets and comparing the effectiveness with and without this stage. Experimental data included images of fresh and diseased salmon, mainly sourced from aquaculture companies and some from the Internet. During the experimental process, input images of various sizes were resized to a fixed size of 600×250 pixels, from which disease points on the fish could be easily identified. The SVM model's performance was recorded with accuracy rates of 91.42% and 94.12% with and without augmentation, respectively. However, the input images must undergo pre-processing techniques to extract features from the images before classification based on the SVM model for better results, resulting in a more elaborate processing pipeline to obtain classification outcomes.

3   Methods

An overview of this section will be shown sequentially. First, Subsect. 3.1 describes how we collect, measure, and analyze data. Next, Subsect. 3.2 briefly describes the input data processing method and lists methods to enhance data usage in the article. We will provide an appropriate solution for the primary processing stages for the overall system architecture in Subsect. 3.3. Figure 1 presents the steps detailed in the following sections.

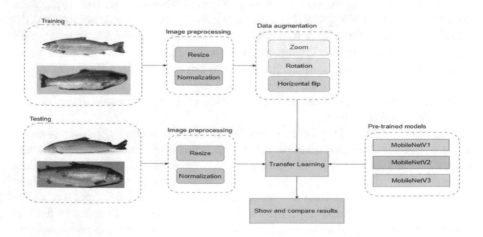

Fig. 1. The proposed workflow for disease diagnosis on fish

3.1   Data Collection

For the fish classification stage, we used the SalmonScan dataset [16, 18] taken from the GitHub data repository, which is an ideal salmon dataset for use in machine learning as well as deep learning, shown in Fig. 2a, images in the dataset include healthy salmon images and infected salmon images shown in Fig. 2a. From the initial dataset consisting of 456 Healthy Salmon images labeled as FreshFish and 752 Infected Salmon images labeled as InfectedFish, with a total of 1208 images, we divide them into 60% training, 20% validation, and 20% testing sets. Table 1 records each directory's detailed number of images.

Table 1. Classes of Salmon fish image and its usage in learning

Salmon Dataset	Train	Validation	Test	Total
FreshFish	273	91	92	456
InfectedFish	451	150	151	752
Total	754	241	243	1208

3.2   Pre-processing and Data Augmentation

The input image described in the MobileNet article has square images. We resize images to a square dimension of 224×224 before training without losing image features to fit the network. The solution is that we use a bilinear interpolation algorithm to standardize image size. This technique calculates the value of an image pixel by interpolating from its four nearest neighbors linearly [19]. Moreover, images are passed through the Normalization layer. According to [20], the layer will perform pre-calculation of the mean and variance of the data. It can shift and scale the input to a distribution centered around 0 with a standard deviation of 1. In this way, the network can stabilize the training process and improve performance. Figure 2b illustrates the whole dataset preprocessing step.

The poor number of images is an issue that directly affects the performance of the deep learning network. In this situation, data augmentation is a savior for deep learning models. The main goal of data augmentation is to enrich the dataset, which helps models improve accuracy and allows better performance on small datasets [21].

To increase the network performance, we implement additional data augmentation techniques to increase the data diversity in quantity and quality visualized in Fig. 2c. First, we scale the pixel values of the image. Next, rotating random images within -10 to $10°$, crop and randomly move images horizontally, and height within -10 to $10°$. Then, we zoom in and out from 80–120% of the original size. Finally, randomly flip images horizontally to match many real-life environmental conditions.

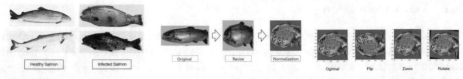

(a) Healthy Salmon and Infected Salmon images

(b) Image after resizing and normalizing

(c) Images after performing data augmentation

Fig. 2. Salmon images and these images after pre-processing and performing data augmentation

3.3   MobileNet and Its Variants for Disease Diagnosis on Images

MobileNet [22] decomposes standard convolution into depthwise convolution used for input filtering and pointwise convolution used to combine the outputs of depthwise convolution, significantly reducing the number of computational operations and the model size. In 2018, MobileNetV2 [23] was introduced, continuing to utilize Depthwise Separable Convolutions from MobileNet. Version 2 was enhanced with Linear Bottlenecks and Inverted Residuals to improve performance and mobility for embedded systems or mobile applications. Shortly after that, in 2019, Google's development team continued to enhance the MobileNets model and introduced MobileNetV3 by adding the NetAdapt algorithm. In this version, MobileNetV3 [24] aimed at tasks such as object detection and semantic segmentation, achieving notably high efficiency while reducing computation time.

In this study, we propose using this architecture with the MobileNetV1 version; we also compare it with two other versions, namely MobileNetV2 and MobileNetV3, to evaluate the performance of the three versions in the fish disease classification task. MobileNetV1 has 28 layers; we apply transfer learning to retain the weights and save model training time. The first is to freeze the weights and change the model's output to fit the problem. Next, create an additional data augmentation layer for the image to pass through before entering the model. After entering the model, the first standard convolutional layers have the effect of extracting features, reducing the size of the image, and increasing the channel; depthwise convolution will continue to extract features, and pointwise convolution will combine information from different channels and increase the number of channels as desired. Finally, we add layers and activation functions to the output to classify the image according to the number of layers of the data. Figure 3 shows the disease fish dataset passing through the MobileNetV1 architecture.

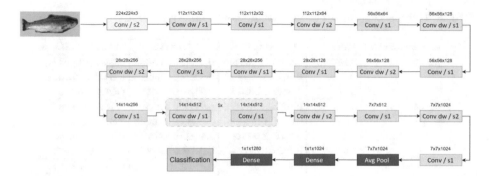

Fig. 3. An illustration of MobileNetV1 architecture to classify diseases fish

4   Experimental Results

4.1   Environmental Settings

To take advantage of available resources on the internet, experiments are performed on the Kaggle virtual machine environment with the following parameters: CPU Intel P100 with two core four thread 2 GHZ, GPU P100-PCIE 16 GB, 29 GB RAM, and 73.1 GB Disk.

4.2   Hyper-Parameter Tuning

Quite a few hyper-parameters are used when training the model, and each hyper-parameter value will affect the results, so we compared the parameters with many different values to get results. Get the results that match our expectations. Here, we choose the hyper-parameters to include in the model: optimizers, batch size, learning rate, weight decay, and dropout. For the number of epochs, use the Early stopping technique to stop training when the validation loss index no longer decreases. As described in Table 2, we record the optimal parameters.

During testing with three versions of MobileNet, we conducted several tests with different settings. We got the following results: The results will displayed as an accuracy index ± standard deviation, and the highest hyper-parameter value will be bolded for easy identification.

Table 2. Hyper-Parameter tuning result (bold values represent the highest results)

Hyper-parameters	Value	MobileNetV1	MobileNetV2	MobileNetV3Large
Optimizers	RMSProp	0.9676 ± 0.0048	**0.9485 ± 0.0033**	0.9444 ± 0.0085
	Adam	**0.9742 ± 0.0031**	0.9436 ± 0.0020	**0.9477 ± 0.0056**
	SGD	0.9419 ± 0.0108	0.9328 ± 0.0017	0.8921 ± 0.0157
Learning rate	0.01	**0.9734 ± 0.0056**	0.9419 ± 0.0026	0.9436 ± 0.0067
	0.001	0.9701 ± 0.0048	**0.9452 ± 0.0048**	**0.9502 ± 0.0037**
	0.0001	0.9568 ± 0.0100	0.9419 + 0.0079	0.9477 + 0.0042
	0.00001	0.9029 ± 0.0374	0.9079 ± 0.0132	0.9311 ± 0.0072
Dropouts	0.1	0.9710 ± 0.0045	0.945 ± 0.0031	0.9369 ± 0.0109
	0.2	**0.9734 ± 0.0050**	0.9378 ± 0.0059	0.9477 ± 0.0062
	0.3	0.9685 ± 0.0081	0.9469 ± 0.0069	**0.9485 ± 0.0033**
	0.4	0.9651 ± 0.0097	**0.9477 ± 0.0056**	0.9477 ± 0.0093
	0.5	0.9602 ± 0.0042	0.9436 ± 0.0089	0.9461 ± 0.0079
Weight decay	0.001	0.9676 ± 0.0048	0.9461 ± 0.0069	0.9436 ± 0.0056
	0.0001	0.9710 ± 0.0079	0.9469 ± 0.0048	**0.9519 ± 0.0033**
	0.00001	**0.9718 ± 0.0066**	0.9452 ± 0.0048	0.9452 ± 0.0031
	0.000001	0.9635 ± 0.0017	**0.9477 ± 0.0056**	0.9461 ± 0.0069
Batch size	16	0.9618 ± 0.0061	0.9419 ± 0.0087	0.9485 ± 0.0085
	32	0.9685 ± 0.0089	**0.9469 ± 0.0041**	0.9477 ± 0.0072
	64	**0.9718 ± 0.0031**	0.9461 ± 0.0037	0.9444 ± 0.0042
	128	0.9701 ± 0.0017	0.9469 ± 0.0055	**0.9485 ± 0.0042**

For the Optimizers, MobileNetV1 runs well with the Adam at 97.42%, MobileNetV2 runs well with the RMSProp at 94.85%, and MobileNetV3 runs well with the Adam at 94.77%. For the Learning rate MobileNetV2 and MobileNetV3, good results are obtained with an index of 0.001 with an accuracy of 94.52% and 95.02%, respectively. MobileNetV1 also produces relatively high results, but the index of 0.01 is slightly better at 97.34%. For batch sizes in order of index 32, 64, 128 are MobileNetV2 94.69%, MobileNetV1 97.18%, MobileNetV3 94.85%. MobileNetV1 gives a weight decay result of 0.00001 with 97.18%, MobileNetV2 gives 94.77% with 0.000001, MobileNetV3 gives 95.19% with 0.0001. For dropout rate, MobileNetV2 archives 94.77% with 0.4, while MobileNetV1 gives 97.34% with 0.2, and MobileNetV3 gives 94.85% with a dropout rate of 0.3. After having the values of the Hyper-parameters that give the highest accuracy for each model, MobileNetV1 (Optimizers: Adam, Learning rate: 0.01, Dropouts: 0.2, Weight decay: 0.00001, Batch size: 64), MobileNetV2 (Optimizers: RMSProp, Learning rate: 0.001, Dropouts: 0.4, Weight decay: 0.000001, Batch size: 32), MobileNetV3 (Optimizers: Adam, Learning rate: 0.001, Dropouts: 0.3, Weight decay: 0.0001, Batch size: 128), we conduct training and have the following results in Table 3, We can see that MobileNetV1, although it is the first model to be created, gives results with the highest accuracy, demonstrating the suitability of the learning data with the model even though there is no need for augmentation. Although MobileNetV2 and MobileNetV3 have more modern improvements, they do not produce results as good as MobileNetV1 in some cases.

Table 3. Results None using data augmentation

Model	Accuracy	F1-score	Precision	Recall
MobileNetV1	0.9547	0.9511	0.9598	0.9445
MobileNetV2	0.9300	0.9258	0.9249	0.9267
MobileNetV3Large	0.9218	0.9143	0.9322	0.9031

4.3 The Efficiency of Data Augmentation

In this scenario, we choose the model with the highest accuracy in Subsect. 4.2 is MobileNetV1 to continue training with the data set with augmentation applied and compare between with and without augmentation.

Table 4. Results between non-using data augmentation and using data augmentation

MobileNetV1	Accuracy	F1-score	Precision	Recall
Using data augmentation	0.9630	0.9473	0.9007	1.0000
Non-using data augmentation	0.9547	0.9511	0.9598	0.9445

As described in Table 4, the MobileNetV1 model has improved slightly in terms of accuracy of 96.30% when used with the dataset with augmentation measures compared to 95.47%, although the index has not changed too much, we still see the importance of data augmentation in model training, the following charts in Fig. 4 track the detailed results of data augmentation.

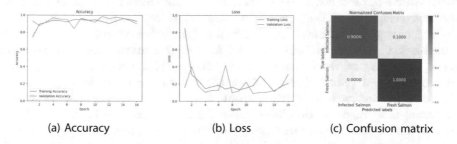

(a) Accuracy　　　　(b) Loss　　　　(c) Confusion matrix

Fig. 4. Accuracy, Loss, and Confusion matrix of MobinetNetV1 after using data augmentation

4.4 Discussion

In this study, we propose the MobileNets architecture, a lightweight CNN model that helps increase efficiency and accuracy when classifying sick fish. The training and evaluation images from the Salmon dataset, including two types of images, healthy salmon and infected salmon, many research groups have used the dataset to perform the classification process to identify diseases. The preprocessing step matches the image to the model input while normalizing the image to deliver better results. In addition, data augmentation is used to avoid overfitting due to poor data, and we also compare results when this method is not used. However, most studies use manual processing techniques with input fish images, using feature extraction to change the image's resolution to clarify the fish's disease spots. This method is complicated, requires specialized knowledge, and takes time to implement.

Simultaneously, the training process is conducted multiple times with adjustments to hyperparameters to observe effectiveness and select the hyperparameter set that optimizes the model's accuracy during experimentation. The experimental results yielded relatively high results; MobileNetV1 achieved the highest values with (96.30%) and (95.47%), respectively, for accuracy with and without data augmentation. The remaining two versions still show effectiveness in classification, with MobileNetV2 accuracy reaching a value of (93.00%), and MobileNetV3 with (92.18%) without data augmentation. These values have shown the classification efficiency when applying the MobileNet architecture to the problem of classifying diseases in fish.

5   Conclusion

Controlling diseases in aquaculture fish helps ensure harvest yield and fish quality supplied to the market. One of the primary solutions is promptly detecting diseased fish and implementing timely measures to protect the aquatic environment and prevent its spread to other individuals. In this study, we conducted disease classification on fish using MobileNet - a variant of CNN characterized by depth-wise separable convolutions enabling compact model size and significantly reduced computation time to achieve high efficiency in disease classification tasks. The support from MobileNet is further demonstrated as it directly extracts features from images without the need for complex manual feature extraction steps. Experimental results show the success of MobileNets with accuracy values of (96.30%) and (95.47%) with and without data augmentation, respectively. These figures indicate relatively high accuracy in disease classification on fish, and using data augmentation helps improve accuracy and avoid overfitting issues due to limited data. In the future, we aim to collect more diverse and specific data on various fish diseases to identify diseases and accurately enhance practicality when applied in aquaculture. The research results show that it can be extended to the internet service environment, an environment that requires low latency and high performance. This requires a model that meets two requirements: fast and lightweight. MoblileNet completely meets the above two requirements.

References

1. Chen, J., Jayachandran, M., Bai, W., Xu, B.: A critical review on the health benefits of fish consumption and its bioactive constituents. Food Chem. **369**, 130874 (2022). https://www.sciencedirect.com/science/article/pii/S030881462101880X
2. Burger, J., Gochfeld, M.: Perceptions of the risks and benefits of fish consumption: individual choices to reduce risk and increase health benefits. Environ. Res. **109**(3), 343–349 (2009). https://www.sciencedirect.com/science/article/pii/S0013935108002715
3. FAO: The state of world fisheries and aquaculture 2022. FAO (2022). https://www.fao.org/3/cc0461en/online/cc0461en.html
4. Ai, L., Ma, B., Shao, S., Zhang, L., Zhang, L.: Heavy metals in Chinese freshwater fish: levels, regional distribution, sources and health risk assessment. Sci. Total Environ. **853**, 158455 (2022). https://www.sciencedirect.com/science/article/pii/S0048969722055541
5. Burić, M., Bavčević, L., Grgurić, S., Vresnik, F., Križan, J., Antonić, O.: Modelling the environmental footprint of sea bream cage aquaculture in relation to spatial stocking design. J. Environ. Manag. **270**, 110811 (2020). https://www.sciencedirect.com/science/article/pii/S0301479720307428
6. Pedrazzani, A.S., Quintiliano, M.H., Bolfe, F., Sans, E.C.d.O., Molento, C.F.M.: Tilapia on-farm welfare assessment protocol for semi-intensive production systems. Front. Vet. Sci. **7** (2020). https://www.frontiersin.org/articles/10.3389/fvets.2020.606388/full
7. Watts, J.E.M., Schreier, H.J., Lanska, L., Hale, M.S.: The rising tide of antimicrobial resistance in aquaculture: sources, sinks and solutions. Marine Drugs **15**(6) (2017). https://www.mdpi.com/1660-3397/15/6/158
8. Abdelsalam, M., Elgendy, M.Y., Elfadadny, M.R., Al, S.S., Sherif, A.H., Abolghait, S.K.: A review of molecular diagnoses of bacterial fish diseases. SpringerLink **31**, 417–434 (2023)
9. Opiyo, M.A., Marijani, E., Muendo, P., Odede, R., Leschen, W., Charo-Karisa, H.: A review of aquaculture production and health management practices of farmed fish in Kenya. Int. J. Vet. Sci. Med. **6**(2), 141–148 (2018). https://doi.org/10.1016/j.ijvsm.2018.07.001, pMID: 30564588
10. Mia, M.J., Mahmud, R.B., Sadad, M.S., Asad, H.A., Hossain, R.: An in-depth automated approach for fish disease recognition. J. King Saud Univ. Comput. Inf. Sci. **34**(9), 7174–7183 (2022). https://www.sciencedirect.com/science/article/pii/S1319157822000672
11. Li, D., Li, X., Wang, Q., Hao, Y.: Advanced techniques for the intelligent diagnosis of fish diseases: a review. Animals **12**(21) (2022). https://www.mdpi.com/2076-2615/12/21/2938
12. Yu, S., Jia, S., Xu, C.: Convolutional neural networks for hyperspectral image classification. Neurocomputing **219**, 88–98 (2017). https://www.sciencedirect.com/science/article/pii/S0925231216310104
13. Phiphiphatphaisit, S., Surinta, O.: Food image classification with improved mobilenet architecture and data augmentation. In: Proceedings of the 3rd International Conference on Information Science and Systems, ICISS 2020, pp. 51–56. Association for Computing Machinery, New York (2020). https://doi.org/10.1145/3388176.3388179
14. Zhao, S., et al.: Application of machine learning in intelligent fish aquaculture: a review. Aquaculture **540**, 736724 (2021). https://www.sciencedirect.com/science/article/pii/S0044848621003860

15. Mamun, M.R.I., Rahman, U.S., Akter, T., Azim, M.: Fish disease detection using deep learning and machine learning. Int. J. Comput. Appl. **185** (2023)
16. Ahmed, M.S., Aurpa, T.T., Azad, M.A.K.: Fish disease detection using image based machine learning technique in aquaculture. J. King Saud Univ. Comput. Inf. Sci. **34**(8, Part A), 5170–5182 (2022). https://www.sciencedirect.com/science/article/pii/S1319157821001063
17. Pisner, D.A., Schnyer, D.M.: Support vector machine. In: Machine Learning, pp. 101–121. Elsevier (2020)
18. Ahmed, M.S.: Salmonscan: a novel image dataset for fish disease detection in salmon aquaculture system (2024). https://data.mendeley.com/datasets/x3fz2nfm4w/1
19. Rukundo, O., Schmidt, S.E.: Effects of rescaling bilinear interpolant on image interpolation quality. In: Dai, Q., Shimura, T. (eds.) Optoelectronic Imaging and Multimedia Technology V, vol. 10817, p. 1081715. International Society for Optics and Photonics, SPIE (2018). https://doi.org/10.1117/12.2501549
20. Ba, J.L., Kiros, J.R., Hinton, G.E.: Layer normalization. arXiv preprint arXiv:1607.06450 (2016)
21. Mumuni, A., Mumuni, F.: Data augmentation: a comprehensive survey of modern approaches. Array **16**, 100258 (2022)
22. Howard, A.G., et al.: MobileNets: efficient convolutional neural networks for mobile vision applications. CoRR abs/1704.04861 (2017). http://arxiv.org/abs/1704.04861
23. Sandler, M., Howard, A.G., Zhu, M., Zhmoginov, A., Chen, L.: Inverted residuals and linear bottlenecks: mobile networks for classification, detection and segmentation. CoRR abs/1801.04381 (2018). http://arxiv.org/abs/1801.04381
24. Howard, A., et al.: Searching for mobilenetv3. CoRR abs/1905.02244 (2019). http://arxiv.org/abs/1905.02244

Methodology to Monitor and Estimate Occupancy in Enclosed Spaces Based on Indirect Methods and Artificial Intelligence: A University Classroom as a Case Study

Alma Mena-Martinez[1(✉)], Joanna Alvarado-Uribe[1,2], Manuel Davila Delgado[3], and Hector G. Ceballos[1,2]

[1] School of Engineering and Sciences, Tecnologico de Monterrey, Monterrey, Mexico
{a00834070,joanna.alvarado,ceballos}@tec.mx
[2] Institute for the Future of Education, Tecnologico de Monterrey, Monterrey, Mexico
[3] Applied Artificial Intelligence (AI) and Business Transformation, Birmingham City University (BCU), Birmingham, UK
manuel.daviladelgado@bcu.ac.uk

Abstract. Despite the indoor occupancy monitoring has been studied over the years to reduce energy and to improve users' comfort, the authors in their works only present a brief explanation of the sensors, algorithms, and models implemented, as well as the results obtained, leaving aside the preview steps related to the sensor selection and deployment. Therefore, this paper proposes a methodology for indoor occupancy monitoring in real-life scenarios in a non-intrusive manner, including data collection aspects, data preprocessing, and models selection. Besides, to evaluate the effectiveness of the methodology proposed, a case study is presented. The experiment was conducted in a classroom at the University of the West of England, Bristol UK deploying four Internet of Things (IoT) environmental sensors to collect the data. The case study showed that the methodology can guide researchers interested in monitoring occupancy in enclosed spaces through non-intrusive sensors and using Artificial Intelligence.

1 Introduction

Real-time monitoring, controlling, and managing of buildings and occupants, appliances, systems, and the environment are crucial for energy savings and effective work-space management, especially considering COVID-19 sanitary measures [13,15].

To measure indoor occupancy there are wide approaches and tools categorized as direct and indirect. Direct methods, like vision systems and radio frequency identification (RFID) tags, offer precise detection but raise privacy concerns. These intrusive methods could gather extremely delicate information

L. Barolli (Ed.): IMIS 2024, LNDECT 214, pp. 213–225, 2024.
https://doi.org/10.1007/978-3-031-64766-6_21

from users [10,12,24]. Conversely, indirect methods such as Passive Infrared (PIR), electricity meters, ultrasonic, and environmental sensors are less intrusive [12,15]. Common environmental sensors include CO_2, temperature, relative humidity (RH), Total Volatile Organic compounds (TVOC), and illuminance sensors [12,15].

On the other hand, occupancy monitoring can be addressed as binary presence detection, counting occupants, and occupancy levels (empty, low, medium, high) [15]. Furthermore, Machine Learning (ML) models, including Support Vector Machine (SVM), Artificial Neural Networks (ANN), and probabilistic models like the Hidden Markov model, have proved to be effective for this occupancy classification task [15,18]. Additionally, the Internet of Things (IoT) facilitates connectivity among diverse devices, information exchange, and privacy/security preservation through wireless communication technologies [13].

Despite the indoor occupancy monitoring has been studied over the years to reduce energy and to improve users' comfort [18], the authors in their works only present a brief explanation of the sensors, algorithms, and models implemented, as well as the results obtained, without describing in detail the steps related to sensor selection and deployment. Therefore, the main objective of this work is to put forward a methodology for indoor occupancy monitoring using indirect methods (environmental sensors), IoT, and ML models. Thereby, the contributions of this work are: (i) A tested methodology for indoor occupancy monitoring in real-life scenarios using indirect methods and AI. (ii) An evaluation of the methodology through a case study. (iii) A generic checklist to prompt the adoption of monitoring systems by practitioners and new researchers.

The rest of this article is organized as follows. Firstly, the related work is presented in Sect. 2. Subsequently, Sect. 3 describes the proposed methodology, whereas the case study is presented in Sect. 4. Finally, Sect. 5 shows the conclusion of this work.

2 Related Work

IoT is a system comprised of Internet-enabled objects, such as sensors and devices, connected over the Internet. It can transfer data over wired and wireless networks, enabling remote communication between these objects [13]. Over the years, various tools and approaches have been used to address indoor occupancy. For example, Zemouri et al. [22] proposed an embedded system using sensors like temperature and RH connected to the IBM Bluemix cloud to detect human presence in an office. Similarly, Zimmermann et al. [25] designed a system capable of sensing CO_2, TVOC, air temperature, and RH. The data were saved on a microSDHC card and the ground truth by a cellphone app. Multiple ML models were training for occupancy detection and counting occupants. On the other hand, Vela et al. [19] proposed a low-cost embedded system focusing solely on measuring air temperature, RH, and barometric pressure to estimate occupancy levels in a living room and an university gym. The sensor data was stored in the cloud and a file was exported to train SVM, k-Nearest Neighbor (k-NN), and

Decision Trees (DT). Finally, Ahmed et al. [3] detected and counted occupants using environmental gaseous parameters, such as CO_2, liquefied petroleum gas, nitrogen dioxide and sulfur dioxide, along with weather parameters like air temperature and RH. The experiment was conducted in a laboratory and data were saved on a Raspberry Pi (RPi) through Serial Communication. Multiple ML models were trained.

The related work presented various perspectives and methodologies that describe their sensors and approaches. Most studies provide a brief explanation of the designed system and preprocessing methods. Likewise, the ML algorithms are presented along with their performance evaluations. However, several studies lack detailed information on the sensor selection and deployment steps.

3 Methodology to Monitor Occupancy in Enclosed Spaces Through Indirect Methods

This work proposes a methodology that allows the deployment or adoption of monitoring systems in real scenarios. It encompasses three general phases constituted by data collection, preprocessing activities, and classification task, as well as insights from practical experience. Figure 1 illustrates the proposed methodology.

3.1 Data Collection

In this initial phase, a robust data collection protocol is developed for ongoing indoor occupancy monitoring. The goal is to gather information without drawbacks during experimentation. The key activities to consider are listed below:

- **Monitoring device:** Considering privacy concerns in indoor occupancy monitoring, it is recommended to use non-intrusive sensors, such as PIR, ultrasonic, CO_2, temperature, RH, and atmospheric pressure. Devices can either be self-developed by integrating sensors with a micro-controller [19, 22, 25] or purchased as commercial sensor devices [11,14,24], depending on the research budget and objectives.
- **Set up the device:** Before deployment, it is advisable to test the monitoring device to ensure accurate data acquisition and detect any special adaptations needed (e.g., electric contacts, WiFi, batteries, additional devices). Additionally, it is necessary to select where to store the data, whether in a cloud database or on a local computer. In case of choosing a cloud storage, the sensor must be correctly linked to the selected cloud platform.
- **Occupancy resolution:** Indoor occupancy can be addressed as a binary problem (occupied or empty) [11,22], counting the number of occupants [24,25], and through occupancy levels (empty, low, medium, high) [14,19]. Defining the occupancy levels requires determining the maximum and minimum capacity of the places to establish proportional intervals between the maximum and minimum number of occupants.

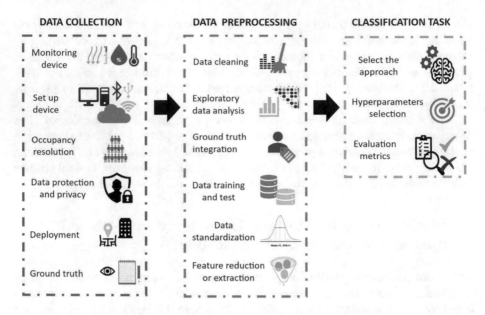

Fig. 1. Proposed experimental methodology.

- **Data protection and privacy:** Always consult the institution's privacy policy where data will be collected. Additionally, it is beneficial to contextualize venue users about the monitoring device's functions and the research objectives so that the device is familiar to them and the research progresses properly.
- **Deployment:** For environmental monitoring it is recommended to place sensors on tables at the room center [15]. Alternatively, it is suggested to install devices on walls at 100–150 cm from the floor [15]. On the other hand, the number of devices depends on the enclosed area, as well as the budget for the experimentation. A linear regression (Eq. 1) was proposed to estimate the number of sensors per area [15].

$$y = 0.0175X + 0.7132 \tag{1}$$

where X represents the area of the enclosed space in m^2.

- **Ground truth:** This data aims to offer accurate occupancy values. To address privacy concerns, a viable method is essential, such as a device with buttons, entrance sensors, or a person taking note of the number of people in a defined time window.

3.2   Data Preprocessing

Before training model, data preparation is crucial. A meticulous and effective pipeline is proposed to ensure information integrity and reliability for subsequent analysis. The following tasks should be carried out:

- **Data cleaning:** For reliability, datasets should be identified by MAC address or device number. Furthermore, to prevent instances with no occupants, data may be selected based on weekdays and work hours [2]. Regarding missing values, if less than 1%, can be removed [1]. Otherwise, suitable methods like interpolation or Rolling Moving Average should be employed for filling missing data [22,25].

- **Exploratory Data Analysis (EDA):** The primary goal is to grasp the database structure, detect outliers, and reveal relationships between variables [10]. Typically, a timeline graph of environmental variables identifies temporal patterns, complemented by scatter plots to explore relationships between variables.

- **Ground truth integration:** Next, to generate labeled data, ground truth related to occupancy is incorporated into the datasets. When occupancy labels are categorical, it is advisable to use a code. For example, for occupancy levels, it is suggested to use the letters "E", "L", "M", and "H" to represent empty, low, medium, and high levels, respectively.

- **Data training and test:** To avoid over-fitting and ensure unbiased model evaluation, the dataset should be randomly divided for training and testing. Typically, splits like 60:40, 70:30, or 80:20 are common (training:testing), but it depends on the dataset size [17].

- **Data standardization:** Many ML algorithms are sensitive to feature scale. Keeping features within a similar scale facilitates importance comparison and expedites algorithm convergence, improving overall performance and training times [9].

- **Feature reduction or extraction:** To reduce the number of features while preserving the most relevant information, two main methods are employed: (i) Feature Selection, which selects a subset of features, and (ii) Feature Extraction, transforming the original features into a more informative and compact set [21].

3.3 Classification Task

In this phase, data classification is performed according to the defined occupancy resolution, requiring model and hyperparameter selection. Key activities include:

- **Select the approach:** To estimate the defined indoor occupancy resolution, models commonly used in the literature include Supervised ML, being the most popular: SVM, RF, and ANN [15,18]. Additionally, probabilistic methods like the Hidden Markov Model are widely explored [15,18].

- **Hyperparameters selection:** To achieve an optimal model that effectively minimizes a predefined loss function on provided test data, careful selection of hyperparameters for each model is necessary [8]. Hyperparameters serve to fine-tune different facets of the learning algorithm, varying significant influence on the resultant model and its performance [8].

- **Evaluation metrics:** Using multiple metrics provides a comprehensive view of model performance, preventing overreliance on a single metric. However, it is crucial to select appropriate metrics based on the trained model. Common metrics for evaluating classification tasks include accuracy, precision, recall, F1-score, confusion matrix, log-loss, and area under the curve (AUC). In contrast, root mean square error (RMSE), mean squared error (MSE), and root relative squared error (RRSE) are employed for regression tasks (predicting numeric scores) [18, 23].

Additionally, a checklist based on the proposed methodology is provided in the Appendix for students and new practitioners interested in researching occupancy monitoring.

4   Case Study and Discussion

The proposed methodology has been applied as the case study in a classroom 5X103 at the University of the West of England (UWE), Bristol, United Kingdom (Fig. 2). Data collection occurred on September 15th, 2022, during a workshop from 13:00 to 16:00 hrs. The classroom, measuring $59.86 \, m^2$ and accommodating 25 occupants, includes four tables with six seats, a speaker's computer desktop, a heater, whiteboards on three walls, four large windows ($100 \times 150 \, cm$), and one access door. During the experiment, the windows remained open and the door was mostly closed.

To carry out data collection, an IoT environmental sensor previously designed by some authors of this paper (patent pending MX/a/2022/016112) was used, which is a low-cost, non-intrusive device that is easy to transport and install onsite. This IoT monitoring device includes air temperature (°C), relative humidity (%), and atmospheric pressure (hPa) sensors (Fig. 3). Sensor specifications are: (i) pressure 300 to 1100 hPa (± 1 hPa accuracy), (ii) temperature -40 to $85 \, °C$ ($\pm 1 \, °C$ accuracy), and (iii) relative humidity 0% to 100% ($\pm 3\%$ accuracy) [19]. However, the IoT device, with a case-insulated sensor, has an approximate $\Delta Temperature$ of $11.5 \, °C$ and $\Delta relative\_humidity$ of 19.5%.

Fig. 2. Case study in Classroom 5X103.

The device was connected to AWS IoT core [4], employing the MQTT protocol for communication. Measurements from the device were sent as JSON documents to a DynamoDB database [5,19]. All traffic to and from AWS IoT is securely transmitted over Transport Layer Security (TLS). Hence, each IoT device requires a unique set of X.509 certificates and private keys for secure communication [4].

Fig. 3. The IoT device is integrated with temperature, relative humidity, and atmospheric pressure sensors.

4.1 Data Collection

During this phase, IoT devices were placed at the center of tables, but complications arose with some devices due to connectivity issues with the WiFi network, causing delays in data transmission. However, cloud storage was efficiently executed.

On the other hand, occupancy levels were assigned for each table: "E" $= 0$, "L" $= 1$–2, "M" $= 3$–4, and "H" $= +5$, focusing solely on the illustrative purpose of the proposed methodology. Regarding data protection and privacy, the University's privacy policy was followed in this experiment. Furthermore, the ground truth was gathered manually, that is, one person counted the occupants without recording specific behaviors or characteristics. The ground truth dataset includes date, start-end hour (indicating the period of unchanged conditions), occupant count, and door/windows status. Furthermore, all data adheres to the (EU) 2016/679 General Data Protection Regulation (GDPR), the Data Protection Act 2018 (or any successor legislation), and relevant privacy laws [20].

4.2 Data Preprocessing

To clean datasets, a Comma-Separated Values (.CSV) file was downloaded from AWS and ground truth data was manually added. Each sensor's dataset is generated considering weekdays (Monday-Friday) and work hours (7 am–7 pm). These individual datasets no contain missing values as they solely consider the timestamp gathered for each sensor, eliminating the need for any filling method. Table 1 provides a description of the datasets generated and used to train the ML models. The resulting datasets are made up of the following variables: (i) **Date:** time in Hour:Minutes and date in Year-Month-Day, (ii) **MAC Address:**

device's identifier, (iii) **Temperature:** air temperature in °C, (iv) **Humidity:** relative humidity in %, (v) **Pressure:** barometric pressure in hPa, (vi) **Occupants:** number of occupants, and (vii) **Label:** occupancy level as E, L, M, and H.

Table 1. Data teaser of the resulting datasets.

Date	ID Sensor	Dataset Size	Training Data	Testing Data	Training Classes			
					E	L	M	H
September 15th 2022 1 pm–4 pm	S03	72	59	13	11	7	34	7
	S10	64	50	14	12	7	31	0
	S12	84	66	18	24	0	42	0
	S09	75	59	16	16	22	21	0

For a comprehensive understanding of the variables, an EDA was performed using a timeline graph (Fig. 4). This graph illustrates the relationship among temperature, RH, pressure, and occupancy over time. The temperature increases as people remain around the table, while the humidity decreases. Pressure changes are notable, especially during transitions from medium to low occupancy levels, showing a decrease, and from medium to high occupancy levels, indicating an increase. The variable behaviors exhibit a lag with respect to the real-time occupancy records, considering the time that the environmental conditions of the place are updated.

Finally, 80% of the dataset was used for training models, and the remaining 20% for testing. Moreover, the data was standardized using the Standard scikit-learn function [16], which centered the variable at zero and standardized the variance to 1, along with Principal Component Analysis (PCA) using three components to reduce noise [3].

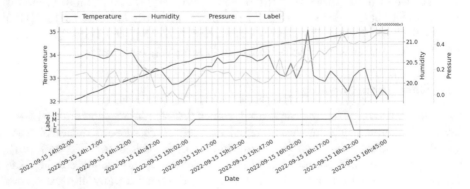

Fig. 4. Temperature (blue), humidity (green), barometric pressure (yellow), and occupancy level (red) of the IoT device S10.

4.3   Classification of Occupancy Levels

SVM, RF, and MLP (supervised ML algorithms) were chosen to classify indoor occupancy levels per table. Their hyperparameters were selected using the Grid-SearchCV function from the scikit-learn library [16], using the values shown in Table 2.

Table 2. Parameters established for the GridSearchCV function.

Algorithm	Parameter	Values
SVM	C	1, 10, 100, 1000
	gamma	1, 0.1, 0.01, 0.001, 0.0001
	kernel	rbf
RF	n estimators	Range(2, 30)
	max depth	Range(2, 15)
	criterion	gini, entropy
MLP	hidden layer	[(range(5, 15), range(5, 15))], [(range(5, 15), range(5, 15), range(5, 15))], [(range(5, 15))]
	activation	tanh, relu, identity, logistic
	solver	adam, sgd, lbfgs
	alpha	10.0 ** -np.arange(1, 10)
	learning rate	constant, adaptative, invscaling

The models were finally trained with the following parameters: SVM with fixed C and Gamma values of 1; RF using entropy criterion, 17 estimators, and a max depth of 7; and MLP tuned with a constant learning rate, tanh activation, lbfgs solver, alpha of 0.1, and a hidden layer of (11,). All experiments were executed on a laptop with an AMD® Ryzen 7 pro 4750u processor, 46.3 GiB RAM, 1.0 TB SSD, and Ubuntu 20.04.4 LTS - 64 bit as the operating system. The models were implemented in Python 3.8.10 using the scikit-learn library.

To assess the model performance, the accuracy and F1-Score metrics and the confusion matrix were chosen. Accuracy [6] represents the number of correctly classified instances over the total number of data instances. Therefore, a result closed to 1 implies an effective classification. Whereas, F1-Score [23] can be interpreted as a weighted harmonic mean of the precision and recall. The score can achieve its highest value at 1 and lowest at 0. The confusion matrix provides a visual summary of correct and incorrect classifications for each class, with each matrix row representing the true label (TL) and each column representing the classifier's prediction [23].

Table 3 presents the results by model and sensor, considering that one sensor was placed per table. In general, the models achieve on average high accuracy (SVM: 0.90, RF: 0.87, MLP: 0.88) and their F1-scores (SVM: 0.67, RF: 0.64,

MLP: 0.65) suggest a trade-off between precision and recall. This may be due to class imbalances. Likewise, the confusion matrix reveals a consistent issue with classifying the "M" level. All models misclassify it as "E" or "L" levels. It is important to highlight that the experiment was not controlled, that is, the room conditions remained unchanged and the capacity was not exceeded. Consequently, some classes were absent. Therefore, based on these results, some recommendations are to increase data collection efforts to capture a more balanced representation of all classes or to apply a class balancing algorithm, such as SMOTE [7]. In addition to considering the trade-off between model performance and hyperparameter tuning time. For instance, although MLP offers similar accuracy, tuning its hyperparameters can take longer than RF.

Table 3. Model evaluation.

ID Sensor	Metric	SVM	RF	MLP	Avg Sensor
S03	Accuracy	0.85	0.85	0.85	0.85
	F1-Score	0.83	0.83	0.83	**0.83**
S10	Accuracy	0.93	0.93	0.93	**0.93**
	F1-Score	0.70	0.70	0.70	0.70
S12	Accuracy	0.94	0.94	0.94	**0.94**
	F1-Score	0.47	0.47	0.47	0.47
S09	Accuracy	0.88	0.75	0.81	0.81
	F1-Score	0.66	0.57	0.61	0.61
Avg	Accuracy	**0.90**	0.87	0.88	
Avg	F1-Score	**0.67**	0.64	0.65	

5   Conclusion

In this paper, a methodology to measure and estimate indoor occupancy is proposed. The methodology is based on IoT environmental sensors (indirect method) for data collection, cloud storage, and supervised ML algorithms for occupancy estimation. Furthermore, a case study in a real-life scenario (classroom) was illustrated. SVM, RF, and MLP models were trained to classify occupancy levels, which were established as Empty, Low, Medium, and High. Finally, the accuracy and F1-Score metrics, as well as the confusion matrix, were used to evaluate the model performance. The results show that the models can classify the occupancy level for each table. The SVM model achieved the best average model accuracy of 90%, while the devices S12 and S10 achieved the best average sensor accuracy.

Thereby, the objective of this work was fulfilled. Therefore, the contributions of this work were: (i) A flexible method to monitor indoor occupancy while avoiding privacy concerns, (ii) a methodology can be applied using different kind of non-intrusive sensors in real-scenarios, (iii) the use of low-cost, homemade, and easy-to-install IoT devices, and (iv) a generic checklist to support the adoption of monitoring systems. Nevertheless, some aspects that emerged during the experimentation need to be analyzed in future work. For instance, determine how much the external weather conditions affect the model performance. Therefore, future work should focus on an analysis of sensor placement to evaluate whether external factors, such as open doors or windows, can affect sensor measurements and impact model accuracy. Furthermore, a data fusion stage should be proposed to integrate sensor fusion information to address missing data and improve model performance. Additionally, the deployment of the classification task should be carried out in real-time instead of running it offline.

Acknowledgements. The authors would like to thank the University of the West of England for its support in carrying out this research project. Furthermore, the first author thanks to Tecnologico de Monterrey and CONAHCYT for the Ph.D. studentship.

Appendix

(See Table 4).

Table 4. Methodology checklist.

	Checklist Item	Comments	Completed
	Data Collection		
1	Monitoring device		
2	Set up the device		
3	Occupancy resolution		
4	Data protection and privacy		
5	Deployment		
6	Ground truth		
	Data Preprocessing		
7	Data cleaning		
8	Exploratory data analysis		
9	Ground truth integration		
10	Data training and test		
11	Data standardization		
12	Feature reduction or extraction		
	Classification Task		
13	Select the aproach		
14	Hyperparameters selection		
15	Evaluation metrics		

References

1. Acuña, E., Rodriguez, C.: The treatment of missing values and its effect on classifier accuracy. In: Banks, D., McMorris, F.R., Arabie, P., Gaul, W. (eds.) Classification, Clustering, and Data Mining Applications. STUDIES CLASS, pp. 639–647. Springer, Heidelberg (2004). https://doi.org/10.1007/978-3-642-17103-1_60
2. Adeogun, R., Rodriguez, I., Razzaghpour, M., Berardinelli, G., Christensen, P.H., Mogensen, P.E.: Indoor occupancy detection and estimation using machine learning and measurements from an IoT LoRa-based monitoring system. In: 2019 Global IoT Summit (GIoTS), pp. 1–5 (2019). https://doi.org/10.1109/GIOTS.2019.8766374
3. Ahmed, S., Kamal, U., Toha, T.R., Islam, N., Alim Al Islam, A.B.M.: Predicting human count through environmental sensing in closed indoor settings. In: Proceedings of the 15th EAI International Conference on Mobile and Ubiquitous Systems: Computing, Networking and Services, pp. 49–58 (2018). https://doi.org/10.1145/3286978.3287021
4. Amazon Web Services, Inc.: AWS IoT Core: Developer Guide (2023). https://docs.aws.amazon.com/iot/latest/developerguide/iot-gs.html
5. Amazon Web Services (AWS): Tutorial: Storing device data in a DynamoDB table (2022). https://docs.aws.amazon.com/iot/latest/developerguide/iot-ddb-rule.html
6. Baratloo, A., Hosseini, M., Negida, A., El Ashal, G.: Part 1: Simple definition and calculation of accuracy, sensitivity and specificity. Emergency (Tehran, Iran) **3**(2), 48–49 (2015). https://www.ncbi.nlm.nih.gov/pmc/articles/PMC4614595/
7. Chawla, N.V., Bowyer, K.W., Hall, L.O., Kegelmeyer, W.P.: SMOTE: synthetic minority over-sampling technique. J. Artif. Intell. Res. **16**, 321–357 (2002). https://doi.org/10.1613/jair.953
8. Claesen, M., De Moor, B.: Hyperparameter search in machine learning (2015). arXiv:1502.02127
9. Galli, S.: Python Feature Engineering Cookbook: Over 70 Recipes for Creating, Engineering, and Transforming Features to Build Machine Learning Models, p. 372. Packt Publishing (2020). ISBN 9781789806311
10. Gudivada, V.N.: Chapter 2 - Data analytics: fundamentals. In: Chowdhury, M., Apon, A., Dey, K. (eds.) Data Analytics for Intelligent Transportation Systems, pp. 31–67. Elsevier (2017). https://doi.org/10.1016/B978-0-12-809715-1.00002-X. ISBN 978-0-12-809715-1
11. Jiang, C., Chen, Z., Png, L.C., Bekiroglu, K., Srinivasan, S., Su, R.: Building occupancy detection from carbon-dioxide and motion sensors. In: 2018 15th International Conference on Control, Automation, Robotics and Vision (ICARCV), pp. 931–936 (2018). https://doi.org/10.1109/ICARCV.2018.8581229.
12. Jin, M., Bekiaris-Liberis, N., Weekly, K., Spanos, C.J., Bayen, A.M.: Occupancy detection via environmental sensing. IEEE Trans. Autom. Sci. Eng. **15**(2), 443–455 (2018). https://doi.org/10.1109/tase.2016.2619720
13. Lawal, K., Rafsanjani, H.N.: Trends, benefits, risks, and challenges of IoT implementation in residential and commercial buildings. Energy Built Environ. **3**(3), 251–266 (2022). https://doi.org/10.1016/j.enbenv.2021.01.009
14. Masood, M.K., Soh, Y.C., Jiang, C.: Occupancy estimation from environmental parameters using wrapper and hybrid feature selection. Appl. Soft Comput. **60**, 482–494 (2017). https://doi.org/10.1016/j.asoc.2017.07.003

15. Mena, A.R., Ceballos, H.G., Alvarado-Uribe, J.: Measuring indoor occupancy through environmental sensors: a systematic review on sensor deployment. Sensors **22**(10), 3770 (2022). https://doi.org/10.3390/s22103770
16. Pedregosa, F., et al.: Scikit-learn: machine learning in Python. J. Mach. Learn. Res. **12**, 2825–2830 (2011). https://scikit-learn.org/stable/modules/preprocessing.html#
17. Raschka, S.: Python Machine Learning, p. 425. Packt Publishing Ltd. (2015). ISBN 9781783555130
18. Saha, H., Florita, A.R., Henze, G.P., Sarkar, S.: Occupancy sensing in buildings: a review of data analytics approaches. Energy Build. **188–189**, 278–285 (2019). https://doi.org/10.1016/j.enbuild.2019.02.030
19. Vela, A., Alvarado-Uribe, J., Davila, M., Hernandez-Gress, N., Ceballos, H.G.: Estimating occupancy levels in enclosed spaces using environmental variables: a fitness gym and living room as evaluation scenarios. Sensors **20**(22) (2020). https://doi.org/10.3390/s20226579
20. Whitbread, J.: Data Protection Policy, V1.1, University of West of England (UWE Bristol), November 2021, Revision Date: December 2022. https://www.uwe.ac.uk/study/it-services/information-security-toolkit/information-security-policies#section-3
21. Zebari, R., Abdulazeez, A.M., Zeebaree, D., Zebari, D., Saeed, J.: A comprehensive review of dimensionality reduction techniques for feature selection and feature extraction. J. Appl. Sci. Technol. Trends **1**, 56–70 (2020). https://doi.org/10.38094/jastt1224
22. Zemouri, S., Magoni, D., Zemouri, A., Gkoufas, Y., Katrinis, K., Murphy, J.: An edge computing approach to explore indoor environmental sensor data for occupancy measurement in office spaces. In: 2018 IEEE International Smart Cities Conference (ISC2), pp. 1–8 (2018). https://doi.org/10.1109/ISC2.2018.8656753
23. Zheng, A.: Evaluating Machine Learning Models, 1st edn. O'Reilly Media, Inc. (2015). ISBN 978-1-491-93246-9
24. Zhou, Y., et al.: A novel model based on multi-grained cascade forests with wavelet denoising for indoor occupancy estimation. Build. Environ. **167** (2020). https://doi.org/10.1016/j.buildenv.2019.106461
25. Zimmermann, L., Weigel, R., Fischer, G.: Fusion of nonintrusive environmental sensors for occupancy detection in smart homes. IEEE Internet Things J. **5**(4), 2343–2352 (2018). https://doi.org/10.1109/jiot.2017.2752134

Estimating Occupancy Level in Indoor Spaces Using Infrared Values and Environmental Variables: A Collaborative Work Area as a Case Study

Angelo Jean Carlo Ovando Franco$^{(\boxtimes)}$, Gerardo Tadeo Pérez Guerra,
Joanna Alvarado-Uribe, and Héctor Gibran Ceballos Cancino

School of Engineering and Sciences, Tecnologico de Monterrey, Monterrey, Mexico
{A00821528,A00832522,joanna.alvarado,ceballos}@tec.mx

Abstract. Improving energy efficiency in indoor spaces is critical to reduce harmful effects of excessive energy consumption worldwide. For this reason, estimating occupancy level of people in indoor spaces has been identified as a significant contributor to improve energy efficiency and space utilization. In this paper, in order to contribute to the solution of this problem, it is proposed to estimate occupancy level of people in enclosed spaces through an indirect approach based on environmental and infrared data, using Machine Learning (ML) techniques. The selected environmental variables are temperature, relative humidity, and atmospheric pressure. In the process, the values of five different workstations from a collaborative work area at Tecnologico de Monterrey were collected to determine the occupancy level of each workstation. To estimate occupancy, supervised ML algorithms were used, obtaining an average accuracy for each workstation of 93%, by using both environmental and infrared data, compared to ground truth counts during occupied hours. Our results show that infrared data plus environmental variables are more accurate than infrared-only sensors for estimating indoor occupancy.

Keywords: smart spaces · machine learning · IoT · indoor occupancy · environmental sensors

1 Introduction

One of the sectors with the highest global energy consumption is the building sector [1], making it a critical area to consider when fighting against climate change and promoting efficient energy use. According to data from World Energy Outlook [2], today's building energy sector accounts for about 30% of global energy demand. In 2018, the demand in this sector increased twice the average rate of the past decade and it is expected to reach 40% of global demand in the next two decades. This will have important environmental impacts, particularly on atmospheric CO_2 levels.

Estimating the occupancy level of people in indoor spaces enhances resource management [3]. For instance, knowing the occupancy level in commercial buildings can

L. Barolli (Ed.): IMIS 2024, LNDECT 214, pp. 226–234, 2024.
https://doi.org/10.1007/978-3-031-64766-6_22

improve the optimization of the energy usage for heating, cooling, and lighting systems. By adjusting resource allocation based on the number of people present, energy waste can be reduced, leading to cost savings and environmental sustainability. Therefore, understanding the occupancy patterns and predicting the occupancy level in enclosed spaces accurately may provide more efficient space planning and utilization [4].

By analyzing occupancy level data, corporations can optimize the layout and design of workspaces, retail areas, and public facilities to improve productivity, customer experience, and overall efficiency [5]. Specifically, it is necessary for ensuring safety and security. Furthermore, it may help security personnel, facility managers, and emergency workers to assess and manage potential risks, for instance, fire hazards, evacuation plans, crowd control, and adherence to building capacity limits. This knowledge of the occupancy level inside the building could help prevent overcrowding, mitigate safety hazards, and optimize resource allocation of people inside it. [6, 7].

Energy in indoor spaces of buildings is often not used efficiently [8], which is why we aim to reduce consumption in these areas by turning off the lights and the AC when they are unoccupied. We want to implement a predictive model based on a presence sensor and considering the following indoor variables: temperature, humidity, pressure. Providing a low-cost sensor that could detect these values and then use Machine Learning (ML) techniques that could help reduce the energy consumption, would be an important benefit for the environment and for the company/organization which is saving money by reducing the amount of energy they consume. There was a previous initiative of Tecnologico de Monterrey bachelor students during the 2019, who intended to detect the amount of people in the gym and locker rooms of the campus [9].

Thus, our hypothesis is that with the installation of a low-cost Internet of Things (IoT) sensor in enclosed spaces that collects data related to temperature, relative humidity (RH), pressure, and infrared values (indirect method), we can have a better estimation of the number/level of people by using supervised ML algorithms, which will help us to non-intrusively monitor the status of the space (occupancy level) and make a better use of the monitored space.

The rest of the manuscript is organized as follows. Section 2 presents the background and related work. Subsequently, the methodology used for this research is presented in Sect. 3. Section 4 provides the results and their discussion. Finally, the conclusion and future work are given in Sect. 5.

2 Literature Review

2.1 Sensors

Environmental sensors are designed to measure a variety of environmental parameters such as temperature, humidity, air quality, pressure, and CO_2, providing essential data for building climate management. On the other hand, motions sensors detect movement based on changes in infrared radiation nearby [19].

In this research, a sensor model BME280 [10] from Bosch was used to collect humidity, temperature, and pressure data. Additionally, a PIR HC-SR501 [11] sensor was used to measure the presence of people. Likewise, an ESP32 controller [12] was used to send the collected data remotely.

2.2 Related Work

Vela et al. [9] performed an indoor occupancy level estimation with the same environmental BME280 sensor considering temperature, relative humidity (RH), and atmospheric pressure in a university gym and a work room. They implemented Support Vector Machine (SVM), k-Nearest Neighbor (kNN), and Decision Trees (DT) as their Machine Learning algorithms and reached an accuracy between 95% and 97%.

Candanedo et al. [13] installed one CO_2 sensor, one temperature sensor, one RH sensor, and one light sensor. Their developed models included Linear Discriminant Analysis (LDA), Classification and Regression Tree (CART), Random Forest (RF), and Gradient Boosting Machines (GBM). Furthermore, the combination of parameters performed by Candanedo et al. obtained an accuracy between 32.68% and 99.33%.

According to Mena et al. [14], regarding the size of the test-bed scenarios, the most common scenarios include offices or apartments of 22 m^2 or laboratories of 186 m^2. The smallest size is of an office of 5 m^2 and the largest size is of a museum of 1196 m^2. Other mostly documented sizes are offices with an area between 11 m^2 and 15 m^2. They are followed by spaces with areas between 16 m^2 and 20 m^2.

Therefore, estimating the occupancy level in enclosed spaces over 186 m^2 using infrared values and environmental variables presents a significate research opportunity.

3 Methodology

The methodology followed in this research was CRISP-DM (Cross-Industry Standard Process for Data Mining). CRISP-DM [15] is a widely used process model for Data Mining and Data Science projects that provides a structured and iterative approach for managing and executing these projects, which ensures that they are well planned, executed, and deliver desired results.

In this research, the general phases are first to fully understand the problem to be solved, in order to select the sensors that best fit the problem. With these, collect the necessary data, clean it, and then apply the ML models.

3.1 Understanding the Problem

Starting the research requires understanding the problem to be solved. In this case, the goal is to improve the prediction of occupancy in enclosed spaces for efficient use of space, lighting, and ventilation. For this reason, the incorporation of an infrared sensor can enhance occupancy level estimations using ML and environmental variables (temperature, humidity, and pressure).

3.2 Data Understanding

3.2.1 Sensor Selection

To achieve this, the quality of the sensors used is extremely important because the data they collect is vital for further analysis and experimentation.

3.2.2 Data Collection

Environmental variables and infrared values were systematically collected using sensors installed at each workstation within the collaborative work area at the university to calculate the occupancy level of the specified areas.

3.3 Data Preparation

Once the data was collected, it was cleaned to avoid significant errors when applying the ML algorithm. The data was also integrated with the labels corresponding to the occupancy level in the collaborative area where the sensors were installed.

3.4 Modeling

After data cleaning, Random Forest algorithm was applied to the data to estimate the occupancy level for each workstation.

3.5 Evaluation and Deployment

Finally, the accuracy was obtained to evaluate the performance of the algorithm.

4 Results and Discussion

For the environmental sensors, it was decided to use the BME280 [10] sensor since it combines three types of sensors into one: a thermometer (temperature), a barometer (pressure), and a hygrometer (humidity). This integration makes it easier to implement in the research. Additionally, the following essential characteristics influenced the selection of this sensor:

1. The temperature range is $-40\,°C$ to $+85\,°C$ with a precision of $\pm 1.5\,°C$, allowing for use in various regions without being affected by the temperature.
2. The sensor measures environmental relative humidity from 0% to 100% with an accuracy of $\pm 3\%$.
3. The sensor has an internal barometer which can measure the atmospheric pressure, with a range of 300 hPa–1100 hPa with an accuracy of ± 1 hPa.
4. Low power consumption as it operates with voltages of 1.2 V–3.3 V with an average current consumption of 1.8 μA–2.8 μA.
5. Small size as it has dimensions of 2.5 mm $\times$ 2.5 mm, which makes it can be easily integrated into the design of electronic devices.
6. Low cost.

For PIR sensors, it was decided to use the HC-SR501 [11] sensor because it offers high sensitivity, high reliability, and ultra-low voltage operation mode. It is widely used in various types of automatic detection equipment. Additionally, it has the following key features:

230   A. J. C. O. Franco et al.

1. Automatic induction, because when a person enters the detection range the sensor output sends a high signal, when the person leaves the range automatically the sensor output sends a low signal.
2. Operating voltage between 4.5 V–20 V.
3. Current consumption is less than 50 μA.
4. Maximum detection distance adjustable between 3 m–7 m.

Sensors were integrated into the device for data collection in the 217 m^2 collaborative work area on the second floor of Centrales Norte. As depicted in Fig. 1, the work area features five workstations, four of them having a length of 6.5 m with capacity for eight people, and one has a length of 5 m with a six-person capacity, totaling a capacity of 38 people. The room is equipped with an HVAC system ventilation and non-operational windows during the test. The collaborative work room has two entrances, a main one through a 1.7 m wide corridor and a secondary external door that leads directly to the outside.

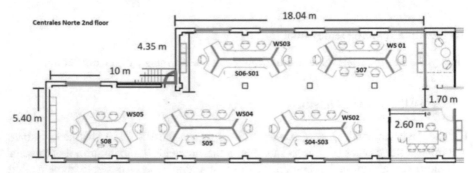

Fig. 1. The monitored collaborative work area.

Due to the size of the work area and the limitations of the sensors, it was decided to take advantage of the modularity of the device for more accurate readings. Therefore, one device was placed on each workstation, as shown in Fig. 2. The PIR sensor was positioned in the center of each workspace to maximize its detection range, aiming to cover more seats and thus enhance people detection. Additionally, the detection range of the PIR sensor was set to 5 m for the first four workstations (WS01–WS04) and 4 m for the last one (WS05) to cover as much space as possible without depending on the presence of people in the work area. Moreover, the triggering type of the PIR sensors was adjusted to single shot with the trigger time set to a minimum of 3 s. This configuration was empirically determined that provides the best results.

Data collection was conducted over 18 days, between October 5, 2023, and November 21, 2023, in sessions averaging 6.5 h with minimum periods of 1 h and maximum periods of 9 h. Data readings were taken every minute, obtaining an approximate of more than 7,000 readings per device, for a total of more than 35,000 readings between the five devices. Due to space and resource limitations, ground-truth collection was done manually by one person on the team responsible for data collection. Occupancy ranges were established and labeled taking proportional intervals between the maximum

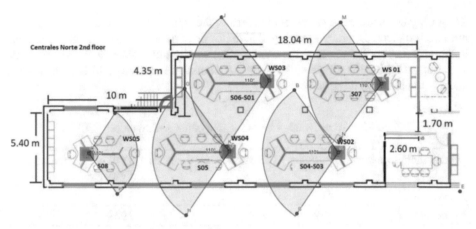

Fig. 2. Sensors position and PIR detection range for each workstation.

and minimum number of people observed. Thus, with the maximum number of people observed being 7, the occupancy levels were established as follows: E (Empty) with 0 people, L (Low) 1–2 people, M (Medium) 3–4 people, and H (High) 5+ people.

Subsequently, to carry out the preprocessing of the data obtained, they were exported from the cloud storage as a CSV file, being a file for each day and put together in one. In turn the ground-truth of the tests was saved as another file. By putting different files together, we found periods of sensor data without ground-truth. Since these instances where there was a lack of ground-truth were of short duration, it was decided to discard them. In order to have a better analysis of the data, it was decided to apply the principal component analysis (PCA) technique [16] to summarize the information contained in a set of variables into a smaller number of variables in terms of the new uncorrelated variables.

On the other hand, due to the randomness of the number of people present in the workroom, an imbalance arose among the data as there were very few times when there were enough people for the different workstations to have high levels (H) of occupancy. This presents problems because the ML models are based on majority classes, thereby, they end up discarding minority classes. To avoid this, it was decided to balance the classes by oversampling the levels with the lowest amount of data in order to avoid this problem. This oversampling was performed using the Adaptive Synthetic (ADASYN) algorithm [17], which generates synthetic data by calculating the number of instances of the minority class to be generated.

Machine Learning Algorithm Application
The Random Forest (RF) algorithm can capture complex, non-linear relationships between features [18], in this case, relative humidity, atmospheric pressure, room temperature, and PIR values, as well as the occupancy level without requiring transformations or assumptions about the data distribution. This is particularly useful in occupancy estimation. In Table 1, it can be seen the accuracy of each workstation to predict the correct label of the occupancy level. As can be seen, the PIR data does not have a good accuracy (17%–41%) for estimating occupancy; a better accuracy (90%–96%) is obtained

using only environmental variables. Therefore, the information of the PIR sensor does not add too much accuracy to the predictive power of a model based on environmental information only [9]. However, the combination of PIR and environmental values could help improve the estimation results, having a maximum accuracy of 96%, in the case of already counting on these sensors.

Table 1. Accuracy of each workstation of the workroom using RF.

Workstation\Accuracy	PIR only	Environmental variables	PIR + environmental variables
Workstation 1	41%	90%	94%
Workstation 2	17%	87%	90%
Workstation 3	30%	85%	93%
Workstation 4	21%	86%	92%
Workstation 5	37%	87%	96%

Finally, to understand the classification errors between classes and workstations, confusion matrices were generated. Figure 3 shows the confusion matrix for workstation 2, which had the lowest accuracy among the five workstations. Similarly, the confusion matrix for workstation 5, which had the highest accuracy of all, can be observed in Fig. 4.

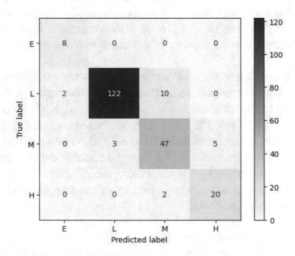

Fig. 3. Confusion matrix of workstation 2

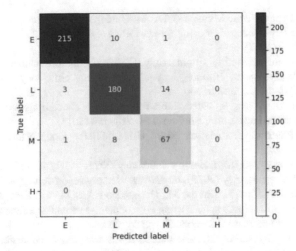

Fig. 4. Confusion matrix of workstation 5

5 Conclusion

This research explored using a PIR sensor, environmental variables, and a Machine Learning algorithm as an indirect method to estimate space occupancy and improve estimation response times. The collected results improved the reviewed literature that has average accuracy of 70%, according to the studies done by Candanedo et al. [13] and were similar to Vela et al. [9] results, which means this proposal could be an effective implementation to reduce energy consumption in indoor spaces and to enable better space planning and management of enclosed spaces in buildings. Thereby, the proposed hypothesis is accepted.

Since the collaborative room is an uncontrolled environment, it was not possible to collect all the occupancy levels in all the workstations, however, the experiment was conducted under real conditions.

Future work will explore the application of other supervised and semi-supervised algorithms to further validate the PIR sensor approach and environmental variables. Additionally, the implementation of an integrated system where sensors are directly connected to the building's HVAC control system to adjust the air conditioning output based on real-time occupancy to optimize energy use. We also recommend to support the development of open interfaces for smart lighting that integrate occupancy levels as input information.

References

1. International Energy Agency (IEA). Breakthrough Agenda Report 2023 (2023). https://www.iea.org/reports/breakthrough-agenda-report-2023
2. International Energy Agency: World Energy Outlook 2019. IEA Publications (2019)
3. European Commission, Joint Research Centre (JRC): Building Occupancy Detection and Management for Energy Efficiency: A Review of Current and Emerging Technologies. Publications Office of the European Union (2019)

4. Office of the Deputy Prime Minister (ODPM): Guidance on the Methodology for Carrying out a Strategic Assessment of Housing Needs. Office of the Deputy Prime Minister (ODPM), United Kingdom (2007)
5. Center for Integrated Facility Engineering (CIFE) at Stanford University. Building for the Occupant: Optimizing Building Layouts for Energy Efficiency and Organizational Performance. http://cife.stanford.edu
6. Hong, T., Yan, D., D'oca, S., Chen, C.F.: Ten questions concerning occupant behavior in buildings: the big picture. Build. Environ. **114**, 518–530 (2017)
7. National Fire Protection Association: NFPA 101: Life Safety Code. National Fire Protection Association (2018)
8. International Energy Agency: Energy Efficiency 2023: Analysis and outlooks to 2030. International Energy Agency (2023). https://www.iea.org/reports/energy-efficiency-2023
9. Vela, A., Alvarado-Uribe, J., Davila, M., Hernandez-Gress, N., Ceballos, H.G.: Estimating occupancy levels in enclosed spaces using environmental variables: a fitness gym and living room as evaluation scenarios. Sensors **20**, 6579 (2020)
10. BME280 Bosch datasheet. https://www.bosch-sensortec.com/products/environmental-sensors/humidity-sensors-bme280/. Accessed Feb 2024
11. Alldatasheet. (n.d.). HC-SR501 Datasheet, PDF - PIR Motion Detector. https://www.alldatasheet.com/view.jsp?Searchword=HC-SR501
12. Esp32 espressiff datasheet. https://www.espressif.com/sites/default/files/documentation/esp32_datasheet_en.pdf. Accessed Feb 2024
13. Candanedo, L.M., Feldheim, V.: Accurate occupancy detection of an office room from light, temperature, humidity and CO2 measurements using statistical learning models. Energy Build. **112**, 28–39 (2016)
14. Mena, A.R., Ceballos, H.G., Alvarado-Uribe, J.: Measuring indoor occupancy through environmental sensors: a systematic review on sensor deployment. Sensors **22**, 3770 (2022). https://doi.org/10.3390/s22103770
15. Studer, S., et al.: Towards CRISP-ML(Q): a machine learning process model with quality assurance methodology. Mach. Learn. Knowl. Extr. **3**, 392–413 (2021). https://doi.org/10.3390/make3020020
16. Kurita, T.: Principal Component Analysis (PCA). Computer Vision: A Reference Guide, pp. 1–4 (2019)
17. He, H., Bai, Y., Garcia, E.A., Li, S.: ADASYN: adaptive synthetic sampling approach for imbalanced learning. In: 2008 IEEE International Joint Conference on Neural Networks (IEEE World Congress on Computational Intelligence), pp. 1322–1328. IEEE, June 2008
18. Parmar, A., Katariya, R., Patel, V.: A review on random forest: an ensemble classifier. In: Hemanth, J., Fernando, X., Lafata, P., Baig, Z. (eds.) ICICI 2018. LNDECT, vol. 26, pp. 758–763. Springer, Cham (2019). https://doi.org/10.1007/978-3-030-03146-6_86
19. Verma, M., Kaler, R.S., Singh, M.: Sensitivity enhancement of Passive Infrared (PIR) sensor for motion detection. Optik **244**, 167503 (2021)

Bird Recognition Based on Mixed Convolutional Neural Network

Feiyu Yao[1], Na Deng[1(✉)], and Xu-an Wang[2]

[1] Hubei University of Technology, Wuhan, China
2010300721@hbut.edu.cn, iamdengna@163.com
[2] Engineering University of Peoples Armed Police, Xi'an, China

Abstract. Bird image recognition is a classic experiment in the fields of artificial intelligence and machine vision. In order to better explore the recognition patterns and performance differences of mainstream convolutional structures nowadays, an experiment based on mixed convolutional neural networks was designed. In the experimental model design section, the hybrid convolutional neural network focuses on dense connections and adopts residual connections in dense connections to ensure that the network is easier to learn. In order to ensure that the network extracts more feature information, an Inception structure based on deep separable convolution is also introduced. Afterwards, an improved channel attention mechanism is used to learn the weights between channels and an adaptive convolution structure is used to learn spatial weights. In comparative experiments conducted on a 525 classified bird dataset, the accuracy of the mixed convolutional neural network experimental model was 94% with fewer parameters, which was superior to models such as Inception V2, ResNet101, DenseNet264, MobileNet V3-large, and EfficientNet. This indicates that it has certain performance advantages in bird feature recognition tasks.

1 Introduction

The automatic identification of birds provides important methodological support for understanding and mastering the distribution, quantity, habitat, habits, migration patterns, and other information of birds, and is of great significance for maintaining biodiversity and protecting the ecological environment. At present, convolutional neural networks are the mainstream method for automatic bird recognition [1–3].

Since LeNet [4] was proposed in 1998, convolutional neural networks have made tremendous progress and evolution in the field of computer vision. With the rise of deep learning and the improvement of computing resources, researchers are constantly introducing more complex and deeper network structures, further improving the performance and application scope of models.

Afterwards, in 2012, AlexNet [5] used the ReLU activation function in the ImageNet large-scale visual recognition competition and achieved significant breakthroughs, verifying its effectiveness; GoogLeNet [6] proposed the Inception module, which improves the computational efficiency and receptive field of the network through multi-scale convolution kernels and pooling layers; Reference [7] proposes batch normalization to

L. Barolli (Ed.): IMIS 2024, LNDECT 214, pp. 235–246, 2024.
https://doi.org/10.1007/978-3-031-64766-6_23

reduce internal covariate shifts between different layers in the network and accelerate network convergence; ResNet [8] introduces a residual structure that allows the network to have very deep layers, solving the problem of gradient vanishing and greatly improving the efficiency and performance of training deep networks; DenseNet [9] adopts the idea of dense connections to enhance feature reuse; MobileNet [10] proposed deep separable convolution, which greatly reduces computational costs and is suitable for lightweight devices and real-time applications; A channel attention mechanism, Squeeze and Excitation (SE) Attention [11], is proposed with the goal of adaptively adjusting the importance of input feature maps by learning effective channel weights. The proposal of these methods has led to the rapid development and widespread application of deep learning in the field of vision.

Nowadays, based on these classic convolutional neural networks, more improved complex convolutional neural networks have been proposed and have achieved good results. Reference [12] introduced residual connections into CNN and standardized residual blocks, and the ResNet structure was able to train deep networks to achieve highly competitive recognition performance; Reference [13] use Convolutional Neural Networks (CNN) to locate salient features in the images. Finally, for the test dataset, the average sensitivity, specificity, and accuracy were 93.79%, 96.11%, and 95.37%, respectively; The network structure of different improvement methods may have inadequate feature extraction capabilities for different datasets, so the convolution size and structure of the network are particularly important.

2   Design of Bird Experimental Models

2.1   Improved Channel Attention Mechanism

The current mainstream channel attention mechanism is usually based on the idea of SE Attention. Channel attention mechanisms typically use global average pooling to process input feature maps and extract features on channels. However, global average pooling reduces the feature map to a single value and loses spatial information. In order to extract channel features more effectively, this paper proposes an improved channel attention mechanism.

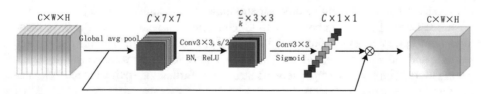

Fig. 1. Schematic diagram of improved channel attention mechanism

By observing the Fig. 1, it can be observed that the only change in the network structure is to replace the original fully connected layer with a convolutional layer. This change has led to an increase in the number of parameters, from the original $2 \times C \times \frac{C}{k}$ to

$2 \times C \times \frac{C}{k} \times 3 \times 3$, which can more effectively extract feature information. Meanwhile, due to the constant computational cost of global average pooling, reducing the size of feature maps does not significantly increase the computational cost of subsequent convolutional layers.

In the case of dense connections in the network, this increased amount of parameters and computation is acceptable. In addition, the improved channel attention mechanism can be more effective in extracting channel features, thereby enhancing the performance and representation ability of the model.

2.2 Adaptive Convolutional Structure

In addition to learning attention weights in the channel dimension, this article also introduces an adaptive attention mechanism in the spatial aspect. Process the spatial features of the input feature map through convolution operations to learn a set of adaptive attention weights for convolution. Unlike global attention, this mechanism applies attention weights to local spatial features to enhance the model's perceptual ability. The effect of this adaptive convolution makes the model more flexible and representative in perceiving and understanding input data.

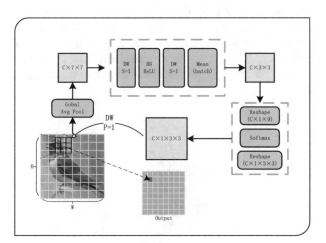

Fig. 2. Adaptive Convolutional Structure Diagram

As shown in Fig. 2, we also first perform global average pooling to $C \times 7 \times 7$, and then output feature maps of size $C \times 3 \times 3$ through depthwise separable convolution (DW) layers. Calculate the average value on the Batch dimension to obtain attention weights of $C \times 1 \times 3 \times 3$ as the generated convolution kernel, and finally perform convolution operations on the input feature map to the output feature map. The convolutional kernel is activated by the Softmax activation function and acts on the last two dimensions, namely 3×3. The formula is as follows:

$$w'_{ij} = \frac{e^{w_{ij}}}{\sum_i \sum_j e^{w_{ij}}} \tag{1}$$

The input feature maps all appear positive after being activated by ReLU(x) = max(0, x) in the previous convolution operation. In this case, using Softmax for weighted summation is a natural choice. By performing local weighted averaging on the input feature map, the network's representation ability can be further improved.

From the perspective of parameters, this adaptive convolution structure mainly consists of two layers of depthwise separable convolutions, with a size of $2 \times C \times 3 \times 3$. This means that the structure has relatively fewer parameters and is suitable for application in resource constrained environments. From the perspective of computational complexity, due to the use of global average pooling to compress feature maps and the use of depthwise separable convolutions, this adaptive convolution structure does not have high computational complexity. Therefore, without sacrificing perceptual ability, this adaptive convolutional structure can locally perceive input features in situations where computing resources are limited.

2.3 Inception

The core idea of the Inception structure is to use multiple convolution kernels of different sizes in the same layer, and then concatenate their outputs. By using multiple convolution kernels of different sizes, Inception structures can capture image details and contextual information in different receptive fields.

Figure 3 shows the structure diagram after introducing depthwise separable convolution into the Inception structure. The advantage of using depthwise separable convolution in Inception structure is that it can more effectively capture the spatial information of input feature maps and reduce resource consumption while maintaining network performance.

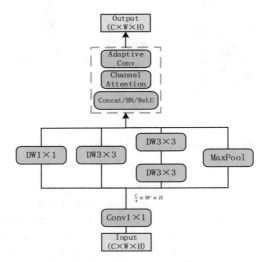

Fig. 3. Transition structure diagram

Compared with the original Inception structure, there is no method of using 1×1 convolution kernels for channel dimensionality reduction for convolution outputs of different scales. On the contrary, this article adopts the same 1×1 convolution kernel for convolution operations. This design can learn features from different receptive fields while maintaining a small number of parameters and calculations. In addition, this article also considers the problem of deep separable convolution not learning the relationships between channels, and introduces an improved channel attention mechanism and adaptive convolution structure. The channel attention mechanism can learn the weight relationship between channels, further improving the representation ability of features; The adaptive convolutional structure can adaptively adjust based on the local information of the input feature map, further enhancing perceptual ability.

2.4 Hybrid Convolutional Neural Network Architecture

This network (MixConvNet) is a convolutional neural network that combines the characteristics of multiple classical networks and makes appropriate improvements. By integrating the advantages and features of different networks, this network has achieved a more powerful and efficient architecture. MixConvNet mainly consists of two parts: Block and Transition. The stacking of multiple blocks is used for feature extraction, while Transition is used to reduce the number of channels and scale size of the feature map.

1) Block

The block structure is shown in Fig. 4. The main components of this structure include a separable convolutional layer, which converts the number of channels in the input feature map into c. Subsequently, the feature map processed by a two-layer Inception structure is concatenated with the input feature map to achieve a dense connection effect. In the process of dense connections, multi-level residual connections are used to solve the problem of gradient vanishing and promote information flow, making the network easier to learn.

During the unidirectional flow of feature map information (indicated by black arrows), the network introduces multi-level residual connections. For example, after undergoing depthwise separable convolution on the input feature map, residual connections are made to the subsequent three blocks (indicated in light blue). To avoid the problem of excessive gradients during backpropagation, we have set parameters $\alpha = 1/3$ and $\beta = 1/2$, limit the coefficient of the residual term to 1 to control the gradient size. If the input of this Block is x_0, and then after feature extraction is x_1, x_2, and x_3, it can be represented as:

$$Input = x_0 \tag{2}$$

$$\begin{cases} x_1 = f_1(x_0) + \alpha x_0 \\ x_2 = f_1(x_1) + \beta x_1 + \alpha x_0 \\ x_3 = f_1(x_2) + x_2 + \beta x_1 + \alpha x_0 \end{cases} \tag{3}$$

$$Output = Concat(x_0, x_3) \tag{4}$$

By stacking multiple blocks, dense connections are achieved, where the current input is associated with all previous outputs. Assuming the size of the input feature map is

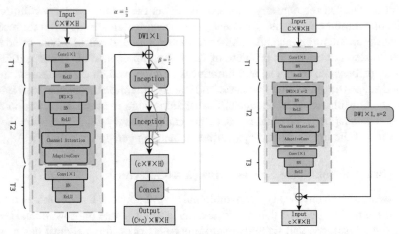

Fig. 4. Hybrid neural network structure diagram (Left: Block, Right: Transition, T1: Raise the dimension to 64, T2: Feature extraction, T3: Reduce the dimension to 64)

$C \times W \times H$, after n blocks of processing, the size of the output feature map will become $(C + n \times c) \times W \times H$.

2) Transition

Compared to the Block structure, the Transition structure only includes the previous feature extraction part and associates the input and output through a layer of residual connections. In addition, a step size of 2 is set in depthwise separable convolution to achieve halving of feature map size. The schematic diagram of the Transition structure is shown in Fig. 4.

The function of the Transition structure is to adjust the channels and scales of the feature map, by reducing the size and number of channels of the feature map, so that subsequent network layers can process feature information more effectively. By introducing residual connections, the Transition structure can maintain the information integrity of input features and play a role in information flow and feature transmission throughout the entire network. In MixConvNet, the Transition structure plays a role in balancing and controlling features between blocks, helping the network adapt to feature changes at different levels and further improving its performance and generalization ability.

3   Experimental Process

3.1   Experimental Data and Preprocessing

We selected the bird dataset on Kaggle as the experimental object, containing 525 bird species, with a total of 84635 training images, 2625 validation images (5 images per species), and 2625 test images (5 images per species). All images are JPG format color images with 3 color channels and a length and width of 224 pixels each. Some bird datasets are shown in Fig. 5. Also, flowchart of bird recognition algorithm is shown in Fig. 6.

Fig. 5. Partial bird experimental data

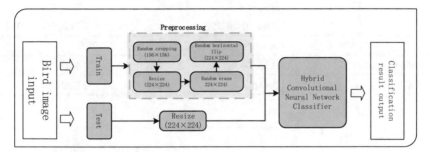

Fig. 6. Flowchart of Bird Recognition Algorithm

3.2 Experimental Model

By combining Block and Transition, we can construct MixConvNet networks of different sizes. The input of the network is an image with a size of $3 \times 224 \times 224$. Firstly, it goes through a convolutional layer with a size of 7×7 and a step size of 2, as well as a max pooling layer with a size of 3×3 and a step size of 2 for feature extraction. Next, we set the number of channels added to each block to $c = 64$, and stack multiple blocks together through dense connections. Finally, the network generates k classification outputs through a fully connected layer. Table 1 shows three different sizes of MixConvNet network structures, including indicators such as network layers, output size, parameter count, and computational complexity.

For fair comparison, all models were trained using the Adam optimization algorithm and a cross entropy loss function with a smoothing parameter of 0.1. The batch size was set to 32, and the initial learning rate was 0.001. After each training round, the learning rate was updated with a decay rate of 0.95, with a total of 50 iterations.

4 Analysis and Discussion of Experimental Results

4.1 Performance Analysis of Attention Mechanism

MixConvNet84 trained different attention mechanisms on the 525 bird dataset, including channel attention mechanism (SE), improved channel attention mechanism (SE/Conv), and adaptive convolutional structure (AdaptiveConv)+SE/Conv. The comparison of accuracy and loss values during these training processes is shown in Fig. 7.

Table 1. MixConvNet structures with different network sizes

network layer	Output size	MixConvNet84	MixConvNet224	MixConvNet444
conv_1	112 × 112	7 × 7, 64, stride 2		
	56 × 56	3 × 3 max pool, stride 2		
conv_2	28 × 28	Block × 3	Block × 3	Block × 7
		Transition(128)	Transition(128)	Transition(256)
conv_3	14 × 14	Block × 2	Block × 6	Block × 12
		Transition(128)	Transition(256)	Transition(512)
conv_4	7 × 7	Block × 2	Block × 12	Block × 24
		Transition(128)	Transition(256)	Transition(1024)
	1 × 1	average pool, 525-d fc, softmax		
Parameter quantity		0.765M	2.578M	6.466M
FLOPs		0.416G	0.658G	1.826G

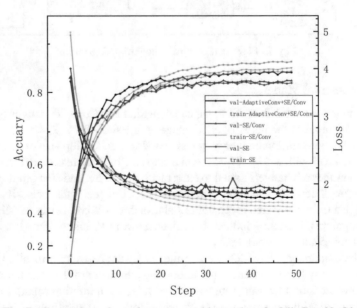

Fig. 7. Comparison chart of accuracy and loss values for MixConvNet84

Based on the analysis of the results in Fig. 7, we can find that using the SE/Conv structure can accelerate the convergence speed of the network. When using the Adaptive Conv structure combined with SE/Conv, compared to using SE/Conv alone, the AdaptiveConv+SE/Conv model achieved significant improvements in both loss value and accuracy. The loss value decreased from 1.90 to 1.83, and the accuracy increased from 82% to 84.5%. This indicates that the introduction of the AdaptiveConv structure is crucial for improving the performance of the model. AdaptiveConv+SE/Conv can significantly improve the performance of networks, not only accelerating convergence speed, but also achieving significant improvements in accuracy.

4.2 Comparative Analysis of Performance of Hybrid Convolutional Neural Networks

By comparing the training results of 8 different networks in the first 4 iterations, we can observe that there are differences in their convergence speeds, as shown in Fig. 8.

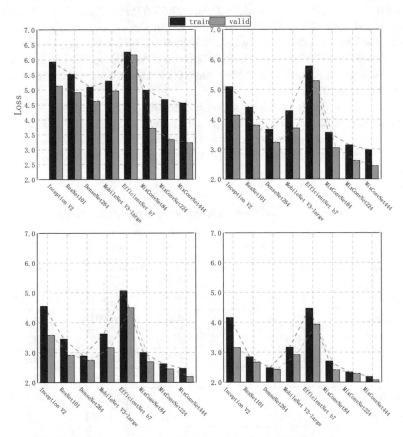

Fig. 8. Training results of the first four iterations of different networks

1) After applying data augmentation to the training set, it was observed that the training loss of the model was initially generally higher than the validation set loss.
2) EfficientNet b7 and Inception V2 have relatively slow convergence speeds. Efficient-Net b7 requires more iterations to achieve lower loss values due to its large model size and large number of parameters. Inception V2, due to the lack of skip connections, makes gradient transfer and network optimization more difficult.

3) MixConvNet has the fastest convergence speed among the three networks, especially
   MixConvNet444 shows the most obvious performance on the validation set.

In Table 2, we compared the accuracy of eight different convolutional neural networks
on the bird dataset. The results showed that MixConvNet444 achieved the best performance on the test set, with an accuracy rate of 94.04%. At the same time, ResNet101 and
EfficientNet b7 have a higher number of parameters and FLOPs, and have also achieved
high accuracy.

According to the results in Table 2, we can measure the complexity of a model by the
number of parameters and FLOPs. ResNet101 and EfficientNet b7 are the models with
the highest number of parameters and FLOPs, with 43.576M and 65.131M, respectively;
MixConvNet84 and MobileNet V3-large are the models with the lowest number of
parameters and FLOPs, respectively, at 0.765M and 0.234G. Furthermore, although the
number of parameters in MixConvNet84 is only 1/6 of MobileNet V3-large, its accuracy
is almost equivalent, indicating that MixConvNet84 can learn more feature information
with fewer parameters.

However, from the perspective of computational speed, MobileNet V3-large still
has significant advantages. Although the FLOPs of the two models are not significantly
different, due to dense connections, the actual training time difference is greater than
what FLOPs describe. Therefore, MobileNet V3-large is still a faster choice in terms of
speed.

Table 2. Comparison of different network prediction results

model	Verification set	Test set	Parameter quantity	FLOPs
Inception V2	0.8278	0.8686	12.518M	1.511G
ResNet101	0.8899	0.9188	43.576M	7.865G
DenseNet264	0.8648	0.8903	19.101M	4.389G
MobileNet V3-large	0.8476	0.8853	4.875M	0.234G
EffcientNet b7	0.8857	0.9238	65.131M	5.346G
MixConvNet84	0.8423	0.8873	0.765M	0.416G
MixConvNet224	0.8899	0.9240	2.578M	0.658G
MixConvNet444	0.9074	0.9404	6.466M	1.826G

MixConvNet demonstrates more efficient features in computing and memory consumption. This indicates that MixConvNet can still achieve excellent performance while
maintaining a relatively small number of parameters and computational complexity. Due
to its excellent performance and small parameter size, its effectiveness in visual tasks
has been confirmed.

5 Conclusion

This article proposes an improved hybrid convolutional neural network model that combines dense connections, residual connections, improved channel attention mechanisms, and adaptive convolutional structures. Through comparative experiments on bird datasets, the hybrid convolutional neural network demonstrated excellent performance with a low number of parameters, achieving an accuracy of 94.04%.

In addition, the experiments in this article were conducted on specific bird datasets, so the conclusions and performance advantages obtained may differ on other datasets. Further research and experiments need to be conducted on a wider and more diverse dataset to verify the generalization ability and robustness of the hybrid convolutional network. Although the performance of this hybrid convolutional network can be evaluated through these comparisons, future research can consider expanding the scope of comparative experiments to include more network architectures and models.

Overall, the improved convolutional neural network model proposed in this article demonstrates significant performance advantages on bird datasets. These results indicate that by combining different convolutional structures and attention mechanisms, as well as appropriate network design and optimization, excellent performance can be achieved with fewer parameters. This is of great significance for further promoting the development of convolutional neural networks, and provides valuable reference and inspiration for solving feature recognition problems in practical applications.

References

1. Xie, J., Zhong, Y., Zhang, J., et al.: A review of automatic recognition technology for bird vocalizations in the deep learning era. Ecol. Inform. **73**, 101927 (2023)
2. Huang, Y.P., Basanta, H.: Bird image retrieval and recognition using a deep learning platform. IEEE Access **7**, 66980–66989 (2019)
3. Ragib, K.M., Shithi, R.T., Haq, S.A., et al.: PakhiChini: automatic bird species identification using deep learning. In: 2020 Fourth World Conference Proceedings on Smart Trends in Systems, Security and Sustainability (WorldS4), pp. 1–6. IEEE, Piscataway (2020)
4. LeCun, Y., Bottou, L., Bengio, Y., et al.: Gradient-based learning applied to document recognition. In: Proceedings of the IEEE, pp. 2278–2324. IEEE, Piscataway (1998)
5. Krizhevsky, A., Sutskever, I., Hinton, G.E.: ImageNet classification with deep convolutional neural networks. Commun. ACM **60**(6), 84–90 (2012)
6. Szegedy, C., Liu, W., Jia, Y., et al.: Going deeper with convolutions. In: Proceedings of 2015 IEEE Conference on Computer Vision and Pattern Recognition, pp. 1–9. IEEE, Piscataway (2015)
7. Ioffe, S., Szegedy, C.: Batch normalization: accelerating deep network training by reducing internal covariate shift. In: Proceedings of the 32nd International Conference on Machine Learning, pp. 448–456. International Machine Learning Society, New York (2015)
8. He, K., Zhang, X., Ren, S., et al.: Deep residual learning for image recognition. In: Proceedings of 2016 IEEE Conference on Computer Vision and Pattern Recognition, pp. 770–778. IEEE, Piscataway (2016)
9. Huang, G., Liu, Z., Van Der Maaten, L., et al.: Densely connected convolutional networks. In: Proceedings of 2017 IEEE Conference on Computer Vision and Pattern Recognition, pp. 4700–4708. IEEE, Piscataway (2017)

10. Howard, A.G., Zhu, M., Chen, B., et al.: MobileNets: efficient convolutional neural networks for mobile vision applications. arXiv preprint arXiv:1704.04861 (2017)
11. Hu, J., Shen, L., Sun, G.: Squeeze-and-excitation networks. In: Proceedings of 2018 IEEE Conference on Computer Vision and Pattern Recognition, pp. 7132–7141. IEEE, Piscataway (2018)
12. Zhou, T., Zhao, Y., Wu, J: ResNeXt and Res2Net structures for speaker verification. In: Proceedings of 2021 IEEE Spoken Language Technology Workshop (SLT), pp. 301–307.IEEE, Piscataway (2021)
13. Li, W., Qi, F., Tang, M., et al.: Bidirectional LSTM with self-attention mechanism and multi-channel features for sentiment classification. Neurocomputing **387**, 63–77 (2020)

Enhancing the Highway Transportation Systems with Traffic Congestion Detection Using the Quadcopters and CNN Architecture Schema

Edy Kristianto[1], Rita Wiryasaputra[1,2], Florensa Rosani Purba[1],
Fernando A. Banjarnahor[1], Chin-Yin Huang[2], and Chao-Tung Yang[3,4]($\boxtimes$)

[1] Department of Informatics, Krida Wacana Christian University, Jakarta 11470, Indonesia
{edy.kristianto,rita.wiryasaputra,florensa}@ukrida.ac.id,
fernando.2017tin017@civitas.ukrida.ac.id
[2] Department of Industrial Engineering and Enterprise Information, Tunghai University, Taichung 407224, Taiwan
huangcy@go.thu.edu.tw
[3] Department of Computer Science, Tunghai University, Taichung 407224, Taiwan
[4] Research Center for Smart Sustainable Circular, Tunghai University, Taichung 407224, Taiwan
ctyang@thu.edu.tw

Abstract. Traffic congestion is a significant issue in urban areas worldwide, impacting mobility, increasing air pollution, and affecting the well-being of city residents. The Unmanned Aerial Vehicles (UAVs) and object recognition technologies such as the YOLO (You Only Look Once) algorithm are engaging solutions to mitigate traffic congestion. This study focuses on using UAVs as a versatile and cost-effective solution for monitoring highway traffic congestion and enhancing traffic management. UAVs are highly manoeuvrable and can navigate around obstacles to capture detailed imagery of congested areas. The flexibility of UAVs allows drones to access challenging terrain or areas with limited road access, providing valuable insights into traffic conditions from different vantage points. YOLOv8n (YOLOv8 nano) and YOLOv8s (YOLOv8 small) as the lightweight models are utilized for congestion detection with a ratio of 70% for data training and 30% for data validation. This study aims to detect congestion and analyze the traffic flow by combining the latest lightweight YOLO architecture and UAV technology. The results indicate that the combination of the UAV and YOLOv8 model for congestion detection is a promising approach for traffic management.

Keywords: congestion · lightweight CNN · monitoring · traffic · UAV

1 Introduction

Traffic congestion is a significant issue in urban areas worldwide. It contributes to the sustainability of transportation development by causing delays, inconveniences, and economic losses to drivers and air pollution [10].

© The Author(s), under exclusive license to Springer Nature Switzerland AG 2024
L. Barolli (Ed.): IMIS 2024, LNDECT 214, pp. 247–255, 2024.
https://doi.org/10.1007/978-3-031-64766-6_24

Traffic monitoring is commonly used as sensor-based and computer-vision-based in [6]. The latest technologies, such as the Unmanned Aerial Vehicle (UAV), the Internet of Things (IoT), edge, and cloud computing, can enhance traffic monitoring for convenience and improve safety in smart cities [2]. Some researchers who have analyzed [2,4,9] smart monitoring traffic using UAVs found that it can improve traffic safety on the highway more efficiently than CCTVs and sensor-based traffic monitoring, especially for monitoring traffic congestion. However, the UAV's limited computational and energy resources necessitate using a lightweight model to ensure efficient energy usage because all of the UAV's components are dependent on its battery [7]. Thus, the lightweight feature and high-accuracy model are essential for combining UAV and AI technology.

This study presents a new method for detecting traffic congestion using the latest lightweight YOLOv8 models, specifically the YOLOv8n and YOLOv8s, in combination with a UAV to overcome the limitation on UAV resources.

The proposed research is structured as follows: Sects. 1 and 2 provide the research background and related works used as the foundation of this research. Section 3 presents the methodology. Section 4 discusses the proposed system's experiment. The last Sect. 5 consists of the research's conclusions and future works.

2   Related Works

Harrou et al. [6] mentioned that traffic monitoring could be classified into a sensor-based and computer-vision-based approach. Sensor-based mostly uses GPS [5] or a piece-wise switched linear traffic (PWSL) [6,17]. The computer vision approaches combine YOLO and CCTV for detecting traffic congestion and have been proposed by some researchers in [3,12,15]. However, fixed CCTV installations or sensor-based systems such as PWSL lack the necessary flexibility, especially for assessing highway congestion. Moreover, implementing CCTVs and sensor-based traffic monitoring is more difficult when maintaining those devices and it also interrupts the traffic, which can cause traffic congestion or endanger traffic workers.

The use of UAVs as surveillance tools is rapidly increasing. The implementation of UAVs offers various advantages, such as improved mobility, coverage, response time, monitoring capabilities, versatility, and cost-effectiveness in comparison to traditional CCTV cameras and sensor-based surveillance systems. This makes UAVs a valuable tool for traffic management authorities and urban planners to address congestion on highways or roads. Some researchers have already combined UAVs with AI in traffic management. Benjdira et al. [1] proposed the combination of UAV and YOLOv3 for monitoring traffic by detecting and counting vehicles from aerial images. This combination offers more flexibility and higher mobility than using the CCTVs. In fact, the UAV produced high-resolution images from different altitudes and locations. Chen et al. [4] and Bisio et al. [2] also found that the merging of YOLO and UAV was more successful in enhancing the traffic surveillance than using the traffic CCTVs.

The UAV has more flexibility for deployment than regular CCTV, although it has limited computation resources. [4] surveyed the YOLO models for the UAV called YOLO-based UAV technology for reducing consumption of the computation resources. [2] reviewed that drones are superior to CCTV or sensor-based traffic monitoring in line with the technological development of smart cities, communication technologies, the Internet of Things (IoT), and artificial intelligence (AI). The AI models that are combined with the UAV are mostly CNN-based deep learning, such as Faster-RCNN and YOLO.

3   Methodology

The proposed research's methodology is as follows:

3.1   UAV's Model

The UAV model generally has four models [7]: single motor, fixed wing, fixed-wing hybrid, and multi-rotor. A quadcopter is a multi-rotor type with four rotors arranged in a square formation, with each rotor equidistant from the center of mass of the quadcopter [13]. This type of UAV is popular because of its user-friendly interface, excellent camera control, and ability to operate within confined spaces. Unlike other types of UAVs, quadcopters are easy to operate and can be used in areas with restricted access.

3.2   Deep Learning Model

Tripathi et al. [14] surveyed that YOLO was mostly used for object detection than the traditional machine learning or CNN models, such as R-CNN or Fast-RCNN because the YOLO only needed an one-stage detector. YOLO also achieved higher accuracy and faster progress in real-time detection. The experiment from Benjdira et al. [1] showed that YOLOv3's accuracy outperformed Faster-RCNN and SVM. We select the YOLO models as the object detection for congestion detection. To detect the vehicle, the images from the dataset were used and also employed the YOLO model for the UAV, which is the most used model in object detection in the survey from [2] and [4]. YOLO is an efficient object detection model, processing images with just one step to detect objects and produce output in the form of bounding boxes, class labels, and confidence. The process involves the division of an image into a grid of cells, of which each cell is responsible for predicting several bounding boxes and object classes. YOLOv8 uses multiple-scale techniques and various architectures to improve detection accuracy. After predictions are made, non-maximum suppression is carried out to address duplication detection [8].

This research used the YOLOv8, the latest model from Ultralytics, due to its improved accuracy and consistent processing speed compared to the previous models[1]. Due to the limited computation resources of the UAV, the lightweight

---

[1] https://docs.ultralytics.com/models/yolov8.

YOLOv8 models were selected. They are YOLOv8n and YOLOv8s, which have a smaller number of parameters at 3.2M and 11.2M, respectively, than the other YOLO models.

A schema of the proposed traffic congestion method is represented in Fig. 1. The steps of the proposed model are as follows:

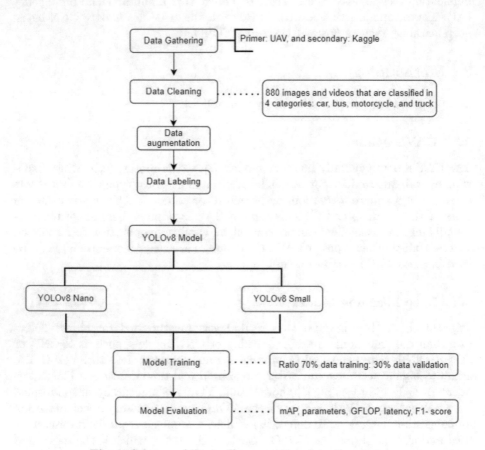

Fig. 1. Schema of the traffic congestion detection system

After collecting traffic images from the UAVid dataset[2] and cleaning the dataset, which consists of 800 images and videos, four classes of vehicles were identified: car (Mobil), truck (truck), bus (bis), and motorcycle (motor). This dataset has a total size of 12GB. However, this dataset is for semantic segmentation and does not have a label. We need to create labels for the classification using the labelling tool from makesense.ai[3]. After the dataset is ready for training and testing, we divide it into data training up to 70% and validation up to

---

[2] https://www.kaggle.com/datasets/dasmehdixtr/uavid-v1.
[3] https://www.makesense.ai/.

30%. Videos in the dataset for testing were also used. During the training stage, we imported the YOLOv8 models. The YOLOv8n and YOLOv8s were selected and then, trained and tested individually. The training was carried out using the standard Google Collabs.

3.3   Congestion Detection

Calculating volume-capacity ratio (V/C ratio) aims to measure traffic congestion. The V/C ratio is the ratio between the traffic volume (V) and the road's capacity (C). It can be calculated as follows:

$$(v/c)_{ratio} = \frac{V}{C} \tag{1}$$

Traffic volume at busy times is obtained according to a one-day survey. The traffic volume at busy times is the highest number of vehicle volumes at a certain time. Meanwhile, road capacity is the maximum capacity or number of vehicles the road can accommodate without experiencing congestion or a significant decrease in speed. Traffic is definitely jammed when the road capacity reaches 90% of vehicle flow.

3.4   Performance Evaluation

YOLO uses mean Average Precision (mAP) for measuring the object detection performance [2]. The mAP score is calculated by varying a detection threshold and computing the area under each class's precision and recall curve. This means that the precision and recall for each class are evaluated, and the resulting scores are averaged to obtain the final mAP score. mAP can be calculated as follows:

$$mAP = \frac{1}{m} \sum_{j=1}^{m} AP_j \tag{2}$$

where m is the number of object detection classes, and AP_j is the Average Precision for class j. Precision ($Prec$) measures the proportion of true positive predictions out of all positive predictions made by the model. Recall (Rec) is the true positive rate, and it is used to measure the proportion of true positive predictions from all actual positive labels in the dataset. The precision and recall use the equation as follows:

$$Prec = \frac{TruePositive}{(TruePositive + FalsePositive)} \tag{3}$$

$$Rec = \frac{TruePositive}{(TruePositive + FalseNegative)} \tag{4}$$

The model prediction performance is measured using the F1-score, the mean from the precision and recall. This performance has the equation:

$$F1 = \frac{2 * Prec * Rec}{Prec + Rec} \tag{5}$$

(a) UAV's detection result

(b) Video frame sample1 result

Fig. 2. The detection results

4   Experiment

The previously mentioned dataset is split into training and validation sets at a
70:30 ratio. The training set is used to develop and tune the prediction models,
while the validation set is kept separate to evaluate the final model.

After we trained the model, it can detect the vehicles as seen in Fig. 2. Our
trained model can detect vehicles in the image (Fig. 2a) and the video as the
sample (Fig. 2b). They can be detected in the bounding boxes.

The accuracy percentage achieved up to 70% as seen in Fig. 3. The cars can
achieve the highest detection. The results of the AP value, parameter numbers,
Giga Floating Point Operations Per Second (GFLOPs), the detection latency,
and the performance can be seen in Table 1.

Table 1. YOLOv8 evaluation

Model	Training	Epoch	mAP	Parameters	GFLOPs	Detection (ms)
YOLOv8n	1	10	0.033	3.006.428	8.1	5.5
	2	126	0.289	3.006.428	8.1	3.7
YOLOv8s	1	128	0.397	11.127.132	28.4	2.9
	2	128	0.421	11.127.132	28.4	2.4

Our model could be trained and learned from the dataset; it could reduce
the loss error during the training, as seen in Fig. 4. The box_loss shows the
predicted bounding box's loss error between the dataset's training and validation. The cls_loss shows the loss error of the correctness between the predicted
classification of the bounding box. The dfl_loss is the distribution focal loss,
demonstrating that the model is trained with the imbalance classes; in our case,
the cars are more dominant.

The model could calculate the flow and contraflow of vehicles based on
whether or not they were facing the UAV camera and vice versa. This is shown
in Fig. 5a and Fig. 5b. The "in" value is for the number of vehicles facing the

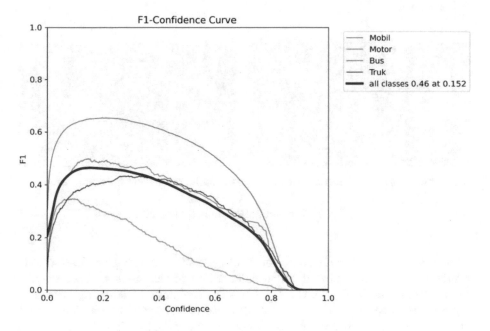

Fig. 3. F1 score curve

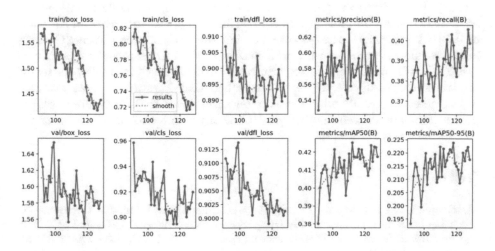

Fig. 4. Loss function

UAV's camera and the "out" value is for the number of vehicles opposite the
UAV's camera.

(a) Traffic congestion detection result 1   (b) Traffic congestion detection result 2

Fig. 5. The congestion detection results

In future work, there is room for improvement concerning the performance
results which requires more detailed calculations of the number of each vehicle
classification. The YOLO in the UAV has the challenge of detecting very small
objects due to the various UAV altitudes. Moreover, traffic congestion is likely
due to the large number of trucks. A comparison of lightweight YOLO models
on our dataset is also necessary, such as the model proposed by [11] and [16].

5   Conclusion

During the experiments, the YOLOv8s model - with 11.2 million parameters,
28.4 GFLOPs, 2.4 ms latency, and trained for 128 epochs - achieved an AP value
of 0.421. Furthermore, it is also found that the performance of the YOLOv8s
model requires a larger dataset. Additionally, the YOLOv8s model outperformed
the YOLOv8n model in the experiments. After testing the YOLOv8s model
for traffic congestion detection, it was found that it provided more accurate
object detection results with a maximum accuracy of 70%. In future work, we
plan to compare different lightweight YOLO object detection methods for traffic
congestion using our dataset.

References

1. Benjdira, B., Khursheed, T., Koubaa, A., Ammar, A., Ouni, K.: Car detection
   using unmanned aerial vehicles: comparison between faster R-CNN and YOLOv3.
   In: 2019 1st International Conference on Unmanned Vehicle Systems-Oman (UVS),
   pp. 1–6. IEEE (2019)
2. Bisio, I., Garibotto, C., Haleem, H., Lavagetto, F., Sciarrone, A.: A systematic
   review of drone based road traffic monitoring system. IEEE Access **10**, 101537–
   101555 (2022)

3. Chakraborty, P., Adu-Gyamfi, Y.O., Poddar, S., Ahsani, V., Sharma, A., Sarkar, S.: Traffic congestion detection from camera images using deep convolution neural networks. Transp. Res. Rec. **2672**(45), 222–231 (2018)

4. Chen, C., et al.: YOLO-based UAV technology: a review of the research and its applications. Drones **7**(3), 190 (2023)

5. D'Andrea, E., Marcelloni, F.: Detection of traffic congestion and incidents from GPS trace analysis. Expert Syst. Appl. **73**, 43–56 (2017)

6. Harrou, F., Zeroual, A., Sun, Y.: Traffic congestion monitoring using an improved KNN strategy. Measurement **156**, 107534 (2020)

7. Iftikhar, S., et al.: Target detection and recognition for traffic congestion in smart cities using deep learning-enabled UAVs: a review and analysis. Appl. Sci. **13**(6), 3995 (2023)

8. Jocher, G., Chaurasia, A., Qiu, J.: Ultralytics YOLOv8 (2023). https://github.com/ultralytics/ultralytics

9. Khan, N.A., Jhanjhi, N., Brohi, S.N., Usmani, R.S.A., Nayyar, A.: Smart traffic monitoring system using unmanned aerial vehicles (UAVs). Comput. Commun. **157**, 434–443 (2020)

10. Kumar, N., Raubal, M.: Applications of deep learning in congestion detection, prediction and alleviation: a survey. Transp. Res. Part C: Emerg. Technol. **133**, 103432 (2021)

11. Sahin, O., Ozer, S.: Yolodrone: improved yolo architecture for object detection in drone images. In: 2021 44th International Conference on Telecommunications and Signal Processing (TSP), pp. 361–365. IEEE (2021)

12. Sonnleitner, E., Barth, O., Palmanshofer, A., Kurz, M.: Traffic measurement and congestion detection based on real-time highway video data. Appl. Sci. **10**(18), 6270 (2020)

13. Sugandi, A.N., Hartono, B.: Implementasi pengolahan citra pada quadcopter untuk deteksi manusia menggunakan algoritma yolo. In: Prosiding Industrial Research Workshop and National Seminar, vol. 13, pp. 183–188 (2022)

14. Tripathi, A., Gupta, M.K., Srivastava, C., Dixit, P., Pandey, S.K.: Object detection using YOLO: a survey. In: 2022 5th International Conference on Contemporary Computing and Informatics (IC3I), pp. 747–752. IEEE (2022)

15. Wu, J.D., Chen, B.Y., Shyr, W.J., Shih, F.Y.: Vehicle classification and counting system using YOLO object detection technology. Traitement du Signal **38**(4) (2021)

16. Yao, Z.B., Douglas, W., O'Keeffe, S., Villing, R.: Faster YOLO-LITE: faster object detection on robot and edge devices. In: Alami, R., Biswas, J., Cakmak, M., Obst, O. (eds.) RoboCup 2021. LNCS (LNAI), vol. 13132, pp. 226–237. Springer, Cham (2022). https://doi.org/10.1007/978-3-030-98682-7_19

17. Zeroual, A., Harrou, F., Sun, Y.: Road traffic density estimation and congestion detection with a hybrid observer-based strategy. Sustain. Urban Areas **46**, 101411 (2019)

Edge AI-Driven Air Quality Monitoring and Notification System: A Multilocation Campus Perspective

Chandra Wijaya[1,2], Anggi Andriyadi[1,3], Shi-Yan Chen[4], I-Jan Wang[1], and Chao-Tung Yang[4,5(✉)]

[1] Department of Industrial Engineering and Enterprise Information, Tunghai University, Taichung 407224, Taiwan
`chandraw@unpar.ac.id`
[2] Informatics Department, Parahyangan Catholic University, Bandung 40141, West Java, Indonesia
[3] Informatics and Bussiness Institute Darmajaya, Bandar Lampung 35141, Lampung, Indonesia
[4] Department of Computer Science, Tunghai University, Taichung 407224, Taiwan
[5] Research Center for Smart Sustainable Circular Economy, Tunghai University, Taichung 407224, Taiwan
`ctyang@thu.edu.tw`

Abstract. With a growing emphasis on environmental health and safety, monitoring and managing air quality in large-scale settings such as campuses are becoming increasingly critical. This research proposes an innovative approach that integrates multilocation Internet of Things (IoT) sensors, edge artificial intelligence (AI), machine learning, and public API integration to create a comprehensive air quality monitoring and notification system for campus environments. Our framework deploys IoT sensors across various locations within the campus to collect real-time data on air quality parameters. Leveraging edge AI capabilities, these sensors process data locally, enabling rapid analysis and anomaly detection without the need for centralized processing. Furthermore, machine learning algorithm is used to analyze the collected data, identify patterns, and predict air quality. To enhance user accessibility and engagement, the system use public APIs to deliver notifications and alerts regarding air quality status.

Keywords: edge ai · air quality monitoring · multinode iot sensors · notification delivery api

1 Introduction

In the research carried out, we built an IoT sensor that can measure air quality and record all the data so that it can be processed quickly at edge node. The

L. Barolli (Ed.): IMIS 2024, LNDECT 214, pp. 256–261, 2024.
https://doi.org/10.1007/978-3-031-64766-6_25

process includes data preprocessing, model training and model evaluation. Edge AI refers to the deployment of machine learning algorithms and models directly on edge devices, such as smartphones, IoT devices, sensors, and other embedded systems, rather than relying on cloud-based processing. There are several benefits to leveraging Edge AI:

– Low Latency: by processing data locally on edge devices, Edge AI reduces the latency associated with transmitting data to centralized cloud servers for processing.
– Privacy and Security: Edge AI enhances data privacy and security by keeping sensitive information local and reducing the need to transmit raw data over networks.
– Bandwidth Efficiency: Edge AI helps reduce the bandwidth requirements by processing data locally and transmitting only relevant insights or aggregated results to the cloud.
– Scalability and Cost: Edge AI distributes computing resources across multiple edge devices, allowing for scalable and cost-effective deployment of AI-powered solutions.
– Real-Time Insights: Edge AI enables the generation of real-time insights and actionable intelligence directly on edge devices.
– Reduced Dependence on Cloud Services: Edge AI reduces the dependency on cloud services and internet connectivity, making applications more robust and resilient to fluctuations in cloud service availability or performance.

Overall, Edge AI offers numerous benefits in terms of latency reduction, privacy and security enhancement, bandwidth efficiency, scalability, cost-effectiveness, and real-time insights, making it a compelling approach for deploying AI-powered solutions in various domains and applications.

2   Related Works

Endah et al. [1] has deployed iSEC, a sophisticated system integrating sensors, cloud, and edge architecture. This system effectively assesses the performance of edge computing in analyzing air quality data and conducting object detection tasks. Cho and Baek [2] proposed a smart system aimed at monitoring and purifying air quality within school premises. This system allows for concurrent monitoring of outdoor and indoor air quality, providing the flexibility to utilize natural ventilation through window opening. Sholahudin et al. [3] conducted a comprehensive analysis of indoor air quality monitoring and automation through a literature review. They found that 55% of the studies examined lacked automated systems for monitoring indoor air quality. Furthermore, the predominant microcontrollers used were integrated on a single development board, and the primary parameters measured did not align with WHO guidelines. Janarthanan et al. [4] proposed an air pollution monitoring approach utilizing sensors and IoT technology. This method offers a safe and straightforward solution for real-time environmental monitoring. The air pollution monitoring package aims to assist individuals in identifying, monitoring, and assessing indoor air quality at specific locations to alert them and safeguard the environment from harmful gases such as CO, CO2, C2H4, NO, NO2, SO2, and CH4.

Various machine learning techniques may be employed for outlier detection in Internet of Things (IoT) and Wireless Sensor Networks (WSN). Given that sensors are pivotal in generating raw data and detecting environmental changes within IoT and WSN frameworks, the detection of outliers is essential for analyzing error-free sensor data. [6] The techniques include random forest, different types of Support Vector Machines (SVMs), visualization methods such as Principal Component Analysis (PCA) and Kernel PCA (KPCA), multi-agent-based approaches, and statistics-based methods. [5] Anomaly detection at the edge devices holds promise for delivering faster response times and improved quality of service in IoT settings. Real-time anomaly detection is essential for handling the continuous data streaming characteristic of IoT applications. [7] Zhang et al. [8] introduces the Isolation Forest algorithm as an unsupervised machine learning technique. Unlike traditional methods that rely on distance or density calculations, the Isolation Forest detects abnormal data in a sample without such computations. It assesses the distribution and density of sample data to accurately identify isolated points.

3 Methodology

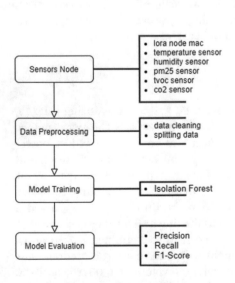

Fig. 1. Research Methodology

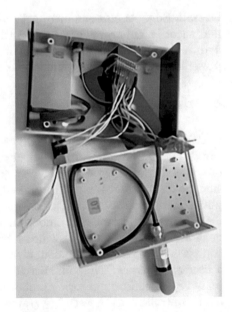

Fig. 2. Sensor Node using Arduino D1 Mini

Figure 1 depicts the overall research stages. The edge node is developed using an Arduino D1 mini microcontroller equipped with various sensors including temperature, humidity, PM2.5, TVOC, and CO2 sensors. Data from each sensor is collected by the microcontroller every 30 s. This data is then transmitted to the gateway, utilizing an Nvidia Jetson Orin Nano device as an edge AI tool. At the gateway, preprocessing tasks such as data cleaning, normalization, feature scaling, and segmentation are performed on the incoming data from multiple sensor nodes. After preprocessing, the Jetson device employs the Isolation Forest algorithm to build a machine learning model aimed at detecting anomalous conditions identified by the sensor nodes. The effectiveness of the developed model is subsequently evaluated for precision using metrics such as Precision, Recall, and F1-Score.

4   Implementation

Figure 2 illustrates the deployed sensor node implementation. These nodes are strategically positioned at three distinct locations within our campus. They transmit data to the gateway via Wi-Fi.

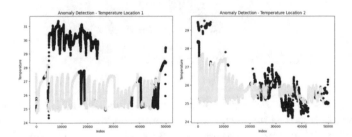

Fig. 3. Temperature Comparison in Location 1 and Location 2

Upon acquiring data from sensor nodes across multiple locations, we conduct data cleaning and segmentation procedures prior to initiating the machine learning algorithm. Employing the Isolation Forest algorithm, we set parameters as follows: n_estimator = 100, contamination = auto, max_features = 1.0. The algorithm being executed aims to detect anomalies within the acquired data. Anomalies are labeled as -1, whereas normal data is labeled as 1. Each data index that has been stored receives this designated label. In Fig. 3, the results of anomaly detection for temperature value are illustrated for both location 1 and location 2. In Fig. 4, the results of anomaly detection for humidity are depicted. Normal data points are represented by yellow dots, while anomalous data points are illustrated in purple.

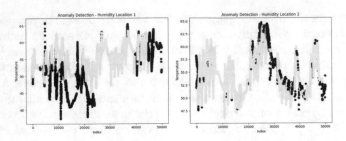

Fig. 4. Humidity Comparison in Location 1 and Location 2

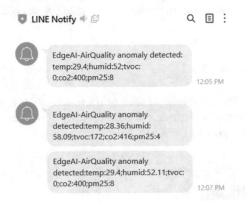

Fig. 5. Line notify about anomalous sensors value

In Fig. 5, shown the result of notification sending about the anomalous value from the temperature, humidity, co2, tvoc and pm2.5.

Recall, Precision, and F1-score serve as standard metrics for assessing the effectiveness of classification models, particularly in binary classification scenarios. We employed these metrics to evaluate the performance of the developed machine learning model. The evaluation results are outlined in Table 1. According to the table, the model achieves an average Recall of 97.91%, Precision of 97.21%, and F1-score of 97.55%. These figures indicate a notably high level of accuracy.

Table 1. Performance Metric Evaluation using Recall, Precision and F1-Score

Location	Recall	Precision	F1-Score
1	0.9875981	0.9836186	0.9855140
2	0.9922893	0.9816494	0.9869293
3	0.9715785	0.9735712	0.9724279
4	0.9732688	0.9798096	0.9765230
5	0.9710328	0.9417012	0.9561038
average	0.9791535	0.9720700	0.9754996

5   Conclusion

Following the implementation involving multiple sensor nodes with Arduino, edge nodes with Jetson, the construction of a machine learning model using isolation forest, and the subsequent performance evaluation, it can be affirmed that the system development has been fruitful. The sensor nodes effectively capture sensor values and transmit the data to the edge, where it undergoes processing utilizing a machine learning algorithm to discern normal and anomalous data. Upon detecting anomalies, the edge node sends notifications via the Line Notify API.

Acknowledgement. This work was sponsored by the National Science and Technology Council (NSTC), Taiwan, under Grant No. 112-2622-E-029-003, 112-2621-M-029-004, 112-2221-E-126-004, 112-2811-E-029-003, and 110-2221-E-029-002-MY3.

References

1. Kristiani, E., Yang, C.-T., Huang, C.-Y., Ko, P.-C., Fathoni, H.: On construction of sensors, edge, and cloud (iSEC) framework for smart system integration and applications. IEEE Internet Things J. **8**(1), 309–319 (2021). https://doi.org/10.1109/JIOT.2020.3004244
2. Cho, H., Baek, Y.: Design and implementation of a smart air quality monitoring and purifying system for the school environment. In: 2022 IEEE International Conference on Consumer Electronics (ICCE), Las Vegas, NV, USA, pp. 1–4 (2022). https://doi.org/10.1109/ICCE53296.2022.9730505
3. Sholahudin, Damey, Y., Fauji, W., Yudono, M.A.S.: Indoor air quality monitoring system with automation: a review. In: 2023 IEEE 9th International Conference on Computing, Engineering and Design (ICCED), Kuala Lumpur, Malaysia, pp. 1–6 (2023). https://doi.org/10.1109/ICCED60214.2023.10425129
4. Janarthanan, A., Paramarthalingam, A., Arivunambi, A., Vincent, P.M.D.R.: Real-time indoor air quality monitoring using the Internet of Things. In: 2022 Third International Conference on Intelligent Computing Instrumentation and Control Technologies (ICICICT), Kannur, India, pp. 99–104 (2022). https://doi.org/10.1109/ICICICT54557.2022.9917990
5. Dwivedi, R.K., Rai, A.K., Kumar, R.: A study on machine learning based anomaly detection approaches in wireless sensor network. In: 2020 10th International Conference on Cloud Computing, Data Science and Engineering (Confluence), Noida, India, pp. 194–199 (2020). https://doi.org/10.1109/Confluence47617.2020.9058311
6. Ghosh, N., Maity, K., Paul, R., Maity, S.: Outlier detection in sensor data using machine learning techniques for IoT framework and wireless sensor networks: a brief study. In: 2019 International Conference on Applied Machine Learning (ICAML), Bhubaneswar, India, pp. 187–190 (2019). https://doi.org/10.1109/ICAML48257.2019.00043
7. Sharma, B., Sharma, L., Lal, C.: Anomaly detection techniques using deep learning in IoT: a survey. In: 2019 International Conference on Computational Intelligence and Knowledge Economy (ICCIKE), Dubai, United Arab Emirates, pp. 146–149 (2019). https://doi.org/10.1109/ICCIKE47802.2019.9004362
8. Zhang, L., Liu, L.: Data anomaly detection based on isolation forest algorithm. In: 2022 International Conference on Computation, Big-Data and Engineering (ICCBE), Yunlin, Taiwan, pp. 87–89 (2022). https://doi.org/10.1109/ICCBE56101.2022.9888169

Peer Selection for Reliability Improvement in P2P Networks by Fuzzy-Based and Ns-3 Simulation Systems

Yi Liu[1]([✉]), Shinji Sakamoto[2], and Leonard Barolli[3]

[1] Department of Computer Science, National Institute of Technology, Oita College, 1666, Maki, Oita 870-0152, Japan
y-liu@oita-ct.ac.jp
[2] Department of Computer and Information Science, Kanazawa Institute of Technology, 7-1, Ohgigaoka, Nonoichi, Ishikawa 921-8501, Japan
shinji.sakamoto@ieee.org
[3] Department of Information and Communication Engineering, Fukuoka Institute of Technology, 3-30-1 Wajiro-Higashi, Higashi-Ku, Fukuoka 811-0295, Japan
barolli@fit.ac.jp

Abstract. In Peer-to-Peer (P2P) networks, the peers should collaborate with trustworthy neighboring peers. However, it should be noted that in real applications, the peer may be faulty, which may make a wrong decision. For safe communication, the peers should be reliable. The reliability can be evaluated by the interaction of a peer with other peers but this procedure needs many parameters. In this paper, in order to make a qualitative evaluation, we implement a Fuzzy-based system and for quantitative evaluation we build another simulation system by taking into account the correlation coefficients and ns-3 simulator. Three parameters Packet Loss (PL), Delay Time (DT) and Throughput (TG) are considered for the evaluation of both systems. For Fuzzy-based system, the output is Peer Selection Decision (PSD). From the simulations results of FL-based system, we conclude that when PL and DL are increasing, the PSD is decreased. While, when TG is increased, the PSD is increased. By ns-3 simulator results, we found that by using the value of correlation coefficients is possible to select the best reliable peers.

1 Introduction

The Peer-to-Peer (P2P) paradigm is a good approach for the development of decentralized applications because the computation can be distributed to peer nodes [1,2]. In P2P systems, the peers share resource and collaborate together to exchange information with trustworthy neighboring peers. However, it should be noted that in real applications, the peer may be faulty, which may make a wrong decision [3–6].

For safe communication, the peers should be reliable. The reliability can be evaluated by the interaction of a peer with other peers but this procedure

© The Author(s), under exclusive license to Springer Nature Switzerland AG 2024
L. Barolli (Ed.): IMIS 2024, LNDECT 214, pp. 262–271, 2024.
https://doi.org/10.1007/978-3-031-64766-6_26

needs many parameters. In this study, for a qualitative evaluation, we implement a system based on Fuzzy Logic (FL) considering three input parameters: Packet Loss (PL), Delay Time (DT), Throughput (TG) for deriving Peer Selection Decision (PSD). While for quantitative evaluation, we implement another system taking into account the correlation coefficients and ns-3 simulator. We consider the same parameters as in the case of Fuzzy-based system: PL, DT and TG. The best peers are decided by calculating the minimum sum of the parameters and the best peer is selected in each simulation [7–11].

The structure of this paper is as follows. In Sect. 2, we introduce P2P and FL approaches. In Sect. 3, we present the proposed simulation systems. In Sect. 4, we discuss the simulation results. Finally, conclusions and future work are given in Sect. 5.

2   P2P and FL Approaches

In this section, we describe P2P technology and give a short description of FL.

2.1   P2P Technology

The P2P group model is shown in Fig. 1, where the load can be distributed among peers. In P2P networks, the clients are also servers and routers, nodes are autonomous, network is dynamic, and nodes collaborate directly with each other. The P2P networks have many benefits such as efficient use of resources, scalability, reliability, and ease of administration.

There are several types of P2P systems such as: Pure P2P, Hybrid P2P and Super Node Type P2P. The Pure P2P has only peers, which connect and share information between each other. While, in Hybrid P2P, there are both: peers and servers, but different from client-server model, clients exchange information mutually. While, the servers store data about clients. Finally, in Super Node Type P2P, several nodes with stable communication and good processing power are selected as super peers and share the information with other super nodes.

The P2P networks have many advantages. One advantage is that they are able to make distributed data management, where the load can be distributed among the peers. Also, they can realize zero downtime by distributing the load and increasing the processing speed. Another advantage is the anonymity ensurance because the data is distributed over the network and is difficult to grasp the information of all nodes.

There are different application of P2P technology in social networks such as Skype and LINE. In LINE application, the data files and other information can be shared among users without using servers. Skype has a hierarchical overlay network, where skype peers connect directly to each other. The super nodes are peers with special functions and the overlay network is used to locate clients.

Also, Bitcoin uses P2P approach. The Bitcoin network protocol allows full nodes (peers) to collaboratively maintain a P2P network for block and transaction exchange. It uses P2P technology to facilitate instant financial transactions and removes the need for third-party involvement.

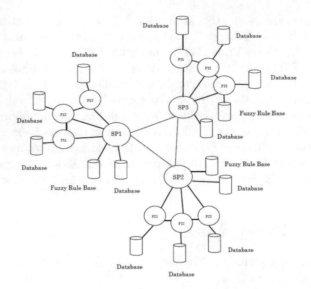

Fig. 1. P2P group-based model.

As a P2P distributed computing platform, we can mention SETI@home, which searches for Extraterrestrial Intelligence (EI). This platform has a central site, which collects radio telescope data and the data is divided into work chunks of 300 Kbytes. The user obtains client, which runs in background and the peer sets up TCP connection to central computer and downloads the chunk. This is not a P2P communication, but the platform exploits the peer computing power.

2.2 FL Outline

The FL have many application because it can model gradual properties or soft constraints whose satisfaction is matter of degree. Also, FL can process the information provided with imprecision and uncertainty. The FL can be used in industrial applications such as appliance control, process control and automotive systems.

The FC systems consider the expert knowledge in the form of fuzzy rules in order to make appropriate action. These rules are put in a Fuzzy Rule Base (FRB) in a form of "if ... then ..." expression, just like in the case of expert systems. If we consider input parameters as x and y, and the output is z, we can build the rules as follows:

> If x is small and y is big, then z is medium;
> If x is big and y is medium, then z is big.

An important concept in the application of FL is the linguistic variable, which may be considered as a form of data compression. One linguistic variable may represent many numerical variables. This form of data compression

can be referred as granulation. The granulation is more general than quantiza-
tion because it can mimic the humans interpretation of linguistic values. Also,
the linguistic value is gradual rather than abrupt, resulting in continuity and
robustness.

3   Fuzzy-Based and Ns-3 Simulation Systems

3.1   Fuzzy-Based System

In this work, for the design of FL-based system, we consider three input parameters: PL, DT and TG. The output parameter is PSD. The membership functions
are shown in Fig. 2. While, in Table 1, we show the FRB, which consists of 27
rules.

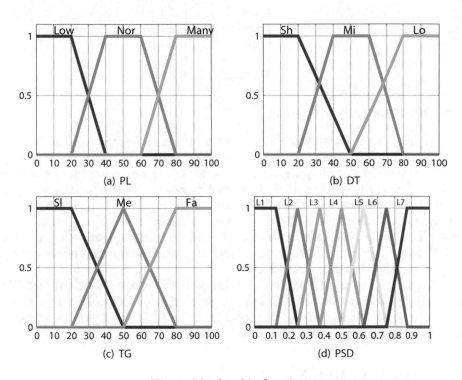

Fig. 2. Membership functions.

The term sets of *PL*, *DT* and *TG* are defined respectively as:

$$PL = \{Many,\ Normal,\ Low\}$$
$$= \{Many,\ Nor,\ Low\};$$
$$DT = \{Long,\ Middle,\ Short\}$$
$$= \{Lo,\ Mi,\ Sh\};$$
$$TG = \{Fast,\ Medium,\ Slow\}$$
$$= \{Fa,\ Me,\ Sl\}.$$

While, the term set for the output *PSD* is defined as:

$$PSD = \begin{pmatrix} Level1 \\ Level2 \\ Level3 \\ Level4 \\ Level5 \\ Level6 \\ Level7 \end{pmatrix} = \begin{pmatrix} L1 \\ L2 \\ L3 \\ L4 \\ L5 \\ L6 \\ L7 \end{pmatrix}.$$

3.2   Ns-3 Simulation System

For quantitative evaluation, we implement a simulation system considering the correlation coefficients and ns-3 simulator. We consider the same parameters as in the case of FL-based system: PL, DT and TG. These parameters are very important parameters for reliability evaluation. For instance, a high throughput of the peer indicates that the communication environment is good. While, in the case of network failure, it can happen the packet loss and communication delay. The packet loss may occur during sudden disasters, hardware problems and network congestion. While, the communication delays may occur during long processing time and link failure.

In order to prevent such failures, it is important to construct a highly reliable network. For this reason, we implement a simulation system using ns-3 simulator and correlation coefficients as shown in Table 2. To construct the network topology, we used the parameters in Table 3.

4   Simulation Results

In this section, we present the simulation results of our proposed FL-based and ns-3 simulation systems.

4.1   Simulation Results of FL-Based System

We show the relation between PSD and PL, DT and TG in Fig. 3. In these simulations, we consider the TG and DT as constant parameters. In Fig. 3(a),

Table 1. FRB.

Rule	PL	DT	TG	PSD
1	Many	Lo	Sl	L1
2	Many	Lo	Me	L1
3	Many	Lo	Fa	L3
4	Many	Mi	Sl	L1
5	Many	Mi	Me	L2
6	Many	Mi	Fa	L4
7	Many	Sh	Sl	L2
8	Many	Sh	Me	L3
9	Many	Sh	Fa	L5
10	Nor	Lo	Sl	L1
11	Nor	Lo	Me	L3
12	Nor	Lo	Fa	L4
13	Nor	Mi	Sl	L2
14	Nor	Mi	Me	L4
15	Nor	Mi	Fa	L5
16	Nor	Sh	Sl	L3
17	Nor	Sh	Me	L5
18	Nor	Sh	Fa	L6
19	Low	Lo	Sl	L5
20	Low	Lo	Me	L4
21	Low	Lo	Fa	L6
22	Low	Mi	Sl	L3
23	Low	Mi	Me	L7
24	Low	Mi	Fa	L6
25	Low	Sh	Sl	L4
26	Low	Sh	Me	L6
27	Low	Sh	Fa	L7

Table 2. Values of correlation coefficients.

Correlation Coefficient	Values	
High Negative Correlation	−1.000	−0.600
Medium Negative Correlation	−0.599	−0.400
Low Negative Correlation	−0.399	−0.200
Uncorrelation	−0.199	+0.199
Low Positive Correlation	+0.200	+0.399
Low Medium Correlation	+0.400	+0.599
High Positive Correlation	+0.600	+1.000

Table 3. Parameters for ns-3.

Items	Values
Number of Peers	10, 20, 30, 40, 50
Wifi	IEEE 802.ac
Model of Propagation Loss	Log-distance Loss Model
Model of Propagation Delay	Steady-state Speed Propagation Model
Peer Location Method	Random
Time for Simulation	150 s

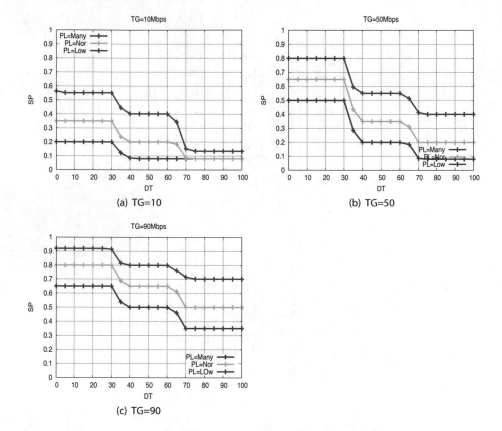

Fig. 3. Simulation result for different TG using FL-based system.

we consider TG value 10 Mbps. We change the PL from Low to Many. When DT increases, the PSD is decreased. Also with the increase of PL, the ASD is decreased. In Fig. 3(b) and (c), we increase TG values to 50 and 90 Mbps, respectively. We see that when the TG increases, the PSD is increased.

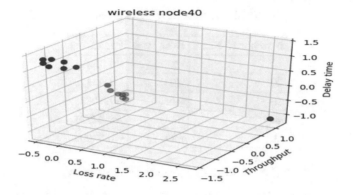

(a) Number of Peers = 40

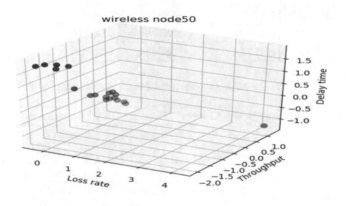

(b) Number of Peers = 50

Fig. 4. Simulation results by ns-3.

4.2   Simulation Results of Ns-3 System

The simulation results by ns-3 are shown in Fig. 4. We consider different number of peers. In our previous work, we considered 10, 20 and 30 peers. While, in this work, we consider also 40 and 50 peers. In Fig. 4(a), there are 40 peers, while in Fig. 4(b) 50 peers. We show the result of best peers in Table 4.

From the results of Table 5, we can see that in case of 10, 20 and 30 peers, n7, n16, n30 are the best peers, respectively. While, for 40 and 50 peers, n32 and n25 are the best peers in group. In Tables 6 and 7, we show the average values and the values of correlation coefficients for each simulation. By using these data for each peer, the best peer in a group of peers can be selected.

Table 4. Results of best peers.

Number of Peers	Best Peer	Peer Data
10	n7	−1.1856
20	n16	−1.55139
30	n16	−1.20736
40	n32	−0.57221
50	n25	−0.69002

Table 5. Parameter values for best peers.

Best Peer	Delay Time (s)	Packet Loss	Throughput (Mbps)
n7 (10)	14.413	2.596679	23.7963
n16 (20)	6.07638	0.490722	18.012
n16 (30)	12.4762	0.889434	17.7486
n32 (40)	8.16726	1.727781	21.5602
n25 (50)	6.41535	0.889434	13.5007

Table 6. Average parameter values for each simulation.

Number of Peers	Delay Time (s)	Packet Loss	Throughput (Mbps)
10	1.146	3.460255	19.242432
20	0.744592	1.172904	17.590646
30	0.9177045	4.9384475	11.350821
40	0.417791468	14.58425002	10.47284063
50	0.5097237368	7.981087206	8.820908947

Table 7. Values of correlation coefficients for each simulation.

Number of Peers	10	20	30	40	50
Delay Time and Throughput	0.68402642	−0.489	−0.235	−0.433	−0.358
Packet Loss and Delay Time	−0.681	−0.406	−0.332	−0.411	−0.332
Throughput and Daley Time	−0.995	0.988	0.983	0.994	0.982

5   Conclusions and Future Work

In this study, for a qualitative evaluation, we implemented a FL-based system considering three input parameters: PL, DT and TG for deriving PSD output value. While for quantitative evaluation, we implemented another system considering the correlation coefficients and ns-3 simulator.

From the simulations results of FL-based system, we concluded that when PL and DL are increased, the PSD is decreased. While, when TG is increased, the PSD is increased. By using the value of correlation coefficients and ns3 was possible to select the best reliable peers.

In the future work, we will investigate other parameters and carry out extensive simulations to evaluate the implemented systems.

References

1. Xhafa, F., Fernandez, R., Daradoumis, T., Barolli, L., Caballé, S.: Improvement of JXTA protocols for supporting reliable distributed applications in P2P systems. In: International Conference on Network-Based Information Systems (NBiS-2007), pp. 345–354 (2007)
2. Barolli, L., Xhafa, F., Durresi, A., De Marco, G.: M3PS: A JXTA-based multi-platform P2P system and its web application tools. Int. J. Web Inf. Syst. **2**(3/4), 187–196 (2007)
3. Aikebaier, A., Enokido, T., Takizawa, M.: Reliable message broadcast schemes in distributed agreement protocols. In: International Conference on Broadband, Wireless Computing, Communication and Applications (BWCCA-2010), pp. 242–249 (2010)
4. Spaho, E., Sakamoto, S., Barolli, L., Xhafa, F., Ikeda, M.: Trustworthiness in P2P: performance behaviour of two fuzzy-based systems for JXTA-overlay platform. Soft. Comput. **18**(9), 1783–1793 (2014)
5. Watanabe, K., Nakajima, Y., Enokido, T., Takizawa, M.: Ranking factors in peer-to-peer overlay networks. ACM Trans. Auton. Adapt. Syst. **2**(3), 1–26 (2007). Article 11
6. Terano, T., Asai, K., Sugeno, M.: Fuzzy Systems Theory and Its Applications. Academic Press Professional Inc., Cambridge (1992)
7. Liu, Y., Sakamoto, S., Matsuo, K., Ikeda, M., Barolli, L., Xhafa, F.: Improvement of JXTA-overlay P2P platform: evaluation for medical application and reliability. Int. J. Distrib. Syst. Technol. (IJDST) **6**(2), 45–62 (2015)
8. Liu, Y., Barolli, L., Spaho, E., Ikeda, M., Caballe, S., Xhafa, F.: A fuzzy-based reliability system for P2P communication considering number of interactions, local score and security parameters. In: International Conference on Network-Based Information Systems (NBiS-2014), pp. 484–489 (2014)
9. Spaho, E., Matsuo, K., Barolli, L., Xhafa, F., Arnedo-Moreno, J., Kolici, V.: Application of JXTA-overlay platform for secure robot control. J. Mob. Multimedia **6**(3), 227–242 (2010)
10. Liu, Y., Sakamoto, S., Matsuo, K., Ikeda, M., Barolli, L.: Improving reliability of JXTA-overlay platform: evaluation for e-learning and trustworthiness. J. Mob. Multimedia **11**(2), 34–50 (2015)
11. Liu, Y., Sakamoto, S., Matsuo, K., Ikeda, M., Barolli, L., Xhafa, F.: Improving reliability of JXTA-overlay P2P platform: a comparison study for two fuzzy-based systems. J. High Speed Netw. **21**(1), 27–45 (2015)

A PFCP Protocol Fuzz Testing Framework Integrating Data Mutation Strategies and State Transition Algorithms

Xiaoyang Feng[1], Wei Tan[1]([✉]), Tao Qiu[1], Wenxiao Yu[2], Zixuan Zhang[2], and Baojiang Cui[2]

[1] The 30th Research Institute of CETC, Chengdu, China
vivifishtl@163.com
[2] Beijing University of Posts and Telecommunications, Beijing, China
{ywx20010107,zzx18731967365,cuibj}@bupt.edu.cn

Abstract. Various standardization organizations have formulated security mechanism designs in the field of 5G security. However, there is a lack of security analysis and testing. This article provides an in-depth analysis of the Packet Forwarding Control Protocol (PFCP) in the 5G core network. It proposes an innovative fuzz testing framework that combines existing network protocol fuzz testing methods and designs mutation strategies and state transition algorithms based on the field characteristics of the PFCP protocol to achieve comprehensive coverage testing of the PFCP protocol. By deploying the tool in the free5GC open-source environment and comparing it to a traditional exhaustive algorithm, the efficiency improvement is about 2^300 times.

1 Introduction

Since 2019, hailed as the "commercial year" of 5G, countries worldwide have embraced 5G as a pivotal driver of progress [4]. Compared to 4G LTE, 5G significantly boosts communication capabilities, achieving millisecond-level end-to-end latency and connecting millions of devices per square kilometer [1]. As 5G technology matures, major operators are gearing up for network evolution and upgrades. 5G not only enhances communication but also deepens the Internet of Things [8, 14], serving as a foundation for digital transformation. It fosters advancements in connectivity, big data, and artificial intelligence [15, 20], influencing various aspects of life and the economy, spurring innovation across industries.

The Packet Forwarding Control Protocol (PFCP) in the 5G core network connects the Session Management Function (SMF) and the User Plane Function (UPF). With features like edge computing and network slicing, along with new technologies such as cloud-native deployment [11, 17], ensuring the security of the PFCP protocol is vital to prevent network attacks.

Specifically, our main contributions are as follows:

1. We introduce a tailored fuzz testing framework for the PFCP protocol to effectively analyze its security with broader coverage.

2. Our framework includes a specialized data mutation strategy for the PFCP protocol, utilizing a state transition algorithm to achieve full-state coverage.
3. Testing in an open-source 5G core network experimental environment validates the effectiveness of our framework in identifying vulnerabilities in the PFCP protocol.

We first review relevant work, then introduces the fuzzing testing scheme and security analysis of PFCP in the background section. Subsequently, we introduce the fuzzing testing framework and algorithm proposed in this article, followed by experimental validation of the framework's effectiveness. Finally, we summarize our contributions.

2 Related Work

Fuzz involves testing a system by providing input data to observe its behavior without focusing on its internal implementation. Hu et al. [13] injects malformed data into the system to uncover vulnerabilities. Ananda et al. [3] demonstrated fuzz testing's importance in identifying vulnerabilities in protocols like HTTP, Modbus, and FTP, crucial for IoT systems.

Data mutation modifies input data to generate diverse test cases. Han et al. [12], Wu et al. [6] and Cao et al. [7] proposed mutation-based fuzz testing methods for different protocols.

State transition ensures fuzz testing covering various system scenarios. Yu et al. [19] and Banks et al. [5] presented stateful fuzz testing tools for protocol testing. Gorbunov et al. [10] proposed a scalable fuzz testing framework that learns protocol syntax.

Combining data mutation and state transition enhances fuzz testing. Pham et al. [16] introduced a gray-box fuzz testing tool combining mutation strategies with state feedback. Song et al. [18] introduced a stateful protocol fuzz testing framework with flexible mutation strategies.

Based on existing works, we propose a novel fuzz testing framework tailored for the PFCP protocol, integrating data mutation and state transition strategies for improved efficiency and accuracy.

3 Background

3.1 Fuzz Testing Schemes

Common fuzz testing tools like Peach and AFL [9] employ various mutation strategies for packets of network protocols: Bit/Byte Flipping, Addition, Subtraction, Special Value, String Overwrite, Insertion, Deletion, Transposition and Coverage. These strategies can be combined in various ways, including:

1. Sequential Test: Applying mutation strategies one by one to all fields needing mutation.
2. Random Test: Randomly selecting fields and applying multiple mutation strategies to each

3.2 Security Analysis of the PFCP Protocol

The PFCP protocol facilitates communication between UPF and SMF. However, if exploited, it could result in severe attacks like UE traffic hijacking, network resource depletion, and billing fraud, while still complying with protocol standards.

1. Node Scanning: Attackers exploit the PFCP heartbeat functionality to detect active UPF and SMF elements in the core network, facilitating subsequent attacks.
2. Malicious Correlation Release Operation: Attackers can impersonate the SMF to disrupt the core network by terminating UPF connections, potentially leading to internet access and billing issues for users [2].
3. PFCP Message Forgery: Inadequate handling of abnormal PFCP messages can lead to resource release anomalies and transmission issues. The protocol's absence of encryption and integrity protection exposes it to attacks like replay, tampering, and forgery [2].

4 The Proposed Approach

4.1 Fuzz Testing Architecture for 5G Core Network Protocols

The fuzz testing process for the 5G core network's PFCP protocol involves analyzing protocol messages, establishing mutation strategies, generating compliant samples, interacting with the network, analyzing results, and optimizing strategies based on feedback to enhance testing efficiency and accuracy as illustrated in Fig. 1. This iterative process ensures thorough testing and identification of anomalies for further investigation and improvement.

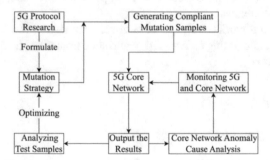

Fig. 1. Fuzz Testing Process.

The fuzz testing tool for the 5G core network PFCP protocol consists of several modules: message capture and analysis, mutation strategy generation, message transmission and reception, and result analysis and feedback. These modules work together to capture, parse, generate, send, and analyze protocol messages, aiming to optimize mutation strategies and improve testing efficiency and accuracy through iterative feedback loops.

1. The Message Capture and Analysis Module captures 5G core network PFCP protocol messages using tools like Wireshark, filters out irrelevant messages, and parses them

to identify message headers and fields. This information is used to generate accurate message formats for mutation strategies and message transmission.

2. The Mutation Strategy Generation Module generates compliant mutation samples based on message formats and variable field characteristics, ensuring coverage of various scenarios and compliance with protocol specifications for fuzz testing.

3. The Message Transmission and Reception Module sends mutation sample messages to interact with the 5G core network and receives response results. It establishes connections, sends messages in protocol-specified formats, listens for responses, parses and processes them, and records results for analysis and feedback.

4. The Result Analysis and Feedback Module analyzes output results from the core network, identifies anomalies, conducts in-depth analysis to pinpoint causes, and optimizes mutation strategies based on feedback to improve fuzz testing efficiency and accuracy. It also records analysis results for generating test reports or logs for future reference and optimization.

4.2 Data Mutation Strategies Based on PFCP Protocol

For the PFCP protocol, its unique packet format and information elements require a specific fuzz testing approach. Randomly mutating the entire packet may produce numerous invalid packets, hindering effective security testing of the core network. Hence, this paper proposes a customized data mutation strategy tailored for the PFCP protocol.

The PFCP protocol comprises a variable-length message header and a protocol data part. The header includes Version, MP flag, S flag, message type, message length, optional SEID field, and sequence number. Node-related messages have an 8-byte header, while session-related messages have a 16-byte header with an optional SEID field. The protocol data part contains Information Elements (IEs), categorized as mandatory, conditional, conditionally optional, and optional. This paper focuses mainly on mandatory IEs.

To ensure diverse and compliant mutation, certain fields like Node ID are preserved to prevent errors, while others like length are recalculated post-mutation to prevent parsing issues. For freely mutable fields, BOOLEAN fields can be set randomly, integers to boundary values, bit strings to random content and length, and sequences to varying numbers of random sequences with recursive element mutation.

The paper introduces five mutation strategies:

1. Multi-bit Mutation: Mutations applied at different bit lengths based on captured protocol data.

2. Protocol Step Mutation: Mutations applied to captured protocol data of a specific process and state.

3. Single-State Multi-bit Mutation: Mutations applied to captured protocol data of a specific process and state, with different bit lengths.

4. Single-State Protocol Field Mutation: Mutations applied to captured protocol data of a specific process and state at the field level.

5. Approximate Full Interaction State Field Mutation: Mutations applied to captured protocol data of all processes and states based on fields after reaching a certain state.

The specific steps for field mutation are as follows:

1. Protocol identification fields remain unchanged.
2. Associated fields are recalculated after mutating others.
3. Other fields undergo mutation using methods like bit flipping, addition/subtraction, replacement, deletion, swapping, and fuzzing strategies like Sequential, Random, and Havoc.

4.3 Protocol Fuzzing Based on State Transition

For multi-state protocols, individual states are interconnected, requiring multiple steps of message interaction to complete the protocol process. Fuzzing a single state alone won't suffice for comprehensive testing. Hence, we've designed a feature allowing users to specify complete security testing for a particular state, enabling more comprehensive testing in a shorter time. All mutation samples are tested against the selected protocol state for comprehensive security testing.

To effectively test a specific state, the protocol state machine must transition to that state. This is accomplished by selecting the protocol transition module to reach the desired single state, enabling comprehensive testing of that state.

The PFCP protocol state transitions, involve processes such as session establishment, modification, and deletion. These transitions occur when the UE initiates requests for data session establishment or closure. Upon completion of these processes, the protocol enters states indicating successful session establishment or deletion.

When testing network protocols, it's crucial to consider the various states of network elements and their interactions. Sending a packet that doesn't match the current state of a network element may lead to packet discard or errors. Proper packet parsing is essential to detect protocol vulnerabilities through mutated fields. Thus, a method should be designed to ensure that network elements are in the appropriate state to process constructed data packets during testing.

5 Experiment

In the open-source free5GC environment, we simulate 5G core network operations, test PFCP protocol functional processes, and capture normal data packets. We use Wireshark to analyze packet formats, specifically capturing UPF-to-SMF data packets in different SMF states, simulating UPF sending normal data packets to SMF, and observing specific feedback from SMF under correct packet conditions. Captured packets are organized and encapsulated into binary files for fuzz testing.

As shown in Fig. 2, our tool conducted fuzz testing on the PFCP protocol within the open-source core network environment. It triggered a protocol vulnerability in 4.9 s, while exhaustive testing had not triggered any vulnerabilities even after 25 s. We compared statistical data to analyze and draw conclusions.

Our tool triggered a vulnerability at the 312th bit. Based on similar sending rates, exhaustive testing would require around 2^{311} data packets to trigger it. With an estimated sending rate of 641 packets per second for exhaustive testing, it would take about 2^{302} s. Therefore, our tool's efficiency in discovering this vulnerability is approximately 2^{300} times that of exhaustive testing.

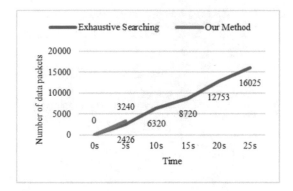

Fig. 2. Comparing with the exhaustive search method.

After debugging and source code analysis, the PFCP protocol was found to have the logical flaws shown in the Fig. 3 below. When parsing the FQDN data of the PFCP protocol, its content is directly copied into memory without checking whether its length matches the subsequent data. An attacker can take advantage of this to construct a packet containing shellcode, which causes the Open5GS system to execute the attacker's commands and open the shell port, and the attacker then logs into the shell to gain full control.

```
int ogs_fqdn_parse(char *dst, char *src, int length)
{
    int i = 0, j = 0;
    uint8_t len = 0;

    do {
        len = src[i++];
        memcpy(&dst[j], &src[i], len);

        i += len;
        j += len;

        if (i < length)
            dst[j++] = '.';
        else
            dst[j] = 0;
    } while (i < length);

    return j;
}
```

Fig. 3. UPF vulnerability core code.

After our construction of the packet, sending a packet such as the one shown in Fig. 4 will cause the UPF network element to crash.

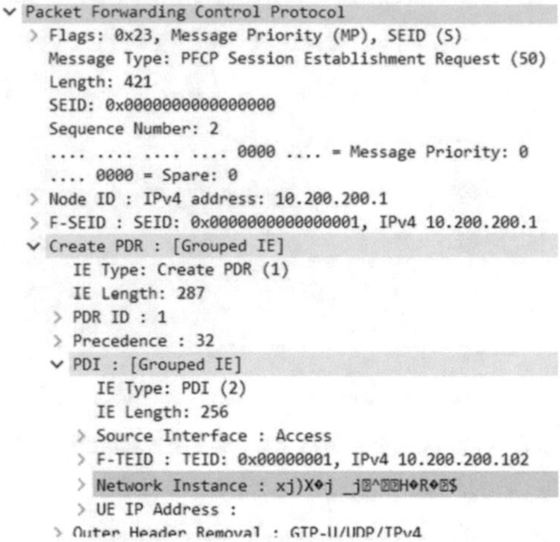

```
∨ Packet Forwarding Control Protocol
  > Flags: 0x23, Message Priority (MP), SEID (S)
    Message Type: PFCP Session Establishment Request (50)
    Length: 421
    SEID: 0x0000000000000000
    Sequence Number: 2
    .... .... .... .... 0000 .... = Message Priority: 0
    .... 0000 = Spare: 0
  > Node ID : IPv4 address: 10.200.200.1
  > F-SEID : SEID: 0x0000000000000001, IPv4 10.200.200.1
  ∨ Create PDR : [Grouped IE]
      IE Type: Create PDR (1)
      IE Length: 287
    > PDR ID : 1
    > Precedence : 32
    ∨ PDI : [Grouped IE]
        IE Type: PDI (2)
        IE Length: 256
      > Source Interface : Access
      > F-TEID : TEID: 0x00000001, IPv4 10.200.200.102
      > Network Instance : xj)X◆j _jⅷ^ⅷⅷH◆R◆ⅷ$
      > UE IP Address :
      > Outer Header Removal : GTP-U/UDP/TPv4
```

Fig. 4. Packet triggering UPF crash.

6 Conclusions

We investigate 5G network security, with a focus on the PFCP protocol. We propose a targeted fuzz testing strategy, analyzing its unique elements and operational flow. This strategy encompasses mutation tactics for various fields and a network element state transition algorithm.

Our framework was utilized to create a security testing tool for the PFCP protocol. Tested within the free5GC environment, it demonstrated significantly enhanced efficiency in generating malformed packets compared to exhaustive algorithms. The fuzz testing tool surpassed other general-purpose tools in analyzing the PFCP protocol.

In the future, we will be testing in real environments and delving into more information elements of the PFCP protocol.

References

1. Agiwal, M., Roy, A., Saxena, N.: Next generation 5G wireless networks: a comprehensive survey. IEEE Commun. Surv. Tutor. **18**(3), 1617–1655 (2016). https://doi.org/10.1109/COMST. 2016.2532458
2. Amponis, G., et al.: Threatening the 5G core via PFCP DoS attacks: the case of blocking UAV communications. EURASIP J. Wirel. Commun. Netw. **2022**(1), 124 (2022). https://doi. org/10.1186/s13638-022-02204-5
3. Ananda, T.K., Simran, G.T., Sukumara, T., Sasikala, D., Kumar, R.P.: Robustness evaluation of cyber physical systems through network protocol fuzzing. In: 2019 International Conference on Advances in Computing and Communication Engineering (ICACCE), pp. 1–6 (2019). https://doi.org/10.1109/ICACCE46606.2019.9079995
4. Andrews, J.G., et al.: What will 5G be? IEEE J. Sel. Areas Commun. **32**(6), 1065–1082 (2014). https://doi.org/10.1109/JSAC.2014.2328098

5. Banks, G., Cova, M., Felmetsger, V., Almeroth, K., Kemmerer, R., Vigna, G.: SNOOZE: toward a Stateful NetwOrk prOtocol fuzZEr. In: Katsikas, S.K., López, J., Backes, M., Gritzalis, S., Preneel, B. (eds.) ISC 2006. LNCS, vol. 4176, pp. 343–358. Springer, Heidelberg (2006). https://doi.org/10.1007/11836810_25
6. Biao, W., Chaojing, T., Bin, Z.: FFUZZ: a fast fuzzing test method for stateful network protocol implementation. Inn: 2021 2nd International Conference on Computer Communication and Network Security (CCNS), pp. 75–79 (2021). https://doi.org/10.1109/CCNS53852.2021.00023
7. Cao, Y., Chen, Y., Zhou, W.: Detection of PFCP protocol based on fuzz method. In: 2022 2nd International Conference on Computer, Control and Robotics (ICCCR), pp. 207–211 (2022). https://doi.org/10.1109/ICCCR54399.2022.9790271
8. Chettri, L., Bera, R.: A comprehensive survey on internet of things (IoT) toward 5G wireless systems. IEEE Internet Things J. 7(1), 16–32 (2020). https://doi.org/10.1109/JIOT.2019.2948888
9. Fioraldi, A., Mantovani, A., Maier, D., Balzarotti, D.: Dissecting American fuzzy lop: a FuzzBench evaluation. ACM Trans. Softw. Eng. Methodol. 32(2) (2023). https://doi.org/10.1145/3580596
10. Gorbunov, S., Rosenbloom, A.: AutoFuzz: automated network protocol fuzzing framework (2010). https://api.semanticscholar.org/CorpusID:18430752
11. Gupta, A., Jha, R.K.: A survey of 5G network: architecture and emerging technologies. IEEE Access 3, 1206–1232 (2015). https://doi.org/10.1109/ACCESS.2015.2461602
12. Han, X., Wen, Q., Zhang, Z.: A mutation-based fuzz testing approach for network protocol vulnerability detection. In: Proceedings of 2012 2nd International Conference on Computer Science and Network Technology, pp. 1018–1022 (2012). https://doi.org/10.1109/ICCSNT.2012.6526099
13. Hu, Z., Pan, Z.: A systematic review of network protocol fuzzing techniques. In: 2021 IEEE 4th Advanced Information Management, Communicates, Electronic and Automation Control Conference (IMCEC), pp. 1000–1005 (2021). https://doi.org/10.1109/IMCEC51613.2021.9482063
14. Li, S., Xu, L.D., Zhao, S.: 5G internet of things: a survey. J. Ind. Inf. Integr. 10, 1–9 (2018). https://doi.org/10.1016/j.jii.2018.01.005
15. Morocho Cayamcela, M.E., Lim, W.: Artificial intelligence in 5G technology: a survey. In: 2018 International Conference on Information and Communication Technology Convergence (ICTC), pp. 860–865 (2018). https://doi.org/10.1109/ICTC.2018.8539642
16. Pham, V.-T., Böhme, M., Roychoudhury, A.: AFLNET: a greybox fuzzer for network protocols. In: 2020 IEEE 13th International Conference on Software Testing, Validation and verification (ICST), pp. 460–465 (2020). https://doi.org/10.1109/ICST46399.2020.00062
17. Rost, P., et al.: Mobile network architecture evolution toward 5G. IEEE Commun. Mag. 54(5), 84–91 (2016). https://doi.org/10.1109/MCOM.2016.7470940
18. Song, C., Yu, B., Zhou, X., Yang, Q.: SPFuzz: a hierarchical scheduling framework for stateful network protocol fuzzing. IEEE Access 7, 18490–18499 (2019). https://doi.org/10.1109/ACCESS.2019.2895025
19. Yu, Y., Chen, Z., Gan, S., Wang, X.: SGPFuzzer: a state-driven smart graybox protocol fuzzer for network protocol implementations. IEEE Access 8, 198668–198678 (2020). https://doi.org/10.1109/ACCESS.2020.3025037
20. Zhang, C., Ueng, Y.-L., Studer, C., Burg, A.: Artificial intelligence for 5G and beyond 5G: implementations, algorithms, and optimizations. IEEE J. Emerg. Sel. Top. Circuits Syst. 10(2), 149–163 (2020). https://doi.org/10.1109/JETCAS.2020.3000103

An Efficient Smart Contracts Event Ordering Vulnerability Detection System Based on Symbolic Execution and Fuzz Testing

Yitao Li, Baojiang Cui[✉], Dongbin Wang, Yue Yu, and Can Zhang

School of Cyberspace Security, Beijing University of Posts and Telecommunications, Beijing, China

`{liyitao,cuibj,dbwang,yuyue_999,zhangcan_bupt}@bupt.edu.cn`

Abstract. Event Ordering Vulnerability, which may lead to fund losses or system paralysis, can be detected using tools based on symbolic execution and fuzz testing techniques. However, the path explosion problem is unavoidable for these tools. Thus, in this paper, we first introduce advanced path pruning techniques and then adopts optimized parallel algorithms to enhance analysis efficiency. Finally, the experimental evaluation verifies the efficiency and effectiveness of the proposed scheme.

Keywords: Smart Contract · Fuzz · Event Ordering Vulnerability · Concurrency

1 Introduction

A contract can be initiated with multiple transactions concurrently at a certain moment, and the specific processing ordering of these transactions is undetermined. Since contracts are stateful, the state can be altered by transaction calls. Therefore, the outcome of multiple concurrent transactions may vary depending on the specific transaction ordering. Developers often assume a certain sequence for executing a transaction, thus often neglecting this event ordering vulnerability.

As we know, there were many incidents of asset loss due to various recognized security risks in smart contracts [1]. Researchers have introduced symbolic execution and fuzz testing technologies to detect contract vulnerabilities [2–6]. While previous related work has mostly focused on analyzing and detecting various types of vulnerabilities for individual events. The study of vulnerabilities caused by the above event ordering vulnerability in a group began only after 2019 [7]. However, analyzing multiple events in a contract leads to a steep increase in the analysis of state space [8], resulting in longer testing times and difficulty in addressing the continuously growing number of smart contracts.

Thus, in this paper, we investigate an improvement detection system for smart contract vulnerability caused by concurrent execution. We adopted trick path pruning techniques for symbol execution and optimized parallel computing techniques for fuzz testing so as to alleviate the path explosion problem.

© The Author(s), under exclusive license to Springer Nature Switzerland AG 2024
L. Barolli (Ed.): IMIS 2024, LNDECT 214, pp. 280–287, 2024.
https://doi.org/10.1007/978-3-031-64766-6_28

2 Background

2.1 Atomicity of Contract Execution

Smart contracts have unique addresses. When a user creates and signs a transaction, and then submits it to the Ethereum network, miners will process it at an appropriate time. This process involves executing computations based on specified smart contract rules, updating the contract's state, and storing the new state on the blockchain [9].

While we discussed how concurrent transaction execution may lead to vulnerabilities, it doesn't necessarily mean vulnerabilities arise from transactions executing concurrently and interleaving. Transaction execution within a contract is atomic. Once started, a transaction continues until it either completes normally or exits abnormally [10].

2.2 Event Ordering Vulnerability

A smart contract contains variables representing its state and function interfaces. It supports concurrent invocation, with the order determined by miners [7].

Contracts find it difficult to identify the true intentions of the caller, which sometimes results in executions not occurring in the expected order. As illustrated in Fig. 1, this is a real contract named *Goochain* deployed on the Ethereum mainnet, which is a customized token system [11]. To briefly illustrate the event ordering vulnerability, the contract code is simplified, focusing on two key functions: *transferFrom* and *approve*. Now, consider a specific scenario where user A inadvertently initiates a transaction *approve(B, 10000)* and later discovers an input error after the transaction succeeds but cannot be revoked. Consequently, user A initiates another transaction *approve(B, 1000)* to correct the mistake. However, user B is unaware of user A's mishap and proceeds to initiate a transaction *transferFrom(A, B, 1000)* based on their pre-agreed token amount of 1000. Due to the miner's selection, it is possible that user A's second transaction *approve(B, 1000)* is executed after user B's transaction *transferFrom(A, B, 1000)*. Although neither of their transactions resulted in an error or rollback, the smart contract's state has indeed experienced a vulnerability. In other words, account B has transferred 1000 tokens but is still able to transfer another 1000.

```
contract Goochain is owned, token {
    ...
    function transferFrom(address _from, address _to, uint256 _value)
        returns (bool success)  {
        require(!frozenAccount[_from]);
        require(balanceOf[_from] >= _value);
        require(balanceOf[_to] + _value >= balanceOf[_to]);
        require(_value <= allowance[_from][msg.sender]);
        balanceOf[_from] -= _value;
        balanceOf[_to] += _value;
        allowance[_from][msg.sender] -= _value;
        Transfer(_from, _to, _value);
        return true;
    }
    funtion approve(address _spender, uint256 _value)
        return (bool success)  {
        allowance[msg.sender][_spender] = _value;
        return true;
    } ...
}
```

Fig. 1. Simplified Goochain Contract Code.

We surveyed 40 real contracts, as shown in Table 1. Nearly half of the contracts contain the *approve* and *transferFrom* functions, indicating the practical significance of our work.

Table 1. Survey statistics.

Function Name	Quantity	Proportion
approve	19	47.5%
transferFrom	20	50%

2.3 Fuzz Testing Combination Explosion

Fuzz testing detects abnormal outputs and states in smart contracts by using a large number of random inputs. However, fuzz testing is less effective in detecting vulnerabilities in a set of transactions containing multiple transactions compared to single transactions because there are multiple different input values for individual transactions within a set [12]. Thus, performing fuzz testing on a set of transactions requires generating permutations of the transactions, making it a time-consuming process for a complex smart contract [13].

3 The Core Module

Our core module is primarily divided into two parts: dynamic symbolic execution and multi-core parallel fuzz testing. The former involves analyzing executable paths of contracts and gathering information, while the latter focuses on automating vulnerability detection based on the collected results and analyzing the output.

3.1 Dynamic Symbolic Execution

Dynamic symbolic execution uses symbolic values as the required parameters for function execution. When encountering conditional branch evaluations, different constraints are collected for different branches until the program terminates normally. The solver provided by Z3 is then utilized to determine if there exists a solution. If a solution exists, specific input values satisfying the conditions are generated, thereby exploring various possible paths of the program [14]. This module initially utilizes the Ethereum Virtual Machine (EVM) as the core for program execution, using contract bytecode obtained from the main network as the code for simulated execution. On this basis, the handling logic for obtaining input values by the Ethereum Virtual Machine under normal circumstances is modified, enabling successful execution of symbolic execution.

Reducing Event Permutation Numbers. In conventional concurrent programming, most high-level languages offer convenient programming methods to prevent multiple threads from entering a critical section simultaneously, thereby ensuring atomicity of transaction execution. In smart contracts, as transactions are atomic, there are no critical sections. We attempted to emulate the MOPS strategy [15] to identify critical paths. During fuzz testing, we observed a common phenomenon: when the order of two events is swapped, the execution cannot proceed normally. Through manual analysis, we found that this phenomenon often occurs because one of the functions in the swapped events restricts the normal execution of the other. *TextMessage* is a contract on the mainnet for sending messages., calling *pauseContract* followed by *sendText* to send a message throws an exception. Consequently, during fuzz testing, encountering situations leading to exceptions renders the execution of the entire event group incomplete, rendering it ineffective. Our insight is to utilize dynamic symbolic execution to identify as many unique event pairs with sequential validity as possible, thereby reducing the number of input event combinations for subsequent fuzz testing to enhance vulnerability detection efficiency.

3.2   Multi-core Parallel Fuzz Testing

The basic idea of fuzz testing is to pass randomly generated input data to a program and observe its response to detect potential vulnerabilities.

Multi-core Parallelism Mode. Since fuzz testing is a CPU-intensive task, it is difficult and inefficient for a single core to handle the heavy workload of fuzz testing. In experiments, the CPU utilization rate often reaches 100%. To alleviate the pressure of fuzz testing, we attempted to distribute fuzz testing tasks across multiple cores in parallel. Tasks are allocated to multiple CPUs based on the number of sets, with each CPU responsible for fuzz testing tasks for a specific set depth. As a result, the efficiency of fuzz testing is improved, and the testing speed depends on the time consumption of the set with the maximum depth. For a depth n to be tested, the efficiency of the outermost layer of testing is optimized to O(n).

4   Detection System

Our system, which we have named Rilupas, shown in Fig. 2, operates at the Ethereum bytecode level, making it compatible with both private and public chains. Initially, contract bytecode is accessed and preprocessed to reset system parameters, extract key information, and prepare for dynamic symbolic execution.

After preprocessing, the dynamic symbolic detection module identifies the modifiable states for each contract function. It then uses symbolic execution on event pairs that could affect sequential validity, as outlined in Sect. 3.1. Two event orderings are tested to verify these relationships. If confirmed, the data aids in accelerated fuzz testing.

Multi-core parallel fuzz testing begins after collecting all relevant information from symbolic execution, including input information required for testing, a list of event pairs with unique sequential validity, and so forth. This approach allows for simultaneous testing across multiple cores to derive vulnerability analysis results.

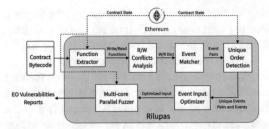

Fig. 2. Main architectures of Rilupas.

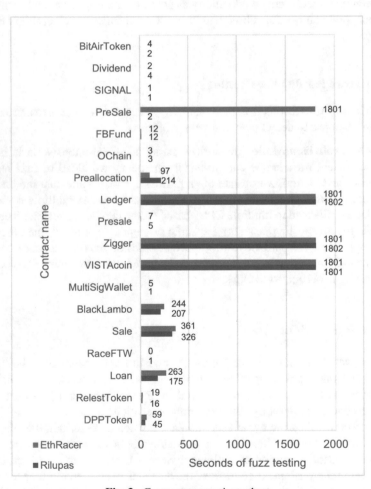

Fig. 3. Contract comparison chart.

5   Evaluation

We tested 100 real contracts from the Ethereum mainnet using our system Rilupas on a Linux server, with 96GB RAM and 12 CPUs Intel(R) Xeon(R) @3.00GHz and detected potential vulnerabilities in 18 contracts after dynamic symbolic execution, prompting fuzz testing. To validate effectiveness, we compared the fuzz testing time with that of EthRacer [7]. As illustrated in Fig. 3, in the testing of the *PreSale* contract, EthRacer erroneously identified state-changing functions, resulting in extended testing durations. However, Rilupas employed a robust strategy for function identification, effectively circumventing such errors. In the case of the *Preallocation* contract, the symbolic execution by Rilupas gathered more information than EthRacer, leading to an unexpected increase in testing time. Nevertheless, for the vast majority of contracts, Rilupas demonstrated superior fuzz testing performance due to enhanced parallelization strategies. Statistical analysis of fuzz testing times across these 18 contracts, as shown in Fig. 4, reveals that our tool, Rilupas, achieved a total time reduction of 22.49% in the fuzz testing phase compared to EthRacer, with an average reduction of 22.61% in the testing time per contract.

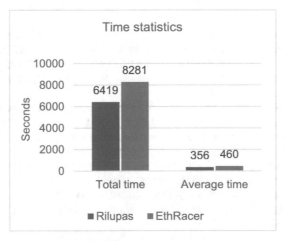

Fig. 4. Time statistics chart.

6   Related Work

Our work has improved the efficiency of detecting this particular type of vulnerability. Most relevant to our work is Kolluri et al.'s EthRacer [7], which we also used as a comparative reference in our evaluation. However, its detection efficiency remains a concern. Currently, most tools primarily focus on Transaction-Ordering Dependency (TOD) vulnerability, i.e., issues related to the order of two transactions, which is a strict subset of event ordering vulnerability. The open-source tool Oyente [16] utilizes symbolic execution to detect TOD errors caused by balance discrepancies. ZEUS [17] also detects

TOD errors, but being closed-source, specific details are unavailable to us. Securify is a static analysis tool [18] that detects TOD errors by defining and checking TOD violation patterns within the dependency graph of smart contracts. ConFuzzius similarly triggers vulnerable transactions by randomly generating transactions [19]. TransRacer explores transaction race conditions in specific states by analyzing function dependencies [20].

7 Conclusion

In this paper, we introduce Rilupas, an efficient system that combines symbolic execution and fuzz testing to detect event-ordering vulnerabilities in Ethereum smart contracts. By leveraging dynamic symbolic execution and constraint solving, Rilupas effectively identifies key event pairs to enhance fuzz testing capability. Utilizing multi-core parallel fuzz testing, Rilupas significantly reduces the time required for contract detection, thereby improving detection efficiency. Comparative experimental evaluations on 100 real-world smart contracts validate the effectiveness and enhancements of Rilupas.

References

1. Dika, A., Nowostawski, M.: Security vulnerabilities in Ethereum smart contracts. In: 2018 IEEE International Conference on Internet of Things (iThings) and IEEE Green Computing and Communications (GreenCom) and IEEE Cyber, Physical and Social Computing (CPSCom) and IEEE Smart Data (SmartData), pp. 955–962 (2018). https://doi.org/10.1109/Cybermatics_2018.2018.00182
2. Tahir, U., Siyal, F., Ianni, M., Guzzo, A., Fortino, G.: Exploiting bytecode analysis for reentrancy vulnerability detection in Ethereum smart contracts. In: 2023 IEEE International Conference on Dependable, Autonomic and Secure Computing, International Conference on Pervasive Intelligence and Computing, International Conference on Cloud and Big Data Computing, International Conference on Cyber Science and Technology Congress (DASC/PiCom/CBDCom/CyberSciTech), Abu Dhabi, United Arab Emirates, pp. 0779–0783 (2023). https://doi.org/10.1109/DASC/PiCom/CBDCom/Cy59711.2023.10361441
3. Siegel, D.: Understanding the DAO attack (2023). https://www.coindesk.com/understanding-dao-hack-journalists
4. Atzei, N., Bartoletti, M., Cimoli, T.: A survey of attacks on Ethereum smart contracts (SoK). In: Maffei, M., Ryan, M. (eds.) POST 2017. LNCS, vol. 10204, pp. 164–186. Springer, Heidelberg (2017). https://doi.org/10.1007/978-3-662-54455-6_8
5. Sun, J., Huang, S., Zheng, C., Wang, T., Zong, C., Hui, Z.: Mutation testing for integer overflow in Ethereum smart contracts. Tsinghua Sci. Technol. **27**(1), 27–40 (2022). https://doi.org/10.26599/TST.2020.9010036
6. Leid, A., Merwe, B., Visser, W.: Testing Ethereum smart contracts: a comparison of symbolic analysis and fuzz testing tools. In: Conference of the South African Institute of Computer Scientists and Information Technologists (2020). https://doi.org/10.1145/3410886.3410907
7. Kolluri, A., Nikolic, I., Sergey, I., Hobor, A., Saxena, P.: Exploiting the laws of order in smart contracts. In: Proceedings of the 28th ACM SIGSOFT International Symposium on Software Testing and Analysis (ISSTA 2019), pp. 363–373. Association for Computing Machinery, New York (2019). https://doi.org/10.1145/3293882.3330560

8. Wüstholz, V., Christakis, M.: Harvey: a greybox fuzzer for smart contracts. In: Proceedings of the 28th ACM Joint Meeting on European Software Engineering Conference and Symposium on the Foundations of Software Engineering (ESEC/FSE 2020), pp. 1398–1409. Association for Computing Machinery, New York (2020). https://doi.org/10.1145/3368089.3417064

9. Ante, L.: Smart Contracts on the Blockchain – A Bibliometric Analysis and Review, 15 April 2020. https://doi.org/10.2139/ssrn.3576393

10. Li, Y.: Finding concurrency exploits on smart contracts. In: 2019 IEEE/ACM 41st International Conference on Software Engineering: Companion Proceedings (ICSE-Companion), Montreal, QC, Canada, pp. 144–146 (2019). https://doi.org/10.1109/ICSE-Companion.2019.00061

11. Etherscan (2024). https://etherscan.io. Accessed 01 Mar 2024

12. Torres, C.F., et al.: Towards Smart Hybrid Fuzzing for Smart Contracts. ArXiv abs/2005.12156 (2020). n. pag

13. Ji, S., Dong, J., Wu, J., Lu, L.: A guided mutation strategy for smart contract fuzzing. In: 2023 IEEE International Conference on Software Maintenance and Evolution (ICSME), Bogotá, Colombia, pp. 282–292 (2023). https://doi.org/10.1109/ICSME58846.2023.00036

14. Tian, Z.: Smart contract defect detection based on parallel symbolic execution. In: 2019 3rd International Conference on Circuits, System and Simulation (ICCSS), Nanjing, China, pp. 127–132 (2019). https://doi.org/10.1109/CIRSYSSIM.2019.8935603

15. Fu, M., Wu, L., Hong, Z., Zhu, F., Sun, H., Feng, W.: A critical-path-coverage-based vulnerability detection method for smart contracts. IEEE Access 7, 147327–147344 (2019). https://doi.org/10.1109/ACCESS.2019.2947146

16. Oyente: Oyente: An Analysis Tool for Smart Contracts (2023). https://github.com/melonproject/oyente

17. Kalra, S., Goel, S., Dhawan, M., Sharma, S.: ZEUS: analyzing safety of smart contracts. In: Proceedings of the 25th Annual Network and Distributed System Security Symposium, NDSS 2018, San Diego, California, USA, 18–21 February 2018 (2018)

18. Tsankov, P., Dan, A.M., Drachsler-Cohen, D., Gervais, A., Buenzli, F., Vechev, M.: Securify: Practical Security Analysis of Smart Contracts. CoRR abs/1806.01143 (2018)

19. Torres, C.F., Iannillo, A.K., Gervais, A., State, R.: ConFuzzius: a data dependency-aware hybrid fuzzer for smart contracts. In: Proceedings of the IEEE European Symposium on Security and Privacy, EuroS&P 2021, Vienna, Austria, 6–10 September, pp. 103–119. IEEE, Vienna (2021)

20. Ma, C., Song, W., Huang, J.: TransRacer: function dependence-guided transaction race detection for smart contracts. I: Proceedings of the 31st ACM Joint European Software Engineering Conference and Symposium on the Foundations of Software Engineering (ESEC/FSE 2023), pp. 947–959. Association for Computing Machinery, New York (2023). https://doi.org/10.1145/3611643.3616281

A Smart Contract Vulnerability Detection System Based on BERT Model and Fuzz Testing

Zhehao Liang, Baojiang Cui[✉], Dongbin Wang, Jie Xu, and Huipeng Liu

School of Cyberspace Security, Beijing University of Posts and Telecommunications, Beijing, China

{liangzhehao,cuibj,dbwang,cheer1107,lhp6666}@bupt.edu.cn

Abstract. Smart contracts have experienced wide and rapid development across various fields due to their decentralization, immutability, and automation advantages. However, vulnerabilities in smart contract have also caused significant losses for contract users and developers. To enhance the accuracy of smart contract vulnerability detection, we combine machine learning pre-training model BERT with improved fuzz testing. We employ the AST tree algorithm to extract crucial information from contracts, converting them into data flow graphs. Then the pre-trained model BERT is utilized to filter contract vulnerabilities. After that fuzz testing is adopted to further classify contracts. Experimental results demonstrate the algorithm's outstanding performance in detecting reentrancy, tx.origin, and timestamp vulnerabilities, with a precision rate of 95.93%, a recall rate of 87.25%, and an F1 score of 91.01% in detecting reentrancy vulnerabilities specifically. Comparison with other vulnerability detection tools confirms the superiority of the proposed scheme.

Keywords: Smart Contract · Fuzz · BERT · Machine Learning

1 Introduction

Smart contracts are computer programs or code based on blockchain technology, designed to execute, control, or coordinate the exchange of digital assets or data without the need for intermediaries. The execution of smart contracts is facilitated by nodes on the blockchain network, which automatically execute pre-written logic, facilitating consensus among the contract participants based on predefined conditions and rules.

BERT [1] (Bidirectional Encoder Representations from Transformers) is a pre-trained language model based on the Transformer architecture, released by Google in 2018. The algorithm utilized in this paper, DeBERTa [2] (Decoding-enhanced BERT with Disentangled Attention), is a pre-trained language model based on BERT proposed by Microsoft Research. DeBERTa improves upon BERT by introducing decoding-enhanced mechanisms and disentangled attention mechanisms, enhancing the performance and efficiency of the model.

Fuzz testing is a software testing technique aimed at discovering vulnerabilities and errors in software programs or systems. It involves inputting large amounts of random,

L. Barolli (Ed.): IMIS 2024, LNDECT 214, pp. 288–295, 2024.
https://doi.org/10.1007/978-3-031-64766-6_29

invalid, or abnormal data (referred to as "fuzzy inputs") into a program, then monitoring the program's behavior to detect potential anomalies or crashes.

Currently, there are several smart contract vulnerability detection tools available, including sFuzzer and ContractFuzzer based on fuzz testing, Securify and Oyente based on symbolic execution, Bhargavan and KEVM based on formal verification, Slither and SmartCheck based on static program analysis, among others. However, these tools often face challenges such as low scalability and low detection accuracy.

The following are the main tasks and contributions we have made:

1. In this paper, during the data processing stage, we increase the weight of correspond-ing key data for different vulnerabilities.
2. In machine learning, we incorporate the natural language processing model BERT to achieve better model training results.
3. We combine machine learning with fuzz testing to quickly filter and analyze vulner-abilities through machine learning and generate a large number of test cases through fuzz testing to increase testing coverage.

2 Related Work

This section mainly introduces the applications of machine learning and fuzz testing in smart contract vulnerability detection.

In 2018, Tann et al. [3] utilized LSTM neural network models for vulnerability detection in smart contracts. In 2019, He et al. [4] combined symbolic execution, fuzz testing, and machine learning to propose ILF. This approach leverages symbolic execu-tion to enhance program branch coverage, followed by machine learning model filtering, and finally validated through fuzz testing. This amalgamation of various technological strengths has provided us with significant inspiration. In 2020, Gao et al. [5] proposed SmartEmbed, a technique based on word embedding and vector space. This method extracts contract symbol flows from AST and then maps them using word embedding algorithms. In 2020, Zhuang et al. [6] proposed a technique for smart contract vulner-ability detection using graph neural networks. This technique consists of three stages: graph generation, graph normalization, and message propagation network. In 2021, Wu et al. [7] proposed Peculiar, a technique based on crucial data flow diagrams and pre-trained models. This technique introduces pre-training technology into smart contract vulnerability detection.

In 2018, Jiang et al. [8] proposed ContractFuzzer, the first smart contract vulnerability detection technique using fuzz testing. In 2018, Liu et al. [9] proposed ReGuard, a solution that compiles smart contracts into C++ code using abstract syntax trees and control flow information, then detects them through fuzz testing. In 2020, Nguyen et al. [10] proposed sFuzz, a feedback-guided adaptive fuzz testing technique. In 2020, Zhang et al. [11] combined taint analysis with fuzz testing, introducing EthPloit. In 2022, Xue et al. [12] proposed xFuzz, a machine learning-guided smart contract fuzz testing framework.

From these past studies, we realize that there are various methods for detecting vulnerabilities in smart contracts. We should organically integrate multiple methods, complementing each other's strengths and weaknesses, to ultimately achieve efficient and satisfactory results.

3   The Structure of the Model and Key Principles

In this chapter, I will introduce the overall architecture of the proposed solution in this article, and then focus on the key technologies used in the solution.

3.1   The Structure of the Model

Figure 1 Display the model's structure. Below are the specific steps of our proposed method based on the pre-trained model BERT and fuzz testing algorithm:

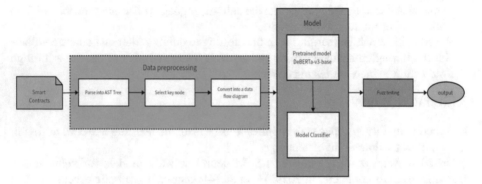

Fig. 1. Vulnerability detect structure

1. We utilized a labeled dataset consisting of 28,388 contracts, all sourced from the Ethereum mainnet. We selected reentrancy vulnerabilities, tx.origin vulnerabilities, and timestamp vulnerabilities respectively for model training, resulting in three trained models.
2. Directly inputting contract source code into the model may lead to poor performance due to excessive irrelevant elements in source code. To amplify smart contract features and reduce interference from irrelevant data, we first constructed abstract syntax trees (ASTs) for smart contracts to extract key information. We used treesitter [13] to parse contract source code into ASTs, and as treesitter does not support the Solidity language, the most commonly used language for writing smart contracts, we utilized JoranHonig's Solidity syntax [14].
3. From the ASTs, we extracted key nodes according to different types of vulnerabilities. For detecting reentrancy vulnerabilities, we identified nodes capable of external calls, such as nodes invoking the 'call' or 'call.value' commands, along with corresponding parameter nodes. For detecting tx.origin vulnerabilities, the crucial nodes include 'tx.origin' itself, along with nodes assigned to or compared with it. For detecting timestamp vulnerabilities, the key nodes are 'block.timestamp' or 'now', along with data nodes assigned to or compared with it.
4. After identifying key nodes, we generated corresponding data flow graphs based on the key node set, ASTs, and source code. We referenced Hongjun Wu's [7] method for data flow graph generation, which involves going through key nodes to find nodes with

assignment and computation relationships, adding these nodes to the key node set, and establishing directed edges representing data relationships until no new elements are added to the set.

5. We encoded the data flow graphs using the tokenizer of DeBERTa-v3 [15] and input them into the model. The model comprises a BERT pre-trained model and a classifier. We chose the DeBERTa-v3-base pre-trained model as it was most suitable for our algorithm among various BERT models.

6. After model training, we devised fuzz testing schemes based on model evaluation results. We initially filtered the test data through the model, then selected 228 contract data for fuzz testing. Finally, we compared the results with smart contract detection tools SmartCheck and Mythril.

3.2 Key Principles

Improved DFG Generation Algorithm
After obtaining the key nodes set V, the algorithm parses the nodes directly related to the nodes in the set from the AST. If a node is not yet included in the set, it is added to set V, and the relationships between nodes are stored in the directed edge set E. This process is repeated until set V no longer changes.

DeBERTa
DeBERTa (Decoding-enhanced BERT with Disentangled Attention) is a pre-trained language model based on BERT, which introduces decoding enhancement and disentangled attention mechanisms to improve model performance. Here are the key algorithms of DeBERTa:

1. Decoding Enhancement: DeBERTa introduces a decoding enhancement mechanism, which utilizes both decoder and encoder simultaneously to improve the model's performance in generative tasks. Decoding enhancement not only effectively handles generative tasks but also increases the model's flexibility and generalization ability.

2. Disentangled Attention: Attention mechanism is a significant feature of BERT, aiming to better grasp crucial information. DeBERTa employs a disentangled attention mechanism, separating content and positional information in the input. Disentangled attention enables the model to better capture both local and global information in the input sequence, thereby enhancing the model's performance.

Formula (1) represents attention disentanglement in DeBERTa:

$$A_{ij} = \{H_i, P_{i|j}\} \times \{H_j \times P_{j|i}\}^T \tag{1}$$

where H represents the content vector, and P represents the relative positional vector between i and j. Expanding the formula further, we obtain formula (2):

$$A_{ij} = H_i \times H_j^T + H_i \times P_{j|i}^T + P_{i|j} \times H_j^T + P_{i|j} \times P_{j|i}^T \tag{2}$$

It can be observed that the attentions between a pair of input tokens is decomposed into four parts, namely content-content matrix, content-position matrix, position-content matrix, and position-position matrix. However, since the position-position matrix has no practical value, it is generally omitted during attention computation, retaining only the first three terms.

Formula (3) for weighting the attention disentanglement matrices into the input data is as follows:

$$H_o = softmax\left(\frac{A}{\sqrt{3d}}\right) \times V_c \qquad (3)$$

where A represents the attention disentanglement matrix, d is the hidden state dimension, and V_c is the value-content matrix in the attention mechanism. By employing the attention disentanglement approach, it effectively reduces information cross-interference and enhances generalization capability.

4   Results and Discussion

In the test dataset, we selected 228 smart contract data, among which 24 smart contracts had reentrancy vulnerabilities, 18 contracts had tx.origin vulnerabilities, and 68 contracts had timestamp vulnerabilities. For these smart contracts, we first conducted model checks on the data and then conducted secondary checking through fuzz testing. Table 1 shows the detection results.

Table 1. Outcomes of fuzz testing, model and algorithm proposed in this paper

	Reentrancy			Tx.origin			Timestamp		
	Precision	Recall	F1	Precision	Recall	F1	Precision	Recall	F1
Fuzz	90.79	66.42	72.23	85.02	79.60	82.03	83.13	58.51	57.21
Model	97.88	81.25	87.38	76.65	94.39	83.23	92.61	95.82	93.95
Model + Fuzz	95.93	87.25	91.01	98.38	80.55	87.11	92.61	95.82	93.95

From the table, we can see that the combination of the BERT model and fuzz testing effectively improves the precision of vulnerability detection, especially in detecting reentrancy vulnerabilities, where there is a significant increase in F1 score. Due to the higher precision but lower recall of fuzz testing, we choose to supplement the results of fuzz testing into the predictions of the model. Experimental results show that the combination of the two methods has a significant effect. Compared to the predictions directly outputted by the model, the F1 scores after combination are generally improved, except for timestamp vulnerabilities where the F1 score remains unchanged, but the model's predictions are already sufficiently accurate.

To demonstrate the superiority of our algorithm, we conducted comparative tests using two smart contract vulnerability detection tools, SmartCheck and Mythril. The experimental results are shown in Fig. 2.

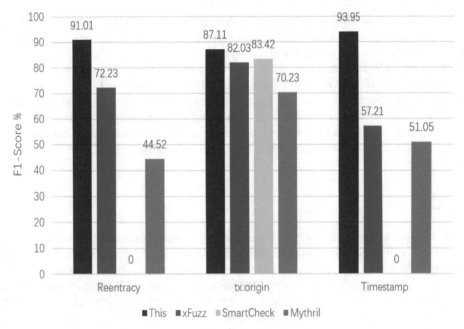

Fig. 2. Comparison of F1-score for Reentracy, tx.origin and Timestamp vulnerabilities detected trough this paper's algorithm, xFuzz, SmartCheck and Mythril

SmartCheck does not support detecting reentrancy vulnerabilities, so its F1 score for reentrancy vulnerabilities is 0. Although SmartCheck can detect timestamp vulnerabilities, the detection results are surprising, as it did not detect any, therefore, SmartCheck's F1 score for timestamp vulnerabilities is also 0.

In detecting reentrancy vulnerabilities, xFuzz detected 9 reentrancy vulnerability contracts, with 1 being a false positive, yielding a precision of 90.79%, but a recall of only 66.42% due to a large number of vulnerable files not being detected. Mythril detected 20 vulnerable contracts, but none of these 20 contracts had reentrancy vulnerabilities, with a precision of 44.20% and a recall of 44.85%. The algorithm proposed in this paper detected a total of 19 reentrancy vulnerability contracts, of which 18 were true positive samples, with 6 vulnerable data not being detected, yielding a precision of 95.93% and a recall of 87.25%, which are superior to other smart contract detection tools.

In detecting tx.origin vulnerabilities, xFuzz detected 23 tx.origin vulnerability contracts, with 9 being false positives, resulting in a precision of 85.02% and a recall of 79.60%. Mythril detected 5 vulnerable contracts, all of which were positive samples, with 13 samples not being detected, resulting in a precision of 97.08% and a recall of 63.88%. SmartCheck detected a total of 11 samples, with one being a false positive, resulting in a precision of 93.61% and a recall of 77.53%. The algorithm proposed in this paper detected a total of 11 vulnerability contracts, all of which were positive samples, with 7 vulnerable data not being detected, resulting in a precision of 98.38% and a recall of 80.55%. Compared to other algorithms, the precision of each algorithm is relatively close, but our algorithm's recall is significantly higher than others.

In detecting timestamp vulnerabilities, xFuzz detected 13 timestamp vulnerability contracts, with a precision of 83.13%, but a recall of only 58.51%, due to a large number of vulnerable files not being detected. Mythril detected 8 vulnerable contracts, with a precision of 79.88% and a recall of 54.83%. The algorithm proposed in this paper detected a total of 78 timestamp vulnerability contracts, of which 67 were true positive samples, with 1 positive sample being missed, resulting in a precision of 92.61% and a recall of 95.82%, which are superior to other smart contract detection tools.

Figure 2 displays the F1-score of this paper's algorithm and other vulnerabilities detection tools. From the figure, it can be seen that the superiority of our algorithm is significant

5   Conclusion

In this paper, we innovatively propose an intelligent smart contract vulnerability detection method that combines the pre-trained machine learning model BERT with fuzz testing. Ultimately, satisfactory results were achieved in detecting reentrancy vulnerabilities, tx.origin vulnerabilities, and timestamp vulnerabilities. We have witnessed the potential of natural language processing models in detecting vulnerabilities in smart contract code, as well as the excellent results achieved by combining them with fuzz testing. In the future, we will continue to expand the detectable types of vulnerabilities, enabling the model to adapt to more complex environments.

References

1. Devlin, J., Chang, M.-W., Lee, K., Toutanova, K.: Bert: pre-training of deep bidirectional transformers for language understanding. arXiv preprint arXiv:1810.04805 (2018)
2. He, P., Liu, X., Gao, J., Chen, W.: DeBERTa: decoding-enhanced BERT with Disentangled Attention. arXiv preprint arXiv:2006.03654 (2021)
3. Tann, W.J.W., Han, X.J., Gupta, S.S., Ong, Y.S.: Towards safer smart contracts: a sequence learning approach to detecting security threats. arXiv:1811.06632 (2019)
4. He, J.X., Balunović, M., Ambroladze, N., Tsankov, P., Vechev, M.: Learning to fuzz from symbolic execution with application to smart contracts. In: Proceedings of the 2019 ACM SIGSAC Conference on Computer and Communications Security. ACM, London, pp. 531–548 (2019). https://doi.org/10.1145/3319535.3363230
5. Gao, Z.P., Jiang, L.X., Xia, X., Lo, D., Grundy, J.: Checking smart contracts with structural code embedding. IEEE Trans. Softw. Eng. 47(12), 2874–2891 (2021). https://doi.org/10.1109/TSE.2020.2971482
6. Zhuang, Y., Liu, Z.G., Qian, P., Liu, Q., Wang, X., He, Q.M.: Smart contract vulnerability detection using graph neural network. In: Proceedings of the 29th International Joint Conference on Artificial Intelligence, IJCAI, Yokohama, pp. 3283–3290 (2021)
7. Wu, H., et al.: Peculiar: smart contract vulnerability detection based on crucial data flow graph and pre-training techniques. In: 2021 IEEE 32nd International Symposium on Software Reliability Engineering (ISSRE), Wuhan, China, pp. 378–389 (2021). https://doi.org/10.1109/ISSRE52982.2021.00047
8. Jiang, B., Liu, Y., Chan, W.K.: ContractFuzzer: fuzzing smart contracts for vulnerability detection. In: Proceedings of the 33rd ACM/IEEE Int'l Conf. on Automated Software Engineering. Montpellier. IEEE, pp. 259–269 (2018)

9. Liu, C., Liu, H., Cao, Z., Chen, Z., Chen, B.D., Roscoe, B.: ReGuard: finding reentrancy bugs in smart contracts. In: Proceedings of the 40th International Conference on Software Engineering: Companion Proceedings Gothenburg. ACM, pp. 65–68 (2018). https://doi.org/ 10.1145/3183440.3183495

10. Nguyen, T.D., Pham, L.H., Sun, J., Lin, Y., Minh, Q.T.: SFuzz: an efficient adaptive fuzzer for solidity smart contracts. In: Proceedings of the ACM/IEEE 42nd International Conference on Software Engineering (ICSE 2020). Association for Computing Machinery, New York, pp. 778–788 (2020). https://doi.org/10.1145/3377811.3380334

11. Zhang, Q.Z., Wang, Y.Z., Li, J.R., Ma, S.Q.: EthPloit: from fuzzing to efficient exploit generation against smart contracts. In: Proceedings of the 27th IEEE International Conference on Software Analysis, Evolution and Reengineering. IEEE, London, pp. 116–126 (2020).https:// doi.org/10.1109/SANER48275.2020.9054822

12. Xue, Y., et al.: xFuzz: machine learning guided cross-contract fuzzing. IEEE Trans. Depend. Sec. Comput. (2022)

13. Gan, T., Lua, et al.: Tree-sitter. https://tree-sitter.github.io/tree-sitter/

14. JoranHonig. tree-sitter-solidity. https://github.com/JoranHonig/tree-sitter-solidity

15. He, P., Gao, J., Chen, W.: DeBERTaV3: improving DeBERTa using ELECTRA-style pre-training with gradient-disentangled embedding sharing. arXiv preprint arXiv:2111.09543 (2023)

An Enhanced Fault Identification Algorithm for PMC-Based Diagnosable Systems

Yuan-Hsiang Teng[1] and Tzu-Liang Kung[2(✉)]

[1] CSIE Department, College of Computing and Informatics, Providence University, Taichung, Taiwan, ROC
[2] CSIE Department, College of Information and Electrical Engineering, Asia University, Taichung, Taiwan, ROC
tlkung@asia.edu.tw

Abstract. Significant advancements in multi-core processors continue to unfold and boost contemporary high-performance computing (HPC) to gain attraction in computational technology. Fault identification plays a critical role in supporting an HPC system's availability. Graph theory models a system's interconnection, whose physical nodes and their communication links are abstractly represented by vertices and edges, respectively. In this paper, an enhanced algorithm is proposed to accomplish fault identification for PMC-based t-diagnosable systems, provided that every fault-free node cannot has no fault-free neighbor, and the total number of faulty nodes is bounded above by $t + 1$.

1 Introduction

Significant advancements in multi-core processors continue to unfold and boost contemporary high-performance computing (HPC) to gain attraction in computational technology. One typical HPC system is the family of multiprocessor systems, also known as parallel computing systems or multi-core systems, which belong to a type of computer architecture where multiple processors (or CPUs) are used within a single computer system. In a multiprocessor system, processors work together simultaneously to execute tasks and process data. In general, multiprocessor systems provide increased computational power and performance compared to single-processor systems, making them suitable for handling complex tasks, parallel processing, and high-performance computing applications. They are commonly used in servers, supercomputers, high-end workstations, and other environments where high computational capacity is required.

Instead of CPUs, which are designed for general-purpose computing, graphics processing units (GPUs) are a suite of advanced integrated circuit optimized to execute specific tasks even faster. Nowadays, GPUs involve a wide range of applications, including gaming, scientific research, machine learning, and artificial intelligence, especially useful for tasks that require a large amount of parallel processing, such as rendering graphics or performing complex calculations.

A modern HPC system comprises numerous processing units, both CPUs and GPUs, interconnected as nodes within an underlying interconnection network [4,6]. The operational reliability of an HPC system hinges on these processing units' dependability; even minor malfunctions can cause significant disruptions. Once faulty units

L. Barolli (Ed.): IMIS 2024, LNDECT 214, pp. 296–305, 2024.
https://doi.org/10.1007/978-3-031-64766-6_30

are detected, they must be replaced with functional ones to ensure the system's normal operation. System-level self-diagnosis refers to a system's comprehensive identification of all faulty units. Preparata et al. [14] introduced the so-called PMC model for system-level self-diagnosis. A *one-step t-diagnosable* system can precisely pinpoint all faulty nodes by means of a single diagnostic process if there exist no greater than t faults in it. A system's *diagnosability* is the maximum positive integer t so as to remains one-step t-diagnosable. Both analytic and algorithmic aspects of system-level self-diagnosis gain the attentions of many researchers [3,5,7–13,16]. This paper's contribution is an enhanced algorithm to accomplish fault identification for PMC-based one-step t-diagnosable systems, provided that every fault-free node cannot has no fault-free neighbor, and the total number of faulty nodes is bounded above by $t + 1$.

The remainder of this paper is organized as follows: Sect. 2 addresses the graph-theory preliminaries to model PMC-based self-diagnosable systems. Section 3 develops an enhanced fault identification method. Finally, Sect. 5 draws some concluding remarks.

2 Preliminary

A system's underlying interconnection is conventionally modeled as an undirected graph [2], whose vertices and edges represent the physical nodes and their communication links. A simple, undirected graph G consists of a vertex set $V(G)$ and an edge set $E(G)$. An undirected edge $e = \{u, v\}$ links two adjacent vertices between u and v; equivalently, u and v are *neighbors* of each other. The *degree* of a vertex $v \in V(G)$ is defined by $deg_G(v) = |\{u \in V(G) \mid \{u, v\} \in E(G)\}|$. The *minimum degree* of G is defined as $\delta(G) = \min\{deg_G(v) \mid v \in V(G)\}$. A *subgraph* S of G is a graph with $V(S) \subseteq V(G)$ and $E(S) \subseteq E(G)$. For any vertex $v \in V(G)$, $N_G(v) = \{u \in V(G) \mid \{u, v\} \in E(G)\}$ denotes the *neighborhood* of v; for any nonempty $C \subseteq V(G)$ and $B \subseteq E(G)$, $G[C]$ and $G[B]$ are *vertex-induced* and *edge-induced* subgraphs of G, respectively. On the other hand, a directed graph (also called digraph) D is comprised of a vertex set $V(D)$ and an arc set $A(G)$. A directed arc $a = (x, y)$ links the adjacent vertices from x to y, where x is an *in-neighbor* of y, and y is an *out-neighbor* of x. The *in-degree* and *out-degree* of a vertex $x \in V(D)$ are defined by $deg_D^-(x) = |\{y \in V(D) \mid (y, x) \in A(D)\}|$ and $deg_D^+(x) = |\{y \in V(D) \mid (x, y) \in A(D)\}|$, respectively. Similar to undirected graphs, for any nonempty $C \subseteq V(G)$ and $B \subseteq A(G)$, $D[C]$ and $D[B]$ are *vertex-induced* and *arc-induced* subgraphs of D, respectively.

For any indirected graph G, the PMC model [14] admits two adjacent vertices to evaluate each other's fault status, and models its *diagnostic test* as a directed graph D with $V(D) = V(G)$ such that $\{(u, v), (v, u)\} \subseteq A(D)$ if and only if $\{u, v\} \in E(G)$. It is noticed that $deg_D^-(v) = deg_D^+(v) = deg_G(v)$ for each $v \in V(G)$. Any arc $a = (u, v)$ indicates that u evaluates v to recognize the fault status of v. For convenience, let σ denote a mapping from $A(D)$ to $\{0, 1\}$ by defining $\sigma(a) = 0$ (respectively, $\sigma(a) = 1$) if u evaluates v as fault-free (respectively, faulty). As convention, the mapping σ is called the *syndrome* of D. Because both u and v might be fault-free or faulty, the complete definition of σ is illustrated in Fig. 1(a). It should be noticed that a faulty vertex makes irregular diagnosis and behaves unpredictable. It is apparent that a given

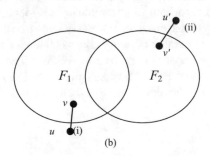

Fault status of u	Fault status of v	$\sigma((u, v))$
0	0	0
0	1	1
1	0	Both 0 and 1 are possible.
1	1	Both 0 and 1 are possible.
0: fault-free 1: faulty		

(a)

(b)

Fig. 1. (a) The definition of diagnostic mapping σ; (b) a sufficient and necessary condition for two distinct fault sets F_1, F_2 to be distinguishable.

fault set $F \subset V(G)$ can cause different syndromes. The set of all possible syndromes that can be caused by F is denoted by $\sigma[F]$. Two distinct fault sets $F_1, F_2 \subset V(G)$ are *distinguishable* if $\sigma[F_1] \cap \sigma[F_2] = \emptyset$; otherwise, F_1 and F_2 are *indistinguishable*. Lemma 1 shows a sufficient and necessary condition for F_1 and F_2 to be distinguishable. Dahbura and Masson [3] further defined that G is called *one-step t-diagnosable* if and only if any two fault sets $F_1, F_2 \subset V(G)$ with $F_1 \neq F_2$, $|F_1| \leq t$ and $|F_2| \leq t$ are always distinguishable, where t is a positive integer.

Lemma 1 [3]. *Let F_1 and F_2 be any two distinct fault sets in a graph G. Then F_1 and F_2 are distinguishable if and only if (i) there exists a vertex $v \in F_1 \setminus F_2$ and a vertex $u \in V(G) \setminus (F_1 \cup F_2)$ with $\{u, v\} \in E(G)$, or (ii) there exists a vertex $v' \in F_2 \setminus F_1$ and a vertex $u' \in V(G) \setminus (F_1 \cup F_2)$ with $\{u', v'\} \in E(G)$. Refer to Fig. 1(b) for illustration.*

Let G be an undirected graph with $v \in V(G)$. Kung and Chen [10] defined that v is *one-step t-identifiable*, $t \geq 1$, if for any two distinct fault sets $F_1, F_2 \subset V(G)$ with $v \in F_1 \triangle F_2$, $|F_1| \leq t$ and $|F_2| \leq t$, they are always distinguishable. It is straightforward that an undirected graph G is one-step t-diagnosable if and only if every vertex of G is one-step t-identifiable. Suppose that the digraph D is the diagnostic test of G. Then an *extending directed star* rooted at v is a subgraph of D, denoted by $EDS_G(v)$, whose vertex set and arc set are

$$V(EDS_G(v)) = \{v\} \cup \{v_{1,i}, v_{2,i} \mid 1 \leq i \leq deg_G(v)\}$$

$$A(EDS_G(v)) = \bigcup_{i=1}^{deg_G(v)} \{(v_{1,i}, v), (v_{2,i}, v_{1,i})\}$$

such that $EDS_G(v)$ includes exactly $2 \times deg_G(v) + 1$ vertices. Figure 2(a) illustrates the tree structure of $EDS_G(v)$, in which arcs $a_{1,i} = (v_{1,i}, v)$ and $a_{2,i} = (v_{2,i}, v_{1,i})$ for $1 \le i \le \beta = deg_G(v)$.

For any $v \in V(G)$, Hsu and Tan [8] developed a novel algorithm to identify the fault status of v based on $EDS_G(v)$. Refer to Algorithm 1.

Algorithm 1. Diagnose$(EDS_G(v))$

Input: $EDS_G(v)$, an extending directed star rooted at v.
Output: The fault status of v, whose value is 0 if v is fault-free, and 1 otherwise.
1: $O \leftarrow |\{1 \le i \le deg_G(v) \mid \sigma(a_{1,i})) = 0, \sigma(a_{2,i}) = 0\}|$;
2: $I \leftarrow |\{1 \le i \le deg_G(v) \mid \sigma(a_{1,i})) = 1, \sigma(a_{2,i}) = 0\}|$;
3: **if** $O \ge I$ **then**
4:     **return** 0;
5: **else**
6:     **return** 1;
7: **end if**

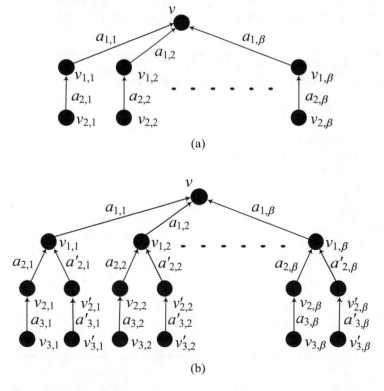

(a)

(b)

Fig. 2. (a) Illustration of an extending directed star rooted at v; (b) illustration of a double-extending directed star rooted at v.

Lemma 2 [8]. *Let G be any undirected graph with $\delta(G) \geq t > 0$. Suppose that F is an arbitrary fault set in G with $|F| \leq t$, and $EDS_G(v)$ appears to be an extending directed star rooted at $v \in V(G)$. Then $v \in F$ if and only if* **Diagnose**$(EDS_G(v))$ *returns 1.*

3   The Enhanced Fault-Identification Algorithm

Let G be an undirected graph with $v \in V(G)$. Suppose that the digraph D is the diagnostic test of G. A *double-extending directed star* rooted at v is a subgraph of D, denoted by $EEDS_G(v)$, whose vertex set and arc set are defined as follows:

$$V(EEDS_G(v)) = \{v\} \cup \{v_{1,i}, v_{2,i}, v'_{2,i}, v_{3,i}, v'_{3,i} \mid 1 \leq i \leq deg_G(v)\}$$

$$A(EEDS_G(v)) = \bigcup_{i=1}^{deg_G(v)} \{(v_{1,i}, v), (v_{2,i}, v_{1,i}), (v'_{2,i}, v_{1,i}), (v_{3,i}, v_{2,i}), (v'_{3,i}, v'_{2,i})\}$$

such that $EEDS_G(v)$ includes exactly $5 \times deg_G(v) + 1$ vertices. Figure 2(b) illustrates the tree structure of $EEDS_G(v)$, in which arcs $a_{1,i} = (v_{1,i}, v)$, $a_{2,i} = (v_{2,i}, v_{1,i})$, $a'_{2,i} = (v'_{2,i}, v_{1,i})$, $a_{3,i} = (v_{3,i}, v_{2,i})$ and $a'_{3,i} = (v'_{3,i}, v'_{2,i})$ for $1 \leq i \leq \beta = deg_G(v)$.

For convenience, let $F(G)$ denote the set of faulty vertices in any graph G.

Algorithm 2. CountSymdrone$(EEDS_G(v))$

Input: $EEDS_G(v)$, a double-extending directed star rooted at v.
Output: A 2×2 array IO.
  1: For $x, y \in \{0, 1\}$, $IO[x][y] \leftarrow 0$;   ▷ Initialization.
  2: **for** $i \leftarrow 1$ to $deg_G(v)$ **do**
  3:     **if** $\sigma(a_{1,i}) = 0$ **then**
  4:         **if** $\sigma(a_{2,i}) = 0$ or $\sigma(a'_{2,i}) = 0$ **then**
  5:             $IO[0][0] += 1$;
  6:         **else**
  7:             $IO[0][1] += 1$;
  8:         **end if**
  9:     **else**
 10:         **if** $\sigma(a_{2,i}) = 0$ or $\sigma(a'_{2,i}) = 0$ **then**
 11:             $IO[1][0] += 1$;
 12:         **else**
 13:             $IO[1][1] += 1$;
 14:         **end if**
 15:     **end if**
 16: **end for**
 17: **return** IO;

Algorithm 3. IdentifyFaultStatus($EEDS_G(v)$)

Input: $EEDS_G(v)$, a double-extending directed star rooted at v.
Output: The fault status of v, whose value is 0 if v is fault-free, and 1 otherwise.
1: **if** $\sigma(a_{3,k}) = 1$ with some k, $1 \le k \le deg_G(v)$ **then**
2:      $S \leftarrow A(EEDS_G(v)) \setminus \{a_{2,i}, a_{3,i}, a'_{3,i} \mid 1 \le i \le deg_G(v)\}$;
3:      **return Diagnose**($EEDS_G(v)[S]$);
4: **end if**
5: **if** $\sigma(a'_{3,k}) = 1$ with some k, $1 \le k \le deg_G(v)$ **then**
6:      $S \leftarrow A(EEDS_G(v)) \setminus \{a'_{2,i}, a'_{3,i}, a_{3,i} \mid 1 \le i \le deg_G(v)\}$;
7:      **return Diagnose**($EEDS_G(v)[S]$);
8: **end if**
9: $IO \leftarrow$ **CountSymdrone**($EEDS_G(v)$);
10: **switch** $IO[0][0] - IO[1][0]$ **do**
11:     **case** $IO[0][0] - IO[1][0] \ge 1$
12:         **return** 0;
13:     **case** $IO[0][0] - IO[1][0] = 0$
14:         **if** $IO[0][0] = IO[1][0] > 0$ **then**
15:             **return** 0;
16:         **else**
17:             **return** 1;
18:         **end if**
19:     **case** $IO[0][0] - IO[1][0] \le -1$
20:         **return** 1;

Lemma 3. *Let v be any fault-free vertex in a graph G, and $EEDS_G(v)$ is a double-extending directed star rooted at v with $|F(EEDS_G(v))| \le deg_G(v) + 1$. If v has at least one fault-free neighbor, then* **IdentifyFaultStatus**($EEDS_G(v)$) *(Algorithm 3) returns 0.*

Proof. Throughout the proof, $|F(EEDS_G(v))|$ does not exceed $deg_G(v) + 1$, and v has at least one fault-free neighbor.

Suppose that $\sigma(a_{3,k}) = 1$ with some k, $1 \le k \le deg_G(v)$. Let $S = A(EEDS_G(v)) \setminus \{a_{2,i}, a_{3,i}, a'_{3,i} \mid 1 \le i \le deg_G(v)\}$. Then, the arc-induced subgraph $EEDS_G(v)[S]$ is an extending directed star rooted at v. Since $\sigma(a_{3,k}) = 1$, $v_{2,k}$ or $v_{3,k}$ is faulty. Thus, $|F(EEDS_G(v)[S])|$ does not exceed $deg_G(v)$. By Lemma 2, **Diagnose**($EEDS_G(v)[S]$) can recognize the fault status of v. Similarly, when $\sigma(a'_{3,k}) = 1$ with some k, $1 \le k \le deg_G(v)$, **IdentifyFaultStatus**($EEDS_G(v)$) returns the correct fault status of v.

In the rest of the proof, suppose that $\sigma(a_{3,i}) = \sigma(a'_{3,i}) = 0$ for every $1 \le i \le deg_G(v)$. According to Algorithm 2, $\sum_{p \in \{0,1\}, q \in \{0,1\}} IO[p][q] = deg_G(v)$. Because v is fault-free, the number of faulty vertices in $EEDS_G(v)$ amounts to at least $3 \times IO[1][0] + IO[0][1] + IO[1][1]$, which is no less than $\sum_{p,q \in \{0,1\}} IO[p][q] + 2 = deg_G(v) + 2$ if $IO[1][0] \ge IO[0][0] + 1$. Therefore, the following cases take into consideration two possibilities for $IO[0][0] - IO[1][0] \ge 0$.

- **Case 1:** $IO[0][0] - IO[1][0] \ge 1$. According to the decision rule of Algorithm 3, **IdentifyFaultStatus**($EEDS_G(v)$) returns 0, which identifies v as fault-free.

- **Case 2:** $IO[0][0] - IO[1][0] = 0$. As $IO[0][0] = IO[1][0]$, the following arguments consider whether the value of $IO[0][0]$ is greater than zero.

  Suppose that $IO[0][0] = IO[1][0] = 0$. Obviously, if $v_{1,i}$ were fault-free for any i, then every of $\{v_{2,i}, v'_{2,i}, v_{3,i}, v'_{3,i}\}$ would be faulty so that $|F(EEDS_G(v))| > deg_G(v) + 1$. That is, every of $\{v_{1,i} \mid 1 \leq i \leq deg_G(v)\}$ is faulty. However, v is fault-free to have at least one fault-free neighbor. By contradiction, $IO[0][0] = IO[1][0]$ must be greater than zero if v is fault-free. According to Algorithm 3, **IdentifyFaultStatus**$(EEDS_G(v))$ returns 0, which identifies v as fault-free.

  The proof is completed. □

Lemma 4. *Let v be any faulty vertex in a graph G, and $EEDS_G(v)$ is a double-extending directed star rooted at v. Then* **IdentifyFaultStatus**$(EEDS_G(v))$ *(Algorithm 3) returns 1 if $|F(EEDS_G(v))| \leq deg_G(v) + 1$.*

Proof. Throughout the proof, $|F(EEDS_G(v))|$ does not exceed $deg_G(v) + 1$.

Suppose that $\sigma(a_{3,k}) = 1$ with some k, $1 \leq k \leq deg_G(v)$. Let $S = A(3EDS_G(v)) \setminus \{a_{2,i}, a_{3,i}, a'_{3,i} \mid 1 \leq i \leq deg_G(v)\}$. Then, the arc-induced subgraph $EEDS_G(v)[S]$ is an extending directed star rooted at v. Since $\sigma(a_{3,k}) = 1$, $v_{2,k}$ or $v_{3,k}$ is faulty. Thus, $|F(EEDS_G(v)[S])|$ does not exceed $deg_G(v)$. By Lemma 2, **Diagnose**$(EEDS_G(v)[S])$ can recognize the correct fault status of v. Similarly, when $\sigma(a'_{3,k}) = 1$ with some k, $1 \leq k \leq deg_G(v)$, **IdentifyFaultStatus**$(EEDS_G(v))$ returns the correct fault status of v.

In the rest of the proof, suppose that $\sigma(a_{3,i}) = \sigma(a'_{3,i}) = 0$ for every $1 \leq i \leq deg_G(v)$. According to **CountSymdrone**$(EEDS_G(v))$, $\sum_{p\in\{0,1\},q\in\{0,1\}} IO[p][q] = deg_G(v)$. Because v is faulty, the number of faulty vertices in $EEDS_G(v)$ amounts to at least $1 + 3 \times IO[0][0] + IO[0][1] + IO[1][1]$, which is no less than $\sum_{p,q\in\{0,1\}} IO[p][q] + 2 = deg_G(v) + 2$ if $IO[0][0] \geq IO[1][0]$ and $IO[0][0] \geq 1$.

By contradiction, it must be $IO[0][0] < IO[1][0]$ or $IO[0][0] = 0$ if v is faulty. According to Algorithm 3, **IdentifyFaultStatus**$(EEDS_G(v))$ returns 1, which identifies v as faulty. □

Theorem 1. *Let v be any vertex in a graph G, and $EEDS_G(v)$ is a double-extending directed star rooted at v. Suppose that $|F(EEDS_G(v))| \leq deg_G(v) + 1$. Then the correct fault status of v matches the returned value of* **IdentifyFaultStatus**$(EEDS_G(v))$.

Proof. Two cases are considered below:

- **Case 1:** v is fault-free. By Lemma 3, **IdentifyFaultStatus**$(EEDS_G(v))$ returns 0 to identify v as fault-free.
- **Case 2:** v is faulty. By Lemma 4, **IdentifyFaultStatus**$(EEDS_G(v))$ returns 1 to identify v as faulty.

As a consequence, this theorem holds. □

4   Applications

Algorithm 3 can apply to many interconnection networks. In this section, we construct instances of double-extending directed stars in the renowned star graphs [1] and hyper-cubes [15].

For $n \geq 2$, the n-dimensional star graph S_n is defined over the permutation set of n distinct identifying digits, for example, $\{1, 2, \ldots, n\}$. Every vertex v is uniquely assigned a permutation $c_1 c_2 \cdots c_n$, and it is adjacent to $(n-1)$ neighbors $(v)^2, (v)^3, \ldots, (v)^n$, which are obtained by a transposition between the first and the kth digit for $2 \leq k \leq n$. Based on the recursive construction of S_n, S_n can be partitioned into n disjoint copies of S_{n-1}, denoted by $S_{n-1}^{(1)}, S_{n-1}^{(2)}, \ldots, S_{n-1}^{(n)}$. Figure 3(a) depicts an instance of $EEDS_{S_4}(1234)$, and Fig. 3(b) depicts an instance of $EEDS_{S_5}(12345)$.

Below we propose a tree construction scheme for $n \geq 5$: Let $v = 1234 \cdots n$ be the vertex in $S_{n-1}^{(n)}$. Suppose that D_n is the diagnostic test of S_n. Denote by T the arc subset of D_n as follows:

$$T = A \left(EEDS_{S_{n-1}^{(n)}} (v) \right) \cup \{((v)^n, v), (((v)^n)^{n-1}, (v)^n), (((v)^n)^{n-2}, (v)^n)\}$$
$$\cup \{(((((v)^n)^{n-1})^{n-2}, ((v)^n)^{n-1}), (((((v)^n)^{n-2})^{n-1}, ((v)^n)^{n-2})\}.$$

Then the arc-induced subgraph $D_n[T]$ is an instance of $EEDS_{S_n}(v)$. Because S_n is vertex-transitive [1] for every $n \geq 2$, the proposed tree construction scheme remains feasible to build a double-extending directed star rooted at each vertex.

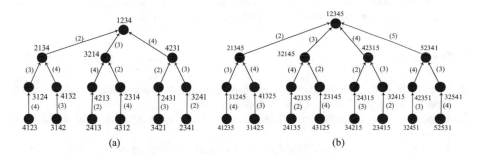

(a)   (b)

Fig. 3. (a) $EEDS_{S_4}(1234)$; (b) $EEDS_{S_5}(12345)$.

For $n \geq 2$, the n-dimensional hypercube Q_n is defined over the binary numbers of n bits, i.e., $\{0, 1\}^n$. Every vertex w is uniquely assigned an n-bit binary numbers, and it is adjacent to n neighbors $[w]^0, [w]^1, \ldots, [w]^{n-1}$, which differs from w only in the ith bit for $0 \leq i \leq n-1$. Figure 4 shows an instance of $EEDS_{Q_6}(000000)$.

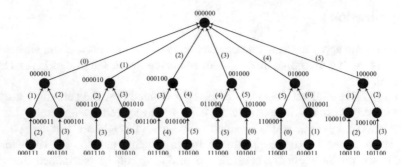

Fig. 4. Illustration of $EEDS_{Q_6}(000000)$.

5 Conclusion

A modern HPC system comprises numerous processing units, both CPUs and GPUs, interconnected as nodes within an underlying interconnection network. Quick fault identification plays a critical role in supporting an HPC system's availability. In this paper, the authors develop an enhanced algorithm to accomplish fault identification for PMC-based one-step t-diagnosable systems, in which every fault-free node has one or more fault-free neighbors, and the total number of faulty nodes is bounded above by $t + 1$. Our future work will be devoted to more efficient methods of self-diagnosable fault identification.

Acknowledgements. This work is supported in part by National Science and Technology Council, Taiwan, ROC, under Grant No. NSTC 112-2221-E-468-006-MY2.

References

1. Akers, S.B., Krishnamurthy, B.: A group-theoretic model for symmetric interconnection networks. IEEE Trans. Comput. **38**(4), 555–566 (1989)
2. Bondy, J.A., Murty, U.S.R.: Graph Theory. Springer, London (2008). https://doi.org/10.1007/978-1-84628-970-5
3. Dahbura, A., Masson, G.: An $o(n^{2.5})$ fault identification algorithm for diagnosable systems. IEEE Trans. Comput. **33**(6), 486–492 (1984)
4. Dally, W.J., Towles, B.: Principles and Practices of Interconnection Networks. Morgan Kaufmann, San Francisco (2004)
5. Das, A., Thulasiraman, K., Agarwal, V.K., Lakshmanan, K.B.: Multiprocessor fault diagnosis under local constraints. IEEE Trans. Comput. **42**(8), 984–988 (1993)
6. Duato, J., Yalamanchili, S., Ni, L.: Interconnection Networks: An Engineering Approach. Morgan Kaufmann, San Francisco (2002)
7. Friedman, A.D., Simoncini, L.: System-level fault diagnosis. Comput. J. **13**(3), 47–53 (1980)
8. Hsu, G.H., Tan, J.J.M.: A local diagnosability measure for multiprocessor systems. IEEE Trans. Parallel Distrib. Syst. **18**(5), 598–607 (2007)
9. Kung, T.L.: Restricted-faults identification in folded hypercubes under the PMC diagnostic model. J. Electron. Sci. Technol. **12**(4), 424–428 (2014)

10. Kung, T.L., Chen, H.C.: Toward the fault identification method for diagnosing strongly t-diagnosable systems under the PMC model. Int. J. Commun. Netw. Distrib. Syst. **15**(4), 386–399 (2015)
11. Lai, P.L., Tan, J.J., Chang, C.P., Hsu, L.H.: Conditional diagnosability measures for large multiprocessor systems. IEEE Trans. Comput. **54**(2), 165–175 (2005)
12. Lin, C.K., Kung, T.L., Tan, J.J.: Conditional-fault diagnosability of multiprocessor systems with an efficient local diagnosis algorithm under the PMC model. IEEE Trans. Parallel Distrib. Syst. **22**(10), 1669–1680 (2011)
13. Lin, C.K., Kung, T.L., Tan, J.J.: An algorithmic approach to conditional-fault local diagnosis of regular multiprocessor interconnected systems under the PMC model. IEEE Trans. Comput. **62**(3), 439–451 (2013)
14. Pretarata, F.P., Metze, G., Chien, R.T.: On the connection assignment problem of diagnosis systems. IEEE Trans. Electron. Comput. **16**(12), 848–854 (1967)
15. Saad, Y., Schultz, M.H.: Topological properties of hypercubes. IEEE Trans. Comput. **37**(7), 867–872 (1988)
16. Somani, A.K., Agarwal, V.K., Avis, D.: A generalized theory for system level diagnosis. IEEE Trans. Comput. **36**(5), 538–546 (1987)

Intelligent Information Transmission Model with Distributed Task Queue Functionality Based on the MQTT Transmission Protocol

Hsing-Chung Chen[1,2(✉)], Yu-Hsien Chou[1], and Wei Lin[1]

[1] Department of Computer Science and Information Engineering, Asia University,
Taichung 413305, Taiwan
cdma2000@asia.edu.tw, shin8409@ms6.hinet.net, {111121019,
109021336}@live.asia.edu.tw
[2] Department of Medical Research, China Medical University Hospital, China Medical
University, Taichung 404327, Taiwan

Abstract. This study examines an intelligent information transmission model that utilizes the MQTT protocol. The primary scenario entails uploading data from MQTT subscription services to a MongoDB database on the backend. To manage multi-user task requests, we introduce a distributed task queue that facilitates collaborative information processing and includes dynamic adjustment capabilities. This new approach, in comparison to traditional models, improves overall efficiency and enhances task management, thereby meeting the increasingly complex requirements of intelligent information systems.

1 Introduction

In the rapidly evolving field of the Internet of Things (IoT), the application spectrum is broadening to include smart cities, smart homes, digital healthcare, transportation, and agriculture [1–4]. IoT devices, encompassing sensors, actuators, and embedded systems, collect and transmit diverse data types—from temperature and humidity to location and motion, and from vital signs like heart rate and blood pressure to energy consumption metrics [5–7]. Data generation may occur in real-time or at predetermined intervals, and data volumes can vary across different application scenarios [1–7]. As the IoT landscape expands, the variety and number of devices proliferate. For instance, a smart city initiative might deploy thousands of sensors throughout the urban area, monitoring various parameters such as traffic flow and air quality. This complexity in data generation intensifies the challenges associated with data utilization.

In response to these challenges, we propose utilizing the Message Queuing Telemetry Transport (MQTT) protocol for data transmission and its brokerage functionalities. MQTT, a lightweight, publish-subscribe communication protocol, is especially suited to IoT applications in environments characterized by low bandwidth and unstable network connections. With MQTT, IoT devices can efficiently transmit data to central servers or peers and subscribe to relevant topics to receive necessary updates, thus facilitating effective data management and transmission. To further enhance system usability, we are implementing a RESTful API as the system's communication interface. REST, or Representational State Transfer, a software architecture style outlined by Roy Thomas and Fielding [9] in 2000, leverages standard HTTP operations. Given the long-term nature of IoT device subscriptions and to prevent prolonged API response times, we incorporate a distributed task queue system to manage asynchronous tasks, thereby improving task efficiency.

Section 2 of this paper reviews related works. Section 3 outlines the designed and implemented system. Section 4 presents experimental results and performance evaluations. Finally, Sect. 5 provides a summary of this research.

2 Related Works

Significant research on MQTT has been conducted across diverse application domains. For instance, in 2020, Biswajeeban Mishra et al. [10] analyzed two decades of Machine-to-Machine (M2M) protocol research [8]—including MQTT, AMQP, and CoAP—highlighting MQTT's prominence and its widespread adoption as the preferred M2M/IoT protocol in various applications.

In the context of the Industrial Internet of Things (IIoT), Michele Amoretti et al. [11] (2020) developed a novel communication framework utilizing MQTT broker bridging. This framework enhances dynamic interoperability among various production lines and industrial sites within IIoT environments, supporting dynamic authentication and authorization while providing heightened isolation and control to improve security. Additionally, Cenk Gündoğan et al. [12] evaluated and contrasted traditional Sensor Network Message Queue Telemetry Transport (MQTT-SN), the IETF's Constrained Application Protocol (CoAP), and the nascent Information-Centric Networking (ICN) methods. Their analysis indicates that deployment context heavily influences outcomes: MQTT-SN is effective for congested links, CoAP manages to coexist with compromised performance, and ICN excels in multi-hop environments.

In the domain of the Internet of Healthcare Things (IoHT), Mehwash Weqar et al. [13] observed network congestion in MQTT-SN medical brokers under high traffic conditions, which impairs network efficiency. To address this, they developed a communication model that adaptively switches between CoAP and MQTT-SN, enhancing transmission reliability and efficiency in IoHT systems. Additionally, in the context of medication transportation, David Samuel Batie et al. [14] addressed the need to maintain critical environmental parameters like temperature and humidity. They proposed a smart medication system utilizing Arduino and MQTT for real-time monitoring, integrated with temperature sensors and portable coolers to ensure the safe transport of drugs and vaccines.

Loom message transmission mechanisms require a high degree of flexibility to meet specific Machine-to-Machine (M2M) communication standards. Yanjun Xiao et al. [15] tackled issues of low efficiency and limited flexibility in current network data transmissions by developing a loom data model categorized by topic. Utilizing the MQTT protocol, they facilitated data transmission across loom networks and optimized the network based on variables such as node RSSI, Quality of Service (QoS) requirements, transmission priority, real-time communication needs, and data volume.

IoT enables cost-effective and reliable ubiquitous computing across contemporary Information and Communication Technologies (ICTs). Manas Pradhan [16] explored MQTT-based integrations through prototype implementations in both military and civilian ICT systems. His work aimed to facilitate lightweight, vendor-neutral, and interoperable message exchanges, leveraging existing information sources and avoiding the creation of pipeline systems.

In IoT frameworks, cloud computing is indispensable, offering substantial computing power and extensive storage capabilities. Lu Hou et al. [17] examined the architecture, implementation, and performance of IoT clouds. They integrated Hypertext Transfer Protocol (HTTP) and MQTT servers within IoT cloud architectures to deliver robust IoT services, while employing diverse technologies to enhance efficiency.

In 2020, Samir Chouali et al. [18] introduced MQTT-CV, a new variant of the MQTT protocol tailored for Connected Vehicles (CVs), enabling local processing by agents to significantly reduce infrastructure demands—achieving an average tenfold improvement over standard MQTT. Additionally, the over-the-air (OTA) update mechanism, a pivotal component of IoT, supports remote firmware/software updates without necessitating device recalls or on-site access. Nian-Zu Wang and Hung-Yu Chien [19] developed and assessed a novel MQTT-based OTA scheme, which leverages end-to-end (E2E) channel establishment to ensure the security and privacy of the transmissions.

The MQTT protocol, widely utilized in IoT, lacks inherent robust security features, posing significant challenges as IoT devices transmit sensitive data across various domains. Eduardo Buetas Sanjuan et al. [20] developed a secure mode for MQTT using encrypted smart cards to enhance authentication and ensure data confidentiality and integrity. In response to issues of data forgery and tampering, Wei Gao et al. [21] introduced a new IoT protocol that incorporates blockchain technology to augment MQTT's security during data transmission. Aleksandar Velinov et al. [22] explored MQTT-based covert channels, identifying seven direct and six indirect channels, and analyzed them within the context of network information hiding techniques. Mohammad Hamad et al. [23] designed the SEEMQTT framework to secure end-to-end publish/subscribe communications for mobile IoT systems. Their approach uses secret sharing schemes with multiple keystores to maintain encrypted message integrity, allowing controlled decryption by authorized subscribers without compromising the Pub/Sub model's decoupling principle. Identity-based encryption safeguards the links between publishers and keystores, ensuring trusted delegation for subscriber access [23].

3 System Design

In this section, Fig. 1 illustrates the architectural diagram of the MQTT distributed task queue system, which is segmented into three main components: MQTT with the distributed task queue, and the MongoDB database. In this system, IoT devices publish sensor data to the MQTT Broker. Users then request subscription services via a RESTful API, prompting the server to publish tasks to the Broker. Workers, upon receiving these tasks, subscribe to a predefined list of topic names managed by the MQTT Broker and forward the data they subscribe to the MongoDB database at the backend.

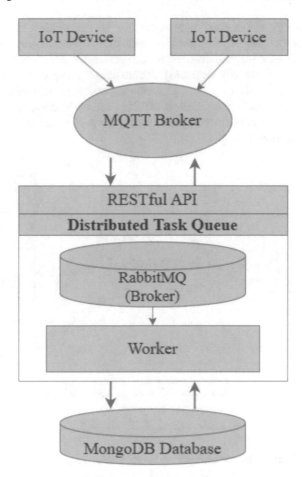

Fig. 1. The architectural diagram of the MQTT distributed task queue system

3.1 MQTT Message Transmission Design

In this subsection, Fig. 2 illustrates the message transmission principle of MQTT, which consists of three key elements: Client, Broker, and Topic. Clients are categorized as either Publishers or Subscribers. Publishers, such as the IoT devices discussed in this paper, send messages to the Broker using the Narrowband Internet of Things (NB-IoT) technology. The Broker, central to the MQTT protocol, then disseminates these messages to Subscribers, who are Clients subscribed to the corresponding Topics. In the context of this paper, Subscribers are defined as Workers that manage subscription tasks on the backend.

The Quality of Service (QoS) mechanism in MQTT's offers three distinct levels to control message transmission: QoS0, QoS1, and QoS2. QoS0, the fastest level, offers no delivery guarantee, potentially leading to data loss. QoS1 guarantees delivery at least once but may result in message duplicates if the Publisher continues to send messages before receiving a PUBACK packet from the Broker. QoS2 ensures exact, once-only delivery; after the Publisher sends a message, the Broker sends a PUBREC packet to acknowledge receipt. The Publisher then responds with a PUBREL packet, prompting the Broker to deliver the message to Subscribers and confirm completion with a PUBCOMP packet. Although QoS2 provides the most reliable delivery by preventing duplicates, it is slower compared to QoS0 and QoS1. For balancing efficiency and reliability, we utilize QoS1, ensuring timely data transmission without excessive delays. Finally, these three distinct levels of QoS have been deployed to testing the performances in the implemented MQTT broker to MongoDB.

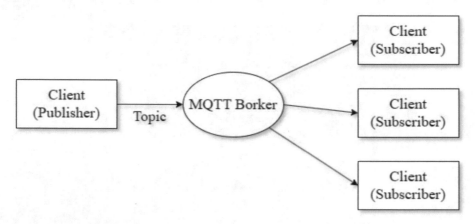

Fig. 2. Flow chart of MQTT message transmission principle

3.2   Distributed Task Queue Design

In addition, Fig. 3 depicts the architecture of the distributed task queue system, segmented into four main components: Broker, Producer, Worker, and Backend. The Producer, which in this study is implemented as a RESTful API, is tasked with publishing tasks and managing the execution messages. The Broker, handled by RabbitMQ, manages the queuing of published tasks. Workers, which are implemented as a single unit in this study, are charged with executing these tasks. Finally, the results of these tasks are transmitted to the Backend, where they are stored and managed using a MongoDB database.

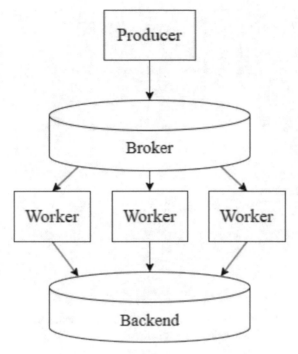

Fig. 3. Flow chart of distributed task queue

4   Experimental Results and Performance Evaluations

Follow the general idea of performance evaluation [24], we have developed an intelligent information transmission mode leveraging the MQTT protocol, with the MQTT Broker and MongoDB database situated in Room H304 at Asia University. The server specifications are as follows: CPU is an Intel(R) Core(TM) i9-10900F at 2.80 GHz; RAM is 32 GB DDR4-2933 MHz; the operating system is Windows 10 Pro 22H2. The RESTful API, implemented using Flask-RESTX—a Python-based extension of Flask—comprises four routes, as illustrated in Fig. 4: Connect, Topic, Subscribe, and Download. 'Connect' configures the connections to both the MQTT Broker and MongoDB database. 'Topic' is designated for storing the names of topics to which subscriptions are required at the MQTT Broker. 'Subscribe' facilitates subscribing to services based on the topics listed under 'Topic' and stores the resultant data in the MongoDB database. 'Download' enables the retrieval of subscribed data from the database.

   To assess the performance and stability of the system, we implemented load testing. Originally designed as a long-term task, the test involved subscribing to services via MQTT. For the purposes of this test, the task was modified to subscribe to ten distinct topics and then conclude. Comparative analysis was also conducted against two alternative systems: the Original API, which lacks asynchronous task capabilities, and a modified API that handles asynchronous tasks utilizing multiple threads. This testing approach allowed us to measure the impact of asynchronous operations and multi-threading on system performance.

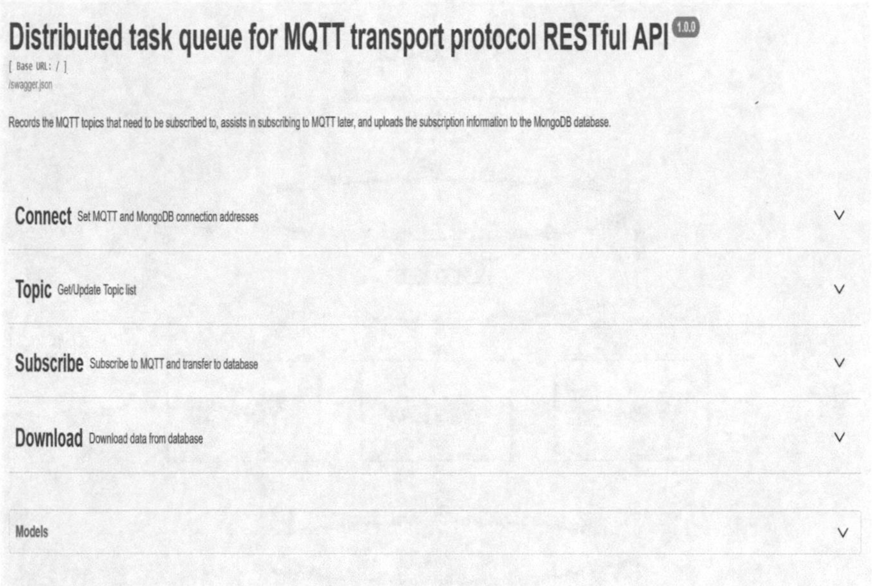

Distributed task queue for MQTT transport protocol RESTful API 1.0.0

[Base URL: /]

/swagger.json

Records the MQTT topics that need to be subscribed to, assists in subscribing to MQTT later, and uploads the subscription information to the MongoDB database.

Connect Set MQTT and MongoDB connection addresses   ∨

Topic Get/Update Topic list   ∨

Subscribe Subscribe to MQTT and transfer to database   ∨

Download Download data from database   ∨

Models   ∨

Fig. 4. The RESTful API documentation generated based on the Swagger tool, which can introduce API usage and interact directly with the API.

The load testing scenario progressively increased the user count from 0 to 100 over the first minute, maintained it at 100 for five minutes, and then reduced it back to 0 within the final minute. This test aimed to evaluate the stability and performance of the proposed API under rapidly increasing user loads. According to the results shown in Table 1, the Original API and Threads API performed faster than the Task Queue API; however, they demonstrated reduced stability. With user requests peaking at 100, the failure rates for these APIs began to increase, as illustrated in Fig. 5. Notably, the Task Queue API maintained a 0% failure rate throughout the testing period, underscoring its reliability for applications demanding long-term stability.

Table 1. The results of the load tests for the three APIs

Measured times API names	API all requests	API requests times	API requests failed	API requests duration (avg)	API requests duration (p90)	API req duration (p95)
Original API	35068	83.42/s	32.4%	20.63ms	31.12 ms	90.91 ms
Threads API	34427	83.79/s	35.6%	39.2 ms	59.72 ms	217.21 ms
Task Queue API	5804	13.8/s	0%	5.23 s	6.35 s	6.46 s

Fig. 5. The failure rate of various APIs under execution time. From left to left are Original API, Threads API, and Task Queue API.

5 Conclusions

This paper proposes a distributed task queue smart information transmission model utilizing the MQTT transmission protocol to address the growing complexity of data in IoT environments. The model enables users to efficiently manage data access by subscribing to sensor-specific topics, such as temperature, humidity, and CO_2 concentration, thereby simplifying the retrieval process. By implementing the system as a RESTful API, we enhance its usability and flexibility, with HTTP methods easing user interaction and integration with existing services. The distributed task queue architecture notably improves the system's stability and scalability, anticipating future increases in task diversity and volume. Furthermore, the system will be deployed on the server in Room H304 (Asia University, Taiwan) for educational use, granting students practical exposure to advanced IoT applications. Looking ahead, we plan to integrate Over-The-Air (OTA) technology, facilitating remote control and dynamic adjustments of IoT devices, thereby advancing our data transmission management capabilities.

Acknowledgments. This work was supported by the Chelpis Quantum Tech Co., Ltd., Taiwan, under the Grant number of Asia University: I112IB120. This work was supported by the National Science and Technology Council (NSTC), Taiwan, under NSTC Grant numbers: 111-2218-E-468-001-MBK, 110-2218-E-468-001-MBK, 110-2221-E-468-007, 111-2218-E-002–037 and 110-2218-E-002-044. This work was also supported in part by Asia University, Taiwan, and China Medical University Hospital, China Medical University, Taiwan, under grant numbers below. ASIA-111-CMUH-16, ASIA-110-CMUH-22, ASIA108-CMUH-05, ASIA-106-CMUH-04, and ASIA-105-CMUH-04.

References

1. Widodo, A.M., Chen, H.-C.: An optimization NPUSCH uplink scheduling approach for NB-IoT application via the feasible combinations of link adaptation, resource assignment and

energy efficiency. Comput. Commun. **218**, 276–293 (2024). https://doi.org/10.1016/j.com com.2024.02.016

2. Chen, H.-C., Widodo, A.M., Lin, J.C.-W., Weng, C.-E.: Reconfigurable intelligent surfaces-aided cooperative NOMA with p-CSI fading channel toward 6G-based IoT system. Sensors **22**(19), 7664, 1–25 (2022). https://doi.org/10.3390/s22197664

3. Chen, H.-C., Putra, K.T., Weng, C.-E.: A Novel predictor for exploring PM2.5 spatiotemporal propagation by using convolutional recursive neural networks. J. Internet Technol. **23**(1), 165–176 (2022). https://doi.org/10.53106/160792642022012301017

4. Putra, K.T., Chen, H.C., Prayitno, Ogiela, M.R., Chou, C.L., Weng, C.E., Shae, Z.Y.: Federated compressed learning edge computing framework with ensuring data privacy for PM2.5 prediction in smart city sensing applications. Sensors **21**(13), 4586, pp. 1–20 (2021). https://doi.org/10.3390/s21134586

5. Chen, H.-C., Yang, W.-J., Chou, C.-L.: An online cognitive authentication and trust evaluation application programming interface for cognitive security gateway based on distributed massive Internet of Things network. Concurr. Comput.-Pract. Exp. **33**(19), 1–14 (2020). https://doi.org/10.1002/cpe.6128

6. Chen, H.-C., Putra, K.T., Tseng, S.-S., Chen, C.-L., Lin, J.C.-W.: A spatiotemporal data compression approach with low transmission cost and high data fidelity for an air quality monitoring system. Future Gener. Comput. Syst. **108**, 488–500 (2020). https://doi.org/10.1016/j.future.2020.02.032

7. Chen, H.-C.: Collaboration IoT-based RBAC with trust evaluation algorithm model for massive IoT integrated application. Mob. Netw. Appl. **24**(3), 839–852 (2019). https://doi.org/10.1007/s11036-018-1085-0

8. Chen, H.-C., You, I., Weng, C.-E., Cheng, C.-H., Huang, Y.-F.: A security gateway application for End-to-End M2M communications. Comput. Stand. Interf. **44**, 85–93 (2016). https://doi.org/10.1016/j.csi.2015.09.001

9. Thomas, R.: Fielding: architectural styles and the design of network-based software architectures. Doctoral dissertation, University of California, Irvine (2000)

10. Mishra, B., Kertesz, A.: The use of MQTT in M2M and IoT systems: a survey. IEEE Access **8**, 202071–202086 (2020)

11. Amoretti, M., Pecori, R., Protskaya, Y., Veltri, L., Zanichelli, F.: A scalable and secure publish/subscribe-based framework for industrial IoT. IEEE Trans. Industr. Inf. **17**(6), 3815–3825 (2021)

12. Gündoğan, C., et al.: The impact of networking protocols on massive M2M communication in the industrial IoT. IEEE Trans. Netw. Serv. Manage. **18**(4), 4814–4828 (2021)

13. Weqar, M., Mehfuz, S., Gupta, D., Urooj, S.: Adaptive switching based data-communication model for internet of healthcare things networks. IEEE Access **12**, 11530–11548 (2024)

14. Bhatti, D.S., Hussain, M.M., Suh, B., Ali, Z., Akobir, I., Kim, K.-I.: IoT-enhanced transport and monitoring of medicine using sensors, MQTT, and secure short message service. IEEE Access **12**, 46690–46703 (2024)

15. Xiao, Y., Pei, E., Wang, K., Zhou, W., Xiao, Y.: Design and research of M2M message transfer mechanism of looms for information transmission. IEEE Access **10**, 76136–76152 (2022)

16. Pradhan, M.: Federation based on MQTT for urban humanitarian assistance and disaster recovery operations. IEEE Commun. Mag. **59**(2), 43–49 (2021)

17. Hou, L.: Internet of Things Cloud: architecture and implementation. IEEE Commun. Mag. **54**(12), 32–39 (2016)

18. Chouali, S., Boukerche, A., Mostefaoui, A., Merzoug, M.A.: Formal verification and performance analysis of a new data exchange protocol for connected vehicles. IEEE Trans. Veh. Technol. **69**(12), 15385–15397 (2020)

19. Wang, N.-Z., Chien, H.-Y.: Design and implementation of MQTT-based over-the-air updating against curious brokers. IEEE Internet Things J. **11**(6), 10768–10777 (2024)

20. Sanjuan, E.B., Cardiel, I.A., Cerrada, J.A., Cerrada, C.: Message Queuing Telemetry Transport (MQTT) security: a cryptographic smart card approach. IEEE Access **8**, 115051–115062 (2020)
21. Gao, W., Zhang, L., Yun, J.: A blockchain-based MQTT protocol optimization algorithm. J. ICT Stand. **11**(2), 135–156 (2023)
22. Velinov, A., Mileva, A., Wendzel, S., Mazurczyk, W.: Covert channels in the MQTT-based Internet of Things. IEEE Access **7**, 161899–161915 (2019)
23. Hamad, M., Finkenzeller, A., Liu, H., Lauinger, J., Prevelakis, V., Steinhorst, S.: SEEMQTT: secure end-to-end MQTT-based communication for mobile IoT systems using secret sharing and trust delegation. IEEE Internet Things J. **10**(4), 3384–3406 (2023)
24. Deng, D.-J., Lien, S.-Y., Lin, C.-C., Gan, M., Chen, H.-C.: IEEE 802.11ba Wake-Up Radio--performance evaluation and practical designs. IEEE Access **8**, 141547–141557 (2020). https://doi.org/10.1109/ACCESS.2020.3013023

Performance Analysis of a DTAG Recovery Method in DTN with Multiple Flows

Shura Tachibana[1], Makoto Ikeda[2]($\boxtimes$) (iD), and Leonard Barolli[2] (iD)

[1] Graduate School of Engineering, Fukuoka Institute of Technology, 3-30-1 Wajiro-Higashi, Higashi-Ku, Fukuoka 811-0295, Japan
mgm23106@bene.fit.ac.jp
[2] Department of Information and Communication Engineering, Fukuoka Institute of Technology, 3-30-1 Wajiro-Higashi, Higashi-Ku, Fukuoka 811-0295, Japan
makoto.ikd@acm.org, barolli@fit.ac.jp

Abstract. In this paper, we introduce the impact of multiple flows on the performance of Dynamic Threshold-based Anti-packet Generation (DTAG) recovery method in Delay/Disruption/Disconnection Tolerant Networking (DTN), which considers grid road maps. Unlike typical recovery methods that use anti-packets, the DTAG approach reduces node storage usage by considering the number of new bundle messages around vehicles. The DTAG has the ability to generate anti-packets at intermediate nodes before the bundle messages reach their end-point. The simulation results have shown that a high delivery rate can be maintained while storage usage and overhead are reduced, even when network flow increases.

Keywords: DTN · Epidemic · DTAG · Multiple Flows · Grid

1 Introduction

In recent years, although cellular networks have improved high-speed handover technology, wireless LANs and ad hoc nodes utilize high frequencies, and their coverage areas are narrow for APs and terminals, making sustained and stable connections challenging. Consequently, communication systems that store messages have also garnered attention. Delay-/Disruption-/Disconnection- Tolerant Networking (DTN) is one such method for message delivery in environments with frequent disconnections, such as terrestrial and space settings. In DTN, nodes continuously replicate the original message to adjacent nodes until the end-point receives it. Because of the proliferation of replicated messages, delivery management becomes critical. The basic architecture and specifications of bundled messages are presented in [7, 15], and there are several existing implementations [2–6, 8–11, 13, 18]. The Epidemic protocol [12, 19] is a typical contact-based DTN protocol.

In [16, 17], Uchimura proposed the Dynamic Threshold-based Anti-packet Generation (DTAG) method, which considers the replication progress of adjacent nodes. DTAG has proven effective in reducing message overhead in grid road models and urban road scenarios. However, the evaluation has only focused on single traffic flow.

© The Author(s), under exclusive license to Springer Nature Switzerland AG 2024
L. Barolli (Ed.): IMIS 2024, LNDECT 214, pp. 316–324, 2024.
https://doi.org/10.1007/978-3-031-64766-6_32

This paper discusses the performance of the proposed DTAG method combined with the Epidemic protocol, considering grid road map scenarios. The simulation results have shown that a high delivery rate can be maintained while storage usage and overhead are reduced, even when network flow increases.

The structure of the paper is organized as follows: Sect. 2 provides an overview of the network management with Epidemic protocol. Section 3 details our approach to integrating the DTAG method with the Epidemic protocol. Section 4 describes the simulation scenario and presents the results. The paper concludes with Sect. 5.

2   Network Management with Epidemic Protocol

The Epidemic protocol disseminates a batch of messages along with two control messages across the network. Each node periodically sends out a Summary Vector (SV) to keep tabs on the messages stored across different nodes. Nodes compare their stored data with the SV, and if they find a message that they do not have, they send out a REQUEST for that message.

However, this method by continuously copying and storing messages from one node to its neighbors does not use efficiently the network's resources and storage capacity. There's a risk that these messages might clog the storage space even after their intended recipients have received them, due to delays in their deletion by recovery methods like anti-packets.

The end-point nodes typically issue anti-packets, which list all bundles they have successfully received. Once other nodes receive and verify the anti-packet, they can delete these messages from their storage. Nonetheless, the use of anti-packets can also strain network resources due to their additional traffic.

3   Message Handling with DTAG Method

The DTAG method can be integrated with DTN's Epidemic and SpW protocols, providing a new recovery function. The flowchart of the DTAG method is shown in Fig. 1. This recovery function involves different processing steps for received packets such as SV, REQUEST, MESSAGE, and anti-packets compared to traditional methods. A notable feature of DTAG is its ability to generate anti-packets at intermediate nodes before the bundle messages reach their end-point. For this generation, nodes count the received SVs without a predetermined threshold in the DTAG procedure to calculate the number of new identical messages. If Eq. (1) is evaluated as true, the DTAG method proceeds to generate an anti-packet. If it is evaluated as false, the message is stored in the node's storage. By executing the process when the conditions of Eq. (2) and Eq. (3) are true, it is possible to avoid handling situations where a new bundle message has not arrived

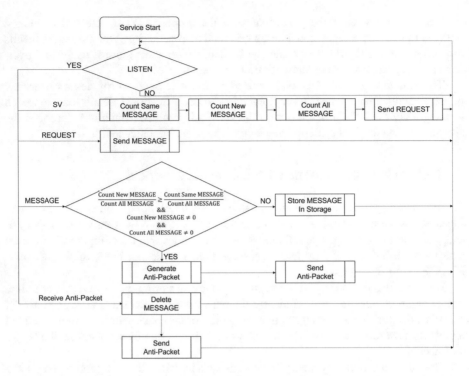

Fig. 1. Flowchart of DTAG method.

at the relay nodes or when a message is undelivered at the relay nodes. We expect this process at relay nodes to reduce the use of resources within the network.

$$\frac{\text{Number of new bundle messages}}{\text{Number of all bundle messages}} \geq \frac{\text{Number of same bundle messages}}{\text{Number of all bundle messages}}, \quad (1)$$

$$\text{Number of new bundle messages} \neq 0, \quad (2)$$

$$\text{Number of all bundle messages} \neq 0. \quad (3)$$

4   Evaluation

To evaluate the effectiveness of the proposed method, we utilize four key metrics: average delivery ratio, delay, overhead and storage usage. These averages are derived from ten unique seeds, representing various vehicle movement patterns. We conduct the evaluation by implementing the DTAG method on the Scenargie platform [14], where it is combined with the Epidemic protocol.

Table 1. Parameters for Simulation Analysis.

Parameter	Value
Area Dimensions	$1,000$ [m] $\times 1,000$ [m]
Vehicle Density	50, 100, 150 [nodes/km^2]
Start-points Count	1, 3 [nodes]
End-points Count	1 [nodes]
Vehicles Speed	8.333 [m/s] to 16.666 [m/s]
Simulation Duration	600 [s]
Recovery Protocols	Conventional Epidemic or DTAG Method
Bundle Message Size	500 [bytes]
Bundle Start and End Time	10 to 400 [s]
Bundle Transmission Interval	10 [s]
Summary Vector Generation Interval	1 [s]

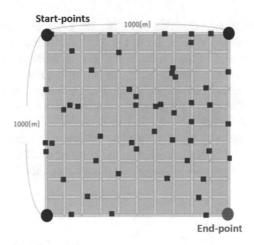

Fig. 2. Scenario with three start-points.

4.1 Setup

To evaluate the proposed approach, we consider buildings with height 10 m within the square blocks between the roads. The grid model is considered with vehicle densities ranging from 50 to 150 nodes/km^2. Table 1 shows the simulation parameters used in the experiment. The road model is illustrated in Fig. 2, with three start-points. In this evaluation, the number of start-points is set to 1 and 3. During the simulation, vehicles move along the roads based on a random way-point mobility model. The Epidemic and the proposed DTAG methods are evaluated in four different cases.

– Epidemic with 1 flow.
– Epidemic with 3 flows.

- DTAG with 1 flow.
- DTAG with 3 flows.

Bundle messages are initially transmitted from the originating node, referred to as the start-point, to an intermediate node, which then relays the message to other nodes until it reaches the end-point. Up to three start-points generate bundle messages. Both the start-points and the end-point are fixed. The simulation duration is set to 600 s. The simulation considers a 5.9 GHz radio frequency and follows the ITU-R P.1411 [1] standard.

4.2 Simulation Results

The simulation results for the delivery success rate are shown in Fig. 3. The Epidemic protocol achieves 100% delivery rate regardless of the number of flows. For DTAG with one flow, the delivery rate decreases as the number of nodes increases, but the reduction is minimal at 2%. The reduction for DTAG with three flows is also very slight, at 1% less than Epidemic.

The results for delay are presented in Fig. 4. The delay increases in all cases as the number of nodes increases, with the largest variations observed in the results for three flows. The proposed method shows a small variation in delay; at 50 nodes, the delay is only 0.1 s. This variation increases to 0.2 s with 100 nodes, and to 0.3 s with 150 nodes.

The results for overhead are shown in Fig. 5. DTAG improves performance regardless of the number of flows. This improvement becomes more significant when the number of vehicles increases.

Figure 6 presents the storage usage results, with the simulation time represented on the horizontal axis. The difference in storage usage between 50 and 150 vehicles for the Epidemic protocol is very small, with slightly higher usage for 150 vehicles.

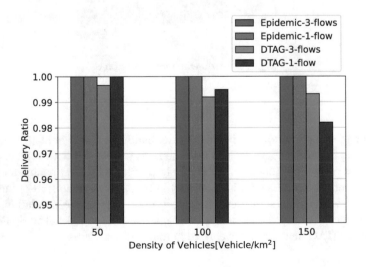

Fig. 3. Delivery ratio.

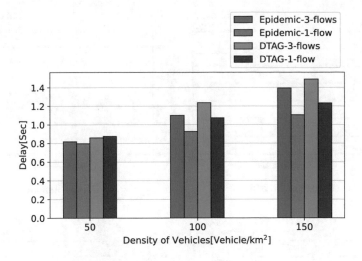

Fig. 4. Delay.

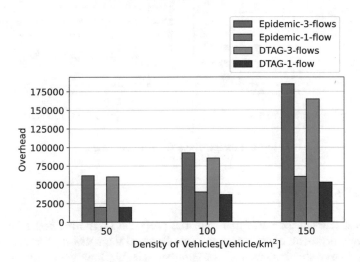

Fig. 5. Overhead.

For DTAG, there is a reduction in storage usage for 150 nodes. When comparing Epidemic and DTAG, the proposed method can reduce storage usage by a factor of 20 in the case of one flow with 50 vehicles and by approximately a factor of 58 in the case of three flows. This effect is further improved when there are 150 nodes in the network. These evaluations show that it is possible to maintain a high delivery rate while reducing storage usage and overhead.

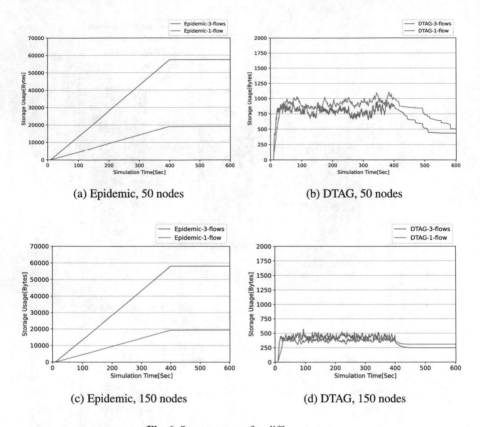

(a) Epidemic, 50 nodes   (b) DTAG, 50 nodes

(c) Epidemic, 150 nodes   (d) DTAG, 150 nodes

Fig. 6. Storage usage for different cases.

5   Conclusions

This paper discussed the performance of the proposed DTAG method combined with
the Epidemic protocol, considering multiple flows in the network. For evaluation, we
considered four evaluation metrics. The simulation results have shown that a high delivery rate can be maintained while storage usage and overhead can be reduced, even when
network flow increases.

In future research, we will investigate other recovery methods and assess various
parameters to enhance the performance of the proposed approach.

References

1. Recommendation ITU-R P.1411-11: Propagation data and prediction methods for the planning of short-range outdoor radiocommunication systems and radio local area networks in
the frequency range 300 MHz to 100 GHz. ITU (2019). https://www.itu.int/rec/R-REC-P.
1411-11-202109-I/en

2. Baumgärtner, L., Höchst, J., Meuser, T.: B-DTN7: browser-based disruption-tolerant networking via bundle protocol 7. In: Proceedings of the International Conference on Information and Communication Technologies for Disaster Management (ICT-DM-2019), pp. 1–8 (2019). https://doi.org/10.1109/ICT-DM47966.2019.9032944
3. Bista, B.B., Rawat, D.B.: EA-PRoPHET: an energy aware PRoPHET-based routing protocol for delay tolerant networks. In: Proceedings of the IEEE 31st International Conference on Advanced Information Networking and Applications (IEEE AINA-2017), pp. 670–677 (2017). https://doi.org/10.1109/AINA.2017.75
4. Burgess, J., Gallagher, B., Jensen, D., Levine, B.N.: MaxProp: routing for vehicle-based disruption-tolerant networks. In: Proceedings of the 25th IEEE International Conference on Computer Communications (IEEE INFOCOM-2006), pp. 1–11 (2006). https://doi.org/10.1109/INFOCOM.2006.228
5. Burleigh, S., Fall, K., Birrane, E.I.: Bundle protocol version 7. IETF RFC 9171 (Standards Track) (2022)
6. Cao, Y., Sun, Z.: Routing in delay/disruption tolerant networks: a taxonomy, survey and challenges. IEEE Commun. Surv. Tutor. 15(2), 654–677 (2013). https://doi.org/10.1109/SURV.2012.042512.00053
7. Cerf, V., et al.: Delay-tolerant networking architecture. IETF RFC 4838 (Informational) (2007)
8. Daly, E.M., Haahr, M.: Social network analysis for information flow in disconnected delay-tolerant MANETs. IEEE Trans. Mob. Comput. 8(5), 606–621 (2009). https://doi.org/10.1109/TMC.2008.161
9. Dhurandher, S.K., Sharma, D.K., Woungang, I., Bhati, S.: HBPR: history based prediction for routing in infrastructure-less opportunistic networks. In: Proceedings of the IEEE 27th International Conference on Advanced Information Networking and Applications (IEEE AINA-2013), pp. 931–936 (2013). https://doi.org/10.1109/AINA.2013.105
10. Leontiadis, I., Mascolo, C.: GeOpps: geographical opportunistic routing for vehicular networks. In: Proceedings of the IEEE International Symposium on a World of Wireless, Mobile and Multimedia Networks 2007, pp. 1–6 (2007). https://doi.org/10.1109/WOWMOM.2007.4351688
11. Lin, H., Qian, J., Di, B.: Learning for adaptive multi-copy relaying in vehicular delay tolerant network. IEEE Trans. Intell. Transp. Syst. 25(3), 3054–3063 (2024). https://doi.org/10.1109/TITS.2023.3292592
12. Ramanathan, R., Hansen, R., Basu, P., Hain, R.R., Krishnan, R.: Prioritized epidemic routing for opportunistic networks. In: Proceedings of the 1st International MobiSys Workshop on Mobile Opportunistic Networking (MobiOpp 2007), pp. 62–66 (2007). https://doi.org/10.1145/1247694.1247707
13. Rhee, I., Shin, M., Hong, S., Lee, K., Kim, S.J., Chong, S.: On the levy-walk nature of human mobility. IEEE/ACM Trans. Netw. 19(3), 630–643 (2011). https://doi.org/10.1109/TNET.2011.2120618
14. Scenargie: Space-time engineering, LLC. http://www.spacetime-eng.com/
15. Scott, K., Burleigh, S.: Bundle protocol specification. IETF RFC 5050 (Experimental) (2007)
16. Tachibana, S., Uchimura, S., Ikeda, M., Barolli, L.: Performance evaluation of DTAG-based recovery method for DTN considering a real urban road model. In: Barolli, L. (ed.) CISIS 2023. LNDECT, vol. 176, pp. 30–37. Springer, Cham (2023). https://doi.org/10.1007/978-3-031-35734-3_4
17. Uchimura, S., Azuma, M., Ikeda, M., Barolli, L.: DTAG: a dynamic threshold-based anti-packet generation method for vehicular DTN. In: Barolli, L. (ed.) AINA 2023. LNNS, vol. 655, pp. 406–414. Springer, Cham (2023). https://doi.org/10.1007/978-3-031-28694-0_39

18. Ulierte, T.A., Timm-Giel, A., Flentge, F.: Enabling quality of service over DTNs for space communications. In: Proceedings of the IEEE International Conference on Pervasive Computing and Communications Workshops and other Affiliated Events (PerCom-2024 Workshops), pp. 249–250 (2024). https://doi.org/10.1109/PerComWorkshops59983.2024.10503494
19. Vahdat, A., Becker, D.: Epidemic routing for partially-connected ad hoc networks. Technical report. Duke University (2000)

An Experiential Learning Platform Adopting PBL and Mix-Reality for Artificial Intelligence Literacy Education

Anthony Y. H. Liao[✉], Shun-Pin Huang, Tomoya Ikezawa, and Kuan-Yu Lin

Department of M-Commerce and Multimedia Applications, Asia University, 500 Liufeng Road, Wufeng, Taichung 41354, Taiwan
dr.tonyliao@gmail.com

Abstract. The objective of this study is to design an artificial intelligence literacy education platform that combines experiential learning and project-based learning and uses AR/VR technology to bring a new experience to artificial intelligence education and learning. The experimental design process is divided into two stages: research on the design of AR online experiential education for observation and learning in AI experience workshop, and research on the design of VR online learning for artificial intelligence education in AI practice workshop for project-based learning. The AR/VR online learning mobile device designed in this study is innovative in two ways. The first is to combine the existing AI experience workshop scenes with AR/VR experiential learning courses to enhance learners' actual experience of AI creativity. The second is to combine the AI practice workshop with project-based learning to conduct virtual operation project development and learning design to confirm whether learners have made progress in artificial intelligence knowledge enhancement and have the capability to apply it.

1 Introduction

With the rapid development of science and technology, the development of AI artificial intelligence has become a key focus of education at all levels in the world to cultivate capabilities [1]. In the university education part, we actively promote university programming education, lay the foundation for university artificial intelligence education, and make university education integrate with industrial AI technology, AI industrialization, etc., and provides universities with multiple innovative teaching models such as micro-courses, problem-oriented teaching, internships and competitions, linking industries for co-creation and co-education, and cultivating artificial intelligence application talents with the ability to integrate multiple technologies.

Therefore, how to lay a solid foundation for artificial intelligence concepts and deeply understand the application of artificial intelligence in innovative fields? This study designs an artificial intelligence literacy education platform that combines experiential learning and project-based learning, using AR/VR technology brings a new experience to education and learning for artificial intelligence, aiming to explore the AR artificial intelligence education learning and observation experience with experiential learning design thinking. The designed AR/VR online learning mobile device not only provides different artificial intelligence application themes (such as smart Home, unmanned factory, smart medical care, etc.), there are also teaching steps guided by artificial intelligence knowledge, as well as explanations of artificial intelligence examples for learners to learn from and interact with, thereby improving the basic capabilities of artificial intelligence knowledge. Based on the above characteristics, the purpose of this study is to achieve three teaching goals. Through online guided teaching with actual scenes and AR/VR, 1. Whether experiential learning can improve learners' artificial intelligence knowledge learning; 2. Whether the integration of virtual and real teaching method is helpful to learners' interest in learning artificial intelligence knowledge; 3. Whether project-based learning is helpful to learners' application of artificial intelligence knowledge.

2   Literature Review

2.1   Experiential Learning

Experiential Learning Theory (ELT) has been widely used in management learning research and practice for more than 35 years. Albort-Morant believes that cultivating experiential learning strategies is conducive to students' understanding of theoretical concepts and achieving excellent results [2]. Holik believes that students have different learning styles, and research shows that they prefer to participate in EL (direct experience) activities to develop their knowledge [3]. Experiential learning theory, based on the foundational work of Kurt Levin, John Dewey and others [4, 5], provides a theory of dynamic learning. Experiential learning is a process of building knowledge in four contexts involving contextual needs [6]. There is a creative tension involved between modes of learning. The basis of this learning model is a learning cycle generated by the four solution methods of experience, reflection, thinking and action [7, 8].

2.2   Project-Based Learning

In 1918, W. H. Kilpatrick had already proposed the concept of "Project-based Instruction and Learning (PBIL)" [9]. PBL is also often translated as "project-based learning", "topic-based learning", "topic-oriented learning", etc. Wang Jinguo believes that the project-based learning method is a "teaching method that allows learners to investigate or respond to real and complex problems or challenges so that students can acquire knowledge and skills." It is different from traditional fixed subject-based learning. It is recommended to let the student find an interesting problem, then collect data, find solutions and the feasibility of experimental methods, revise and modify experiments, and finally share the results of the research with the public. In terms of literacy-oriented curriculum and teaching design, our government has proposed four principles for literacy-transformed curriculum and teaching design: 1. Connect with actual context to make learning meaningful; 2. Emphasize student participation and active learning so that relevant abilities can be used and strengthened; 3. Take into account the content and process of learning to demonstrate that literacy is an integrated ability that includes knowledge, skills, and emotions; 4. Different design focuses can be set for different core literacy projects [10].

Seymour Papert, a professor at MIT, believes that curriculum design is a rigid learning method because students have to follow the curriculum. It suggested that the teaching method should abandon curriculum design and use project-based learning methods. Because this kind of learning method designs courses based on students' interests, which is completely opposite to the traditional concept of course design, he strongly recommends the use of project-based teaching methods [9].

3   Research Methods

The purpose of this study is to design an artificial intelligence literacy education platform for experiential learning and project-based learning. The course design is based on the subject area courses of the Ministry of Education, and is designed to use artificial intelligence, machine learning and other courses so that learners can learn through augmented reality and virtual reality (AR/VR), such as artificial intelligence literacy education through simulated environments, stunning real-life situations through simulation, smart homes, smart factories, smart drones, smart robots, or observation and learning in smart libraries, etc. Finally, the learners' course learning results are analyzed, and learning improvement information feedback is obtained from the learning records. Based on the popularity of mobile devices, we have implemented this tool on mobile devices and adopted an AR/VR touch-based interactive interface design to make the operation more convenient. This design is suitable for beginners who are accustomed to using smartphones.

The research method and experimental process design are to place learners in the real learning scene of artificial intelligence, and cooperate with the AR/VR mobile device learning platform to conduct virtual and real integrated teaching. The learners are in the physical field of AI experience workshop and AI practice workshop. Experiential learning is carried out in the practical application environment of artificial intelligence, and through project-based learning methods, learners can conduct questions and answers on the (AR/VR) virtual education platform to understand the relevant aspects brought by artificial intelligence in the current physical field. Knowledge application allows real-life scenarios to bring real contact and feeling to learners, strengthen learners' perceptions and bring learners to explore problems, and various on-site platform applications allow learners to generate ideas on how to apply artificial intelligence to solve problems. Ideas about the problem, discover errors and propose solutions from actual application operations, and conduct test experiments. From the existing AI experience workshop and AI practice workshop, all the knowledge learned by the learner is used in the process of solving the problem. Ability and skills, so as to discover and learn new knowledge and skills. In the process of hands-on teaching of virtual reality AI artificial intelligence (AR/VR) virtual and real integration courses, learners enjoy the process and results very much, it becomes a learner-centered learning methods which can enhance learners' learning interest and effectiveness.

Therefore, this study uses Unity programming design to create a programming interface for AR/VR reality technology, and designs an artificial intelligence literacy education platform for AR/VR experiential learning and project-based learning used on the Android system. In addition to the AI application scenario learning courses in the AI experience workshop, in order to enhance learners' physical experience of AI creative learning experience, they combine AI practice workshop with project-based learning to carry out virtual operation project development and learning design. The simulation application design of VR is used to simulate the operation process of learners to confirm whether learners have made progress in artificial intelligence knowledge and have the ability to implement it.

3.1 Research Scenario

In this study, 60 students from a university were divided into an experimental group and a control group, with 30 students in each group, to conduct an introduction experiment. Students use the artificial intelligence literacy education platform developed in this study that combines experiential learning and project-based learning for digital learning and teaching. Students in the experimental group learned courses using artificial intelligence (AR/VR) virtual-real integration and physical field in AI experience workshop and AI practice workshop, while students in the control group used the existing AI application scenarios in the AI experience workshop and AI practice workshop for learning and teaching.

3.2 Structure of the Proposed System

This study designs artificial intelligence literacy education platform learning courses using AR/VR experiential learning and project-based learning. The learning categories are divided into real-life simulation learning such as smart homes, smart factories, smart drones, smart robots, and smart libraries. It also introduces the application and technical content of artificial intelligence, including its application in transportation, entertainment, medical care, etc. In addition, there are also "machine learning" principle teaching courses, which include "listening, reading, visual recognition" that simulates human senses, the brain's "reasoning and decision-making, understanding and learning", action-based "movement and action control", etc. artificial intelligence behavioral application. Examples of each type are as follows (see Fig. 1):

(1) Smart Home: Course teaching that introduces how artificial intelligence behavior simulates human sensory "listening, reading, and visual recognition".
(2) Smart Factory: Course teaching that introduces how artificial intelligence behavior simulates the "movement, action control" and "reasoning decision-making, understanding and learning" of the human brain.
(3) Smart UAV: Course teaching that introduces how artificial intelligence behavior simulates the "movement and action control" of the human brain. Learning how AI searches for the best path and the shortest operation, which can be used to speed up computer operations or reduce the burden on the memory, and use artificial intelligence to reason, decide, modify, and other logical inferences about transportation paths.
(4) Intelligent Robots: Course teaching that introduces how artificial intelligence behavior simulates the "movement, action control" and "reasoning and decision-making, understanding and learning" of the human brain. Through the course, learners learn

how smart robots can imitate the structure and function of biological neural networks. They can train with input data to generate corresponding algorithm models, learn to judge specified problems, and perform navigation or explanation applications.
(5) Smart Library: Course teaching that introduces how AI artificial intelligence behavior simulates human "listening, reading, visual recognition" and "movement and action control".

Fig. 1. AR/VR simulated learning scenarios

This study designs (AR/VR) learning courses and classifies the course teaching of each learning category into three levels: basic, advanced, and complex. Each level of learning examples includes at least the following learning functions:

(1) Basic artificial intelligence exercises: Through interesting (AR/VR) virtual reality visual design, the basic knowledge and abilities to be learned are presented.
(2) Advanced practice area for teaching content: Learners can conduct training and learning in various simulated scenarios through learning methods of different learning categories and levels. You can also choose the learning method you want and enter learning training according to your own wishes.
(3) Corresponding to the complex learning mode of designated courses: In (AR/VR) virtual reality, learners complete designated learning actions according to the learning course content of each category and level. After recording and analyzing the learning results, the learning ability level is determined. It is recommended to study proficiently before entering the corresponding learning area.
(4) Simulation test results: Learners can check the learning execution results in the learning behavior record analysis, complete the simulation test and judge for real-machine learning.

3.3 Experimental Design

There were 60 college students, evenly divided into experimental group and control group, participated in the experiment of this study. They used an artificial intelligence literacy education platform that combined experiential learning and project-based learning to conduct a 1 to 4-week experimental course. The course content was AR/VR teaching mode. Course teaching is conducted using AI generative and text description methods. In order to understand the learning effectiveness and satisfaction of learners, the experimental method and process of this study are shown in Fig. 2. The experimental method mainly includes the following six steps:

a. Sixty students were evenly divided into experimental group and control group to conduct experiential learning experiments in the AI experience workshop.
b. Before the experiment, both groups of students received a questionnaire survey (pre-test) to examine whether there were any differences in the learning abilities and learning attitudes of the two groups of students before learning.
c. The experimental group uses artificial intelligence (AR/VR) virtual reality integrated course teaching, and the control group uses AI experience workshop as an experiential learning method.
d. After the experiment, both groups of students reported their learning experience and evaluation of learning results for week one to week four.
e. After the teaching activities, the experimental group and the control group filled out the questionnaire (post-test)
f. Compare the learning effectiveness, attitude, satisfaction and learning results of the two groups.
g. Analyze the difference in the effect of the two groups of learning methods on improving students' learning.

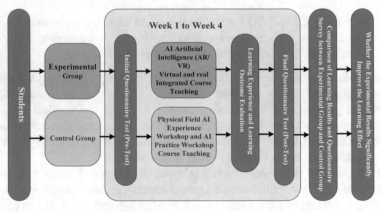

Fig. 2. The process of the experiment

4   Research Results and Performance Analysis

This study uses an artificial intelligence literacy education platform that combines experiential learning and project-based learning to conduct experiments with different teaching materials. It aims to understand the learner effectiveness, attitude, satisfaction and learning of AI-related knowledge and applications before and after the experiment. Are there any significant differences in the results? Therefore, this study used paired samples t-Test and independent samples single-factor covariate analysis for statistical analysis.

4.1   Pre-Test Paired Sample t-Test

In order to understand whether there is any difference in the ability of students in the experimental group and the control group before taking the AI-related knowledge and application knowledge teaching course, we first compared whether the pre-test levels of the two groups of students were the same. Based on the two sets of pre-test results, independent samples analysis was performed to conduct Levene's homogeneity test. Table 1 shows the independent sample results of the pre-test of the two groups of students before the experiment. The results of this t-Test are verified using the F value of the Levene test method. $F = 7.652$, significance (p) $= .008 < .05$, reaching the significant level. Therefore, it can be considered that the two groups of variables are not equal, so look at the t value in the column "Do not assume that the variables are equal", $t = 1.714$, significance (p) $= .093 > .05$, which does not reach the significant level. It means that there is no significant difference in the pre-test scores of students in the experimental group and the control group before the experiment, and there is no statistical difference in the course knowledge learning between the two groups of students. This means that the two groups of students have the same level of knowledge of the course knowledge before the course study.

Table 1. Pre-test paired sample t-Test of two groups

Levene Test for Equal Variance			Equal Mean t-Test		
	F	p-value	t	df	p-value
Assume equal variance	7.652	.008	1.714	58	.092
Do not assume equal variance			1.714	51.537	.093

4.2 Paired Sample t-Test for Test Scores of Two Groups Before and After Learning

This study uses an artificial intelligence literacy education platform teaching course that combines experiential learning and project-based learning. When students learn in different teaching methods in the teaching course, the pre-test of the academic performance test of the two groups of students on understanding AI-related knowledge and applications and post-test to compare the learning effects of students under different teaching methods. Therefore, after the learning activities are completed, the pre-test results and post-test results are analyzed by paired sample T-Test (Matched samples). After analysis and sorting of the experimental group, it can be seen from Table 2 that the t value of the experimental component of the sample $= -6.479$, $p = .000 < 0.050$, reaching the significant level of 0.050. Therefore, it can be seen that there is a significant difference in the results of academic performance between the experimental group's mean pre-test score $M = 85.33$ and post-test score mean $M = 95.03$. Students in the experimental group used the AR/VR digital learning platform developed in this study. Their post-test scores ($M = 95.03$) were significantly better than the pre-test scores ($M = 85.33$), showing that digital teaching methods of AI-related knowledge and applications can significantly help academic performance.

It can be found from Table 3 that the paired sample t-Test results of the control group show that the t value $= -6.530$, $p = .000 < 0.050$, and there is a significant

Table 2. Sample t-Test of experimental group

Variable Name	M	N	SD	t	p-value
Pre-Test Score	85.33	30	7.023	−6.479	.000
Post-Test Score	95.03	30	3.66		

difference in academic performance. Therefore, after the students in the control group used text descriptions for course teaching and digital teaching, their post-test scores (M = 91.83) were significantly better than the pre-test scores (M = 81.46), showing that AR/VR teaching mode uses artificial intelligence (AR/VR) virtual reality integration and the physical field AI experience workshop and AI practice workshop methods have improved the learning effect of course teaching, and have a significant impact on learning performance.

Table 3. Sample t-Test of control group

Variable Name	M	N	SD	t	p-value
Pre-Test Score	81.46	30	10.17	−6.530	.000
Post-Test Score	91.83	30	5.29		

4.3 T-Test of Post-Test Scores

When students use the AR/VR teaching mode to integrate artificial intelligence (AR/VR) virtual reality and physical field AI experience workshop and AI practice workshop in different ways, the academic performance of the two groups of students after the teaching courses has improved. Comparing two groups of students with the same test scores, independent samples t-Test analysis was performed, and Levene's test for homogeneity of variances was performed. Table 4 shows the independent sample t-Test results of the two groups of students before the experiment. The F value test result of Levene's test method, $F = 3.465$, significance (p) $= .068 > .05$, did not reach the significant level. Therefore, it can be considered that the two groups of variables are equal, so looking at the t value in the "Assuming the variables are equal" column, $t = 2.723$, significance (p) $= .009 < .05$, reaching the significance level, as shown in Table 4, indicating that the average post-test score of the experimental group (M = 95.03) was significantly higher than the average post-test score of the control group (M = 91.83).

Table 4. Paired sample t-Test for post-test scores of two groups

Group	N	Mean	SD
Experimental Group	30	95.03	3.662
Control group	30	91.83	5.292

Levene Test for Equal Variance			Equal Mean t-Test		
	F	p-value	t	df	p-value
Assume equal variance	3.465	.068	2.723	58	.009
Do not assume equal variance			2.723	51.595	.009

4.4 Comparison of Learning Attitude Pre-test and Post-test

In this study, for learning AI-related knowledge and application knowledge, the experimental group and the control group were compared to the differences in pre-test and post-test learning status of learning attitudes under different teaching methods. An independent sample t-Test was conducted on learning attitude. The F value of Levene's test for homogeneity test was $= 5.625$ ($p = .021$, $p > 0.050$). There was no significant difference. Therefore, assuming that the variation numbers are equal, the independent sample t-Test F value is $= 1.706$ ($p = .093$, $p > 0.050$), which does not reach a significant difference. It shows that the students in the experimental group and the control group have the same learning attitude before taking the course. As shown in Table 5, the students in the experimental group and the control group have different differences. After learning the teaching methods, the post-test results of students' learning attitude, the homogeneity test Levene's test F value $= 6.334$ ($p = .021$, $p < 0.050$), the difference is significant, so the items without assuming equal variation numbers are compared. The independent sample t-Test F value is 48.885 ($p = .000$, $p > 0.050$), reaching a significant difference. It shows that after students learn through the artificial intelligence literacy education platform that combines experiential learning and project-based learning, the learning attitude of the experimental group was better than that of the control group, as shown in Table 6.

Table 5. The Levene's test of the difference between the two groups in the pre-test of learning attitude

Levene Test for Equal Variance			Equal Mean t-Test		
	F	p-value	t	df	p-value
Assume equal variance	5.625	.021	1.706	58	.093
Do not assume equal variance			1.706	56.609	.094

Table 6. The Levene's test of the difference between the two groups in the post-test of learning attitude

Levene Test for Equal Variance			Equal Mean t-Test		
	F	p-value	t	Df	p-value
Assume equal variance	6.334	.015	5.277	58	.000
Do not assume equal variance			5.277	48.885	.000

4.5 Student Learning Satisfaction Survey

In this study, for learning AI-related knowledge and application knowledge, the experimental group and the control group were compared to the differences in pre-test and post-test learning status of learning satisfaction under different teaching methods. An independent sample t-Test was conducted on learning satisfaction. The F value of Levene's test for homogeneity test was $= .030\,(p = .864, p > 0.050)$. No significant difference was reached. Therefore, it was assumed that the number of variations was equal and the independent sample t-Test was performed. The test value F value $= -.211$ (p $= .833$, p > 0.050), which does not reach a significant difference. It shows that the learning satisfaction of students in the experimental group and the control group before taking the course is the same, as shown in Table 7. After students in the experimental group and the control group learned through different teaching methods, the post-test results of student learning satisfaction, the homogeneity test Levene's test F value $= .397$ (p $= .531, p > 0.050$), the difference is not significant, Therefore, comparing items assuming equal variation, the independent sample t-Test F value $= 57.484$ (p $= .008$, p > 0.050), reaching a significant difference. It shows that after students learn through the artificial intelligence literacy education platform that combines experiential learning and project-based learning, the learning satisfaction of the experimental group was better than that of the control group, as shown in Table 8.

Table 7. The Levene's test of the difference between the two groups in the pre-test of learning satisfaction

Levene Test for Equal Variance			Equal Mean t-Test		
	F	p-value	t	df	p-value
Assume equal variance	.030	.864	-.211	58	.833
Do not assume equal variance			-.211	57.791	.833

Table 8. The Levene's test of the difference between the two groups in the post-test of learning satisfaction

Levene Test for Equal Variance			Equal Mean t-Test		
	F	p-value	t	df	p-value
Assume equal variance	.397	.531	2.735	58	.008
Do not assume equal variance			2.735	57.484	.008

5 Conclusion

This study uses the existing experience environment of a university's AI experience workshop and AI practice workshop to conduct project-based learning and virtual operation project development learning design, and designs and introduces an artificial intelligence literacy education platform that combines experiential learning and project-based learning. The purpose is to design a digital learning environment that is easy for learners to operate and allows learners to learn in an environment with interesting prompts. Through the assistance of AR/VR digital learning, they can practice learning AI-related knowledge and application knowledge. Combining the existing AI experience workshop scenarios with AR/VR experiential learning courses to enhance learners' physical experience of AI creativity and the relevance of artificial intelligence literacy education that combines experiential learning and project-based learning, and confirm learners' understanding of the integration of AI-related knowledge and applications into course knowledge will significantly help students' progress in learning and improve their implementation capabilities. To achieve three teaching goals, through online guided teaching with actual scenes and AR/VR, 1. Experiential learning significantly improves learners' artificial intelligence knowledge learning, 2. The teaching method that integrates virtual and real situations is significantly helpful to learners' interest in learning artificial intelligence knowledge. 3. Project-based learning can significantly help learners apply artificial intelligence knowledge. 4. The experimental group is better than the control group in terms of student learning results.

Acknowledgments. The authors of this paper would like to express their gratitude to the National Science and Technology Council of the Republic of China for its partial grant support to this research. The grant number is NSTC 112-2410-H-468-012.

References

1. Gilbert, J., Watts, M.: Concepts, misconceptions and alternative conceptions: changing perspectives in science education. Stud. Sci. Educ. **10**, 61–98 (1983). https://doi.org/10.1080/03057268308559905
2. Albort-Morant, G.: Promoting innovative experiential learning practices to improve academic performance: Empirical evidence from a Spanish Business School. J. Innov. Knowl. (2017). https://doi.org/10.1016/j.jik.2017.12.001

3. Holik, M.T., Heinerichs, S., Wood, J.: Using experiential learning to enhance student outcomes in a didactic program in dietetics foodservice management course. Int. J. Allied Health Sci. Pract. **19**(1), 15. https://doi.org/10.46743/1540-580X/2021.1921

4. Kolb, D.: Experiential Learning: Experience as the Source of Learning and Development. Englewood Cliffs. Prentice Hall (1984)

5. Kolb, A.Y., Kolb, D.A.: Learning styles and learning spaces: enhancing experiential learning in higher education. Acad. Manage. Learn. Educ. **4**(2), 193–212 (2005)

6. Kolb, A., Kolb, D.: The learning way: meta-cognitive aspects of experiential learning. Simul. Gaming **40**(3), 297–327 (2009)

7. Armstrong, S.J., Fukami, C.V.: The SAGES Handbook of Management Learning, Education and Development. SAGE, pp. 42–68 (2009)

8. Palia, A.P.: Enhancing Experiential Learning via Sustained Student Engagement. Developments in Business Simulation and Experiential Learning, vol. 47 (2020)

9. Retter, H.: The Centenary of William H. Kilpatrick's "Project Method "International Dialogues on Education, Volume 5, Number 2, pp. 10–36 (2018)

10. Banks, J.A.: Published review of Diversity, Transformative Knowledge, and Civic Education: Selected Essays. Schools: Studies in Education, Issue 2, Fall, pp. 332–339 (2020)

Performance Improvement of Multiple Anchors for Three-Dimensional Indoor Positioning Using UWB Wireless Communications

Yung-Fa Huang[1](✉), Guan-Yi Chen[1], and Hsin-Cheng Wu[2](✉)

[1] Department of Information and Communication Engineering, Chaoyang University of
Technology, Taichung, Taiwan
yfahuang@cyut.edu.tw
[2] Department of Information Management, Chaoyang University of Technology, Taichung,
Taiwan
ison.wu@gmail.com

Abstract. With the increasing demand for positioning systems in multi-story buildings, these systems not only need to provide coordinates but also accurately indicate the vertical floor level of objects. This paper explores the performance of multiple anchors in three-dimensional (3D) indoor positioning using Ultra Wide-Band (UWB) communication technology. Experimental results show that the combinations with 4 and 5 anchors yield the smallest errors of 9.8 and 8.6 cm, respectively, while the largest errors are 13.9 and 13.3 cm, respectively. Furthermore, we investigated which anchors contribute the highest improvement in positioning performance. Experimental results indicate that among the proposed anchors, A_4 exhibits the best positioning improvement percentage at 34%, while anchor A_2 has the poorest performance at 1%.

1 Introduction

Location-based services (LBS) are rapidly expanding [1], with indoor positioning technology playing a crucial role in providing necessary support for various service applications, both in military and commercial domains [6]. Popular applications include driving navigation, drone positioning, among others. Currently, the Global Positioning System (GPS) is the most prevalent method for positioning. However, within buildings, its satellite signals are susceptible to obstruction, reflection, and other factors [2]. Hence, many scholars have embarked on research into indoor positioning systems based on short-range signals, such as Bluetooth, Radio-Frequency Identification (RFID), Wireless Local Area Network (Wi-Fi) [3], ZigBee [2], and Ultra Wide-Band (UWB) [4] positioning technologies. However, these methods face challenges such as signal blockage, computational complexity, and poor real-time performance. Among these technologies, UWB technology has garnered significant attention due to its centimeter-level accuracy and precision. For indoor positioning, centimeter-level accuracy is considered the threshold, as errors of a few meters indoors can equate to the distance of an entire room [7].

Currently, the application of 5G technology is gradually expanding into various fields, including smart cities, smart healthcare, intelligent transportation, and more [9].

L. Barolli (Ed.): IMIS 2024, LNDECT 214, pp. 337–347, 2024.
https://doi.org/10.1007/978-3-031-64766-6_34

For instance, in smart cities, 5G communication technology can enable functionalities such as traffic monitoring, smart street lighting, waste management, and intelligent buildings, thereby enhancing urban operational efficiency and quality. The use of mmWave frequency bands in 5G for positioning and tracking systems can be considered analogous to UWB systems due to their wide bandwidth. In recent years, the development of 5G has raised significant concerns regarding spectrum efficiency and coverage range [12]. A vision has been proposed for a vehicular positioning and mapping system utilizing 5G for autonomous driving applications [13]. With the promotion of emerging 5G applications, UWB systems for indoor positioning and tracking now present opportunities for both academia and industry.

UWB is a wireless communication technology used for short-range high-speed data transmission and precise positioning. It utilizes short-pulse signals propagating across the entire bandwidth for data transmission and distance measurement in the time domain. Due to its high temporal resolution, it can provide accurate distance and positioning information. The channel model of UWB describes the effects of attenuation, reflection, diffraction, and scattering during signal propagation [11]. This technology has been widely applied in various indoor positioning methods to meet different application requirements [10]. The DW1000 chip follows the IEEE 802.15.4a standard and operates in the low-frequency portion of the UWB band, ranging from 3.5GHz to 6.5GHz. Utilizing its pulse characteristics and wide bandwidth, it enables accurate real-time location systems (RTLS) [5]. This study employs a modified trilateration method based on distance measurements from tags to anchors to calculate the three-dimensional coordinates of tags. The accuracy of this method is evaluated through experiments.

2 Principle of Positioning

This article employs a method based on a simple linearized position equation. Building upon this linearization, the problem is solved iteratively through the least squares method.

2.1 Principle of Distance Measurement

Based on the DW1000 and ESP32, the ESP32 UWB module functions similarly to a continuous scanning radar, capable of precisely locating another device and establishing communication. When UWB devices come into proximity with each other, they initiate distance measurements using Time of Flight (ToF) distance measurement technology. ToF distance measurement technology calculates the propagation distance of a signal by measuring the round-trip time between the transmitter and receiver, thus estimating the position of the transmitter [8]. According to the measurement principle, the distance d between the Tag and Anchor can be determined as

$$d = c\frac{t_2 - t_1 - \delta - \delta_e}{2} \tag{1}$$

where c is the propagation speed of electromagnetic waves in vacuum; t_1 and t_2 represent the start time of the request clock and the end time of receiving the response, respectively; δ denotes the response clock period, typically a fixed parameter; δ represents the total system time delay.

2.2 Three-Dimensional Localization

The three-dimensional localization method proposed in this paper supports multi-anchor localization, but for UWB anchors, at least four known coordinates are required. After measuring the distance from the tag to each anchor (d_i, $i = 1, 2, 3, ..., i, ..., M$), taking ($M = 4$) as an example, four circles are drawn in Fig. 1. The intersection points of these circles determine the position of the tag, as shown in Fig. 1. The three-dimensional localization method proposed in this paper requires at least four known coordinates for UWB anchors. After measuring the distance from the tag to each anchor (d_i, $i = 1, 2, 3, ..., i, ..., M$), as shown in Fig. 1.

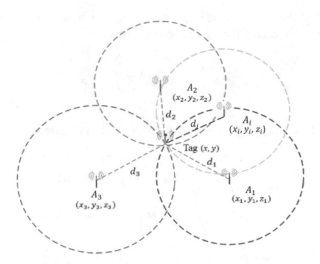

Fig. 1. Three-dimensional positioning diagram.

In Fig. 1, the coordinates of the UWB anchors are represented as A_1 (x_1, y_1, z_1), A_2 (x_2, y_2, z_2), A_3 (x_3, y_3, z_3), ...A_i (x_i, y_i, z_i), ..., A_N (x_M, y_M, z_M). The coordinates of the tag are represented as (x, y, z). Let $\hat{d}_1, \hat{d}_2, \hat{d}_3, ..., \hat{d}_i$ denote the i estimated distances from the tag to each anchor. The localization equation can be organized as

$$\begin{cases} \sqrt{(\hat{x} - x_1)^2 + (\hat{y} - y_1)^2 + (\hat{z} - z_1)^2} = \hat{d}_1 \\ \sqrt{(\hat{x} - x_2)^2 + (\hat{y} - y_2)^2 + (\hat{z} - z_2)^2} = \hat{d}_2 \\ \sqrt{(\hat{x} - x_3)^2 + (\hat{y} - y_3)^2 + (\hat{z} - z_3)^2} = \hat{d}_3 \\ \cdots \\ \sqrt{(\hat{x} - x_M)^2 + (\hat{y} - y_M)^2 + (\hat{z} - z_M)^2} = \hat{d}_M \end{cases} \quad (2)$$

After rearrangement, we obtain the following system of simultaneous equations:

$$
\begin{cases}
2\hat{x}(x_2 - x_1) + 2\hat{y}(y_2 - y_1) + 2\hat{z}(z_2 - z_1) \\
= \hat{d}_1^2 - \hat{d}_2^2 + x_2^2 - x_1^2 + y_2^2 - y_1^2 + z_2^2 - z_1^2 \\
2\hat{x}(x_3 - x_1) + 2\hat{y}(y_3 - y_1) + 2\hat{z}(z_3 - z_1) \\
= \hat{d}_1^2 - \hat{d}_3^2 + x_3^2 - x_1^2 + y_3^2 - y_1^2 + z_3^2 - z_1^2 \\
\quad \cdots \\
2\hat{x}(x_M - x_1) + 2\hat{y}(y_M - y_1) + 2\hat{z}(z_M - z_1) \\
= \hat{d}_1^2 - \hat{d}_M^2 + x_M^2 - x_1^2 + y_M^2 - y_1^2 + z_M^2 - z_1^2
\end{cases}
\tag{3}
$$

By rearranging the above equations and utilizing matrix calculations, the three-dimensional coordinates of the tag can be obtained as

$$
\hat{b} = \begin{bmatrix} \hat{x} \\ \hat{y} \\ \hat{z} \end{bmatrix} = (\mathbf{X}^H \mathbf{X})^{-1} \mathbf{X}^H \mathbf{w}
\tag{4}
$$

where $\mathbf{X}^H$ is the conjugate transpose of matrix $\mathbf{X}$ which is expressed by

$$
\mathbf{X} = \begin{bmatrix}
2(x_2 - x_1) \; 2(y_2 - y_1) \; 2(z_2 - z_1) \\
2(x_3 - x_1) \; 2(y_3 - y_1) \; 2(z_3 - z_1) \\
\cdots \\
2(x_M - x_1) \; 2(y_M - y_1) \; 2(z_M - z_1)
\end{bmatrix}
\tag{5}
$$

and

$$
\mathbf{w} = \begin{bmatrix}
\hat{d}_1^2 - \hat{d}_2^2 + x_2^2 + y_2^2 + z_2^2 - x_1^2 - y_1^2 - z_1^2 \\
\hat{d}_1^2 - \hat{d}_3^2 + x_3^2 + y_3^2 + z_3^2 - x_1^2 - y_1^2 - z_1^2 \\
\cdots \\
\hat{d}_1^2 - \hat{d}_M^2 + x_M^2 + y_M^2 + z_M^2 - x_1^2 - y_1^2 - z_1^2
\end{bmatrix}
\tag{6}
$$

Using the Least Squares method, coordinates of known test points can be determined. Based on the estimated coordinates of the tag and the deployed coordinates of the tag, the measured localization error can be obtained as

$$
\varepsilon = \sqrt{(x - \hat{x})^2 + (y - \hat{y})^2 + (z - \hat{z})^2}
\tag{7}
$$

3   Experimental Results

The experiments in this study were conducted in room 209.1 of the Information Building at Chaoyang University of Technology. During the experiments, Anchors emit signals to the Tag, which then measures distances using ToF and transmits the results to the local machine for subsequent localization experiments. To ensure experimental accuracy, all tests, except for the 8th anchor point, were conducted in Line-of-sight (LOS) channels. As illustrated in Fig. 2, the experiment was conducted in a three-dimensional space measuring $620 \times 783 \times 282$ cm. Eight anchors were used, indicated by the blue dots in Fig. 2. Additionally, one tag was utilized. In this experiment, we selected 27 test points (TPs), represented by the red dots in Fig. 2. The coordinates of the deployed 27 TPs are listed in Table 2. The average 3D positioning error $\overline{\varepsilon_i}$ of the 27 TPs in Table 3 is expressed by

$$\overline{\varepsilon_i} = \frac{\sum_{n=1}^{N} \sqrt{(x_{i,n} - \hat{x}_{i,n})^2 + (y_{i,n} - \hat{y}_{i,n})^2 + (z_{i,n} - \hat{z}_{i,n})^2}}{N} \tag{8}$$

where $x_{i,n}$, $y_{i,n}$ and $z_{i,n}$ represent the actual X, Y, and Z coordinates of TP_i. $N = 1000$ is the total measurement count of TPs. $\hat{x}_{i,n}$, $\hat{y}_{i,n}$ and $\hat{z}_{i,n}$ denote the estimated X, Y, and Z coordinates of TP_i.

This 3D indoor localization is divided into experiments using 4, 5, 6, 7, and 8 anchors, with the anchor addition sequence being A_1-A_8. All experimental results are shown in Table 3. From Table 3, it can be observed that the maximum positioning error of TP_{B2} is 22.7 cm, while the minimum positioning error of TP_{E1} is 6.2 cm. Furthermore, Table 3 also includes the overall average positioning error for different numbers of anchors, providing a comprehensive evaluation of system performance. The values in the "Average" column represent the average positioning errors for 4, 5, 6, 7, and 8 anchors, which are 13.9, 10.5, 8.5, 8.8 and 12.6 cm, respectively, effectively controlling the positioning error (Table 3).

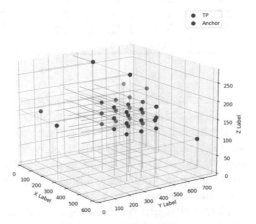

Fig. 2. Anchors and TPs deployment map.

Table 1. The coordinates of anchors (Unit: cm).

Anchor	x coordinate	y coordinate	z coordinate
A_1	613.8	385.5	177.2
A_2	0	131.5	138.2
A_3	0	721.8	175.8
A_4	163.4	357.9	282
A_5	465.8	354.5	282
A_6	307	0	150.4
A_7	122.2	783	173.3
A_8	509.2	783	79.5

Table 2. The coordinates of TPs (Unit: cm).

TP	x coordinate	y coordinate	z coordinate	TP	x coordinate	y coordinate	z coordinate
TP_{A1}	300	300	130	TP_{E3}	400	400	190
TP_{A2}	300	300	160	TP_{F1}	500	400	130
TP_{A3}	300	300	190	TP_{F2}	500	400	160
TP_{B1}	400	300	130	TP_{F3}	500	400	190
TP_{B2}	400	300	160	TP_{G1}	300	500	130
TP_{B3}	400	300	190	TP_{G2}	300	500	160
TP_{C1}	500	300	130	TP_{G3}	300	500	190
TP_{C2}	500	300	160	TP_{H1}	400	500	130
TP_{C3}	500	300	190	TP_{H2}	400	500	160
TP_{D1}	300	400	130	TP_{H3}	400	500	190
TP_{D2}	300	400	160	TP_{I1}	500	500	130
TP_{D3}	300	400	190	TP_{I2}	500	500	160
TP_{E1}	400	400	130	TP_{I3}	500	500	190
TP_{E2}	400	400	160				

Table 3 results indicate that increasing the number of anchors effectively reduces the positioning error. However, it should be noted that the positioning error with 8 anchors is particularly high, primarily due to the deployment of A_8 in a non-line-of-sight (NLOS) environment with respect to the target point (TP). Additionally, the positioning error with 7 anchors is slightly higher than that with 6 anchors, as the position of A_7 is situated at the edge of the entire localization environment, resulting in larger ranging errors due to its larger measuring angle compared to other anchors.

To observe whether other anchors exhibit similar issues, this study removed A_7, A_8, which had larger positioning errors, and investigated the influence of multiple anchors on the positioning performance in different orders. The specific anchor orders are shown in Table 4. In Table 4, "4A" represents selecting four anchors as the basic combination, while "5A" involves adding another anchor for comparison with a different set of added anchors. This was done to test the positioning performance, and the experiment with 6 anchors was not conducted because including all anchors in the positioning does not affect the results. Therefore, we designed a total of six experiments (E_1, E_2, E_3).

Table 3. Comparison of positioning errors $\overline{\varepsilon_i}$ for each TP in different positioning groups (Unit: cm).

TP	4A	5A	6A	7A	8A	Tp	4A	5A	6A	7A	8A
TP_{A1}	17.2	8.0	7.9	11.7	10.4	TP_{E3}	15.7	19.1	17.2	17.5	15.0
TP_{A2}	9.1	6.9	6.7	7.9	11.6	TP_{F1}	7.9	9.5	8.8	9.1	11.5
TP_{A3}	13.6	9.5	7.3	8.9	17.8	TP_{F2}	14.1	11.0	10.6	10.0	11.7
TP_{B1}	14.1	8.7	5.8	3.7	7.6	TP_{F3}	17.2	6.6	6.4	8.9	12.7
TP_{B2}	22.7	11.2	8.6	11.7	15.5	TP_{G1}	9.8	10.3	5.7	7.3	12.2
TP_{B3}	21.6	7.8	6.9	7.2	8.4	TP_{G2}	13.1	10.3	12.7	9.0	10.0
TP_{C1}	13.1	13.0	9.6	7.3	33.6	TP_{G3}	9.1	10.1	6.5	6.7	11.2
TP_{C2}	10.0	4.9	3.9	3.1	18.9	TP_{H1}	15.8	21.3	7.2	6.3	6.6
TP_{C3}	16.3	7.5	4.8	5.4	16.9	TP_{H2}	16.7	9.8	7.5	6.5	5.4
TP_{D1}	7.0	5.0	5.8	4.7	4.9	TP_{H3}	19.5	19.4	14.0	13.1	11.4
TP_{D2}	12.3	12.3	12.4	12.1	8.2	TP_{I1}	15.4	8.7	6.6	7.4	8.8
TP_{D3}	16.3	11.1	9.8	10.3	16.5	TP_{I2}	11.8	14.7	12.5	13.7	11.7
TP_{E1}	6.2	7.0	7.3	11.5	19.1	TP_{I3}	20.5	8.6	7.6	8.2	16.2
TP_{E2}	10.4	11.7	9.8	9.2	7.2	Average	13.9	10.5	8.5	8.8	12.6

Table 4. Experimental design of different positioning groups in multi-anchor positioning systems.

	4A	5A
E_1	A_3, A_4, A_5, A_6	A_1, A_2
E_2	A_1, A_2, A_5, A_6	A_3, A_4
E_3	A_1, A_2, A_3, A_4	A_5, A_6

In Table 5, the experimental results of E_1 are compared, since the results for 4A are all the same, only one set of data is provided. As shown in Table 5, when adding the fifth anchor, the average positioning error decreases for both cases. A comparison between $5A\_A_1$ and $5A\_A_2$ is conducted in Table 5 revealing that the positioning performance at anchor A_1 is superior to that at anchor A_2.

In Table 6, the experimental results of E_2 are compared. It is evident from Table 6 that adding anchors $5A\_A_3$ and $5A\_A_4$ respectively reduces the positioning error of TP. Moreover, from E_2 can be observed that adding anchor A_4 leads to better improvement in positioning performance compared to adding anchor A_3.

Table 5. Comparison of positioning errors $\overline{\varepsilon_i}$ of positioning groups with Experiment E_1 (Unit: cm).

TP	4A	5A_A_1	5A_A_2	Tp	4A	5A_A_1	5A_A_2
TP_{A1}	7.4	7.6	7.9	TP_{E3}	16.1	17.6	15.8
TP_{A2}	8.1	5.2	8.6	TP_{F1}	12.5	8.8	12.8
TP_{A3}	8.3	6.5	9.0	TP_{F2}	11.9	10.8	11.4
TP_{B1}	8.3	5.9	6.5	TP_{F3}	6.7	6.5	6.3
TP_{B2}	8.4	8.3	9.0	TP_{G1}	5.8	5.8	7.8
TP_{B3}	8.8	7.0	7.6	TP_{G2}	15.3	17.2	12.5
TP_{C1}	12.8	9.3	11.6	TP_{G3}	10.3	6.6	9.9
TP_{C2}	3.9	3.5	4.1	TP_{H1}	3.6	4.5	8.3
TP_{C3}	7.6	7.0	4.8	TP_{H2}	10.7	9.2	10.2
TP_{D1}	8.4	4.9	7.6	TP_{H3}	14.9	13.6	14.5
TP_{D2}	12.7	12.3	13.2	TP_{I1}	13.8	6.6	13.3
TP_{D3}	11.6	9.1	12.0	TP_{I2}	13.1	14.5	11.5
TP_{E1}	7.8	6.8	8.5	TP_{I3}	6.2	8.1	6.2
TP_{E2}	10.5	10.1	10.0	Average	9.8	8.6	9.7

Table 6. Comparison of positioning errors $\overline{\varepsilon_i}$ of positioning groups with Experiment E_2 (Unit: cm).

TP	4A	5A_A_3	5A_A_4	Tp	4A	5A_A_3	5A_A_4
TP_{A1}	15.5	13.8	8.6	TP_{E3}	22.0	23.7	15.7
TP_{A2}	11.1	9.6	6.3	TP_{F1}	16.9	10.6	10.1
TP_{A3}	23.4	18.4	8.0	TP_{F2}	10.7	10.7	9.8
TP_{B1}	5.9	5.2	5.5	TP_{F3}	6.8	6.5	8.4
TP_{B2}	6.5	7.0	7.5	TP_{G1}	18.1	8.1	7.3
TP_{B3}	12.8	12.3	10.1	TP_{G2}	8.9	12.2	11.4
TP_{C1}	7.4	9.3	9.5	TP_{G3}	28.9	34.2	6.1
TP_{C2}	12.5	4.0	9.9	TP_{H1}	11.5	11.1	6.5
TP_{C3}	16.1	5.2	11.9	TP_{H2}	13.9	13.0	9.1
TP_{D1}	12.3	6.2	7.0	TP_{H3}	17.9	17.5	14.0
TP_{D2}	12.4	13.3	12.1	TP_{I1}	16.0	6.7	10.2
TP_{D3}	19.2	17.5	9.4	TP_{I2}	17.6	14.4	13.5
TP_{E1}	24.2	6.3	10.5	TP_{I3}	9.0	6.9	8.8
TP_{E2}	10.6	11.6	9.0	Average	14.4	11.7	9.5

Similarly, in Table 7 the experimental results are compared between A_5, A_6 to evaluate the positioning effectiveness. From Table 7, it is observed that adding anchor A_5 leads to a greater improvement in positioning effectiveness,

Table 7. Comparison of positioning errors $\overline{\varepsilon_i}$ of positioning groups in Experiment E_3 (Unit: cm).

TP	4A	5A_A_5	5A_A_6	Tp	4A	5A_A_5	5A_A_6
TP_{A1}	17.2	8.0	16.6	TP_{E3}	15.7	19.1	15.2
TP_{A2}	9.1	6.9	9.0	TP_{F1}	7.9	9.5	8.5
TP_{A3}	13.6	9.5	11.1	TP_{F2}	14.1	11.0	14.1
TP_{B1}	14.1	8.7	14.7	TP_{F3}	17.2	6.6	17.5
TP_{B2}	22.7	11.2	18.4	TP_{G1}	9.8	10.3	8.7
TP_{B3}	21.6	7.8	21.1	TP_{G2}	13.1	10.3	15.7
TP_{C1}	13.1	13.0	13.8	TP_{G3}	9.1	10.1	8.6
TP_{C2}	10.0	4.9	6.7	TP_{H1}	15.8	21.3	12.0
TP_{C3}	16.3	7.5	12.7	TP_{H2}	16.7	9.8	23.6
TP_{D1}	7.0	5.0	7.4	TP_{H3}	19.5	19.4	17.1
TP_{D2}	12.3	12.3	12.3	TP_{I1}	15.4	8.7	15.2
TP_{D3}	16.3	11.1	14.8	TP_{I2}	11.8	14.7	10.0
TP_{E1}	6.2	7.0	6.4	TP_{I3}	20.5	8.6	19.9
TP_{E2}	10.4	11.7	8.9	Average	13.9	10.5	13.3

From the experiments, it can be observed that after removing A_7, A_8, the average error results of all experiments adhere to the principle that the more anchors used, the smaller the error value. Among them, the smallest average error is observed in E_1, with values of 9.8 and 8.6 respectively, while the largest average error is seen in E_6, with values of 13.9 and 13.3 respectively. Additionally, as show in Table 8 the average positioning improvement for each anchor can be defined as

$$e_i = (\frac{\overline{\varepsilon_{4,i}} - \overline{\varepsilon_{5,i}}}{\varepsilon_{4,i}}) \times 100\% \tag{9}$$

where e_i represents the average improvement in positioning error or the i-th anchor, $\overline{\varepsilon_{4,i}}$ denotes the average 3D positioning error $\overline{\varepsilon_i}$ of all TPs in the i-th 4A experiment, and $\overline{\varepsilon_{5,i}}$ denotes the average 3D positioning error $\overline{\varepsilon_i}$ of all TPs in the i-th 5A experiment. In the case of A_1, $\overline{\varepsilon_{4,i}} = 9.8$, and $\overline{\varepsilon_{5,i}} = 8.6$.

From Table 8 can be seen that anchor A_4 exhibits the best average positioning improvement percentage at 34%, while anchor A_2 has the poorest performance at 1%. Through these results, ineffective anchors in our positioning environment can be identified and removed to enhance our positioning performance. These experimental findings are crucial for evaluating the accuracy and reliability of 3D indoor positioning systems under different conditions and scenarios.

Table 8. Average improvement in positioning for different anchors.

A_i	e_i
A_1	12%
A_2	1%
A_3	19%
A_4	34%
A_5	24%
A_6	4%

4 Conclusions

This study aimed to evaluate the impact of multi-anchor positioning on the performance of UWB 3D indoor positioning systems through experiments. The experimental results show that the maximum error is 22.7 cm, while the minimum error is 6.2 cm. Furthermore, when using 4, 5, 6, 7, and 8 anchors for positioning, the errors are effectively controlled within 13.9, 10.5, 8.5, 8.8 and 12.6 cm, respectively. To confirm whether the experimental results adhere to the principle that more anchors lead to smaller errors, we removed problematic distance measurement results and conducted positioning experiments with 4 and 5 anchors in different orders. The results indicate that the average error results of all experiments comply with the principle that more anchors lead to smaller errors. Among them, the smallest average error is observed in E_1, with values of 9.8 and 8.6 cm, while the largest average error is seen in E_6, with values of 13.9 and 13.3 cm. Additionally, anchor A_4 exhibits the best positioning improvement percentage at 34%, while anchor A_2 has the poorest performance at 1%.

Acknowledgments. This research was funded by National Science and Technology Council (NSTC), R.O.C. grant number NSTC 112–2221-E-324–010.

References

1. Chen, C.-M., Huang, Y.-F., Jheng, Y.-T.: An efficient indoor positioning method with the external distance variation for wireless networks. Electronics **10**(16), 1–14 (2021)
2. Akyildiz, I.F., Su, W., Sankarasubramaniam, Y., Cayirci, E.: A survey on sensor networks. IEEE Commun. Mag. **40**(8), 102–114 (2002)
3. Geok, T.K., Aung, K.Z., Aung, M.S., et al.: Review of indoor positioning: radio wave technology. Appl. Sci. **11**(1), 1–44 (2020)
4. Cheng, C.-H., Wang, T.-P., Huang, Y.-F.: Indoor positioning system using artificial neural network with swarm intelligence. IEEE Access **8**(1), 84248–84257 (2020)
5. Bastiaens, S., et al.: Experimental benchmarking of next-gen indoor positioning technologies (unmodulated) visible light positioning and ultra-wideband. IEEE Internet Things J. **9**(18), 17858–17870 (2022)
6. Xu, S.: Optimal sensor placement for target localization using hybrid RSS, AOA and TOA measurements. IEEE Commun. Lett. **24**(9), 1966–1970 (2020)

7. Han, S., Zhang, Y., Meng, W., Li, C., Zhang, Z.: Full-duplex relay assisted macrocell with millimeter wave backhauls: framework and prospects. IEEE Network **33**(5), 190–197 (2019)

8. Wymeersch, H., Seco-Granados, G., Destino, G., Dardari, D., Tufvesson, F.: 5G MmWave positioning for vehicular networks. IEEE Wirel. Commun. **24**(6), 80–86 (2017)

9. Coppens, D., Shahid, A., Lemey, S., Herbruggen, B.V., Marshall, C., Poorter, E. D.: An Overview of UWB standards and organizations (IEEE 802.15.4, FiRa, Apple): interoperability aspects and future research directions. IEEE Access **10**, 70219–70241 (2022)

10. Yang, Z., Wang, Y., Chen, G., Zhang, Y.: Research on high precision indoor positioning technology of UWB. J. Navig. Pos. **2**(4), 31–35 (2014)

11. Decawave: DWM1000 IEEE 802.15.4-2011 UWB Transceiver Module. Technical Report. Decawave (2016)

12. Dotlic, I., Connell, A., Ma, H., Clancy, J., McLaughlin, M.: Angle of arrival estimation using Decawave DW1000 Integrated Circuits. In: 2017 14th Workshop on Positioning, Navigation and Communications (WPNC), pp. 1–6 (2017)

13. Fei, W.: Analysis of UWB indoor positioning accuracy based on TW-TOF. Acad. J. Sci. Technol. **5**(3), 61–64 (2023)

Applying ChatGPT-Based Iterative Improvement Model for Improving Software Maintenance Efficiency

Sen-Tarng Lai[1]([⊠]) and Fang-Yie Leu[2]

[1] Department of Information Technology and Management, Shih Chien University, Taipei 10462, Taiwan
`stlai@mail.usc.edu.tw`
[2] Department of Computer Science, Tunghai University, Taichung 40704, Taiwan
`leufy@thu.edu.tw`

Abstract. Software maintenance takes a lot of manpower and cost. However, it is impossible to effectively improve the quality and efficiency of software maintenance. The biggest problem that maintenance personnel encounter is the lack of correctness, completeness, and consistency of documents. In addition, legacy systems generally lack refactor features of design architecture. These drawbacks are critical factors to cause more time and cost of software maintenance. ChatGPT has a strong analysis capability to provide positive and effective suggestions. Iterative training maintenance concept and technology to chatGPT then requires chatGPT to identify defects of incorrect, inconsistent, and incomplete contents of documents and suggests revision manners of design. With chatGPT, maintenance personnel modify the defects of artifacts and adjust the design architecture quickly and completely. However, it is impossible to completely improve software maintainability in once or twice modifications. Based on chatGPT we propose an Iterative Improvement Model (IIM) with mutation testing to evaluate the improvement effect of software maintainability in this paper. Iteratively feedback and train chatGPT to gradually eliminate the maintenance quality defects and problems. Based on chatGPT, an iterative improvement model can increase software maintainability gradually and improve software maintenance efficiency.

1 Introduction

Software maintenance frequently consumes a lot of resources and effort but cannot effectively improve the quality and efficiency of maintenance. The main reasons include: mutilated documents in the development and documents lack correctness, completeness, and consistency, old versions software systems generally do not have the ability to reconstyruct, and personnel responsible for the maintenance phase are generally assigned to newcomers or those with inexperience or poor skills.[1, 2] These problems result in poor quality and low efficiency of maintenance operations, which greatly affects the market competitiveness of enterprises and organizations. DevOps (Development and Operations) proposed by two Flickr technical employees in 2009 is a development method that promotes cooperation and cultural integration between developers and operators and

L. Barolli (Ed.): IMIS 2024, LNDECT 214, pp. 348–358, 2024.
https://doi.org/10.1007/978-3-031-64766-6_35

effectively improves the market competitiveness of customers, enterprises, and organizations [3, 4]. However, the automated CI/CD process of DevOps combines multiple software tools, which requires a long time to import and the success rate is not high [5].

LLM's chatGPT can assist or even replace multiple tasks of human participation. There are many papers discussing how to use chatGPT to assist in software development, software testing, and resolving software engineering issues [6–9]. Github copilot can serve as a program development partner, assisting programmers in finding coding errors and improving program writing quality and Development speed. ChatGPT can quickly generate the reports according to you needs. However, the reports generated by ChatGPT may have incorrect content, may quote false data, may involve security issues related to personal privacy, may not be able to confirm the time sequence of data generation, and may not be able to effectively analyze visual charts, etc. Therefore, the reports generated by chatGPT must have a verification mechanism and revision method, and are not suitable for direct use, otherwise it will inevitably cause inappropriate consequences. How to make good use of the advantages of chatGPT has become an issue worth discussing.

In this paper, an iterative improvement model (IIM) based on chatGPT is proposed to evaluate the improvement effect of software maintainability by mutation testing. Iteratively feedback undiscovered defects and problems to chatGPT and further interact to gradually eliminate maintenance quality defects and problems. The iterative Improvement Model (IIM) is a method with mutation testing [10] to evaluate the improvement effect of software maintainability. Section 2 will discuss the dilemma of software maintenance operations and the advantages of ChatGPT. Section 3 chatGPT-based IIM key projects and operational processes. Section 4 two java programs as cases to evaluate capability of chatGPT and discuss how to apply chatGPT-based IIM. Section 5 discusses the major tasks in software maintenance operations and the advantages of chatGPT-based IIM. Section 6 describes the contribution of chatGPT-based IIM to software maintenance and draws conclusions for this paper..

2 Phase Documents and ChatGPT for Software Maintenance

2.1 Phase Documents for Software Maintenance Operations

No matter which development model is adopted, the development documents completed in phases in the software development process are the basis for the following phases of work. Phase documents are also the key to interaction and communication between developers [1]. However, most maintenance personnel almost have not participated in the development process. Therefore, to improve the efficiency of maintenance, maintenance personnel must have relevant technology abilities and be good at using tools, need also refer to and quote the development documents at each maintenance. However, the old software systems have many incorrect, incomplete, and inconsistent phase documents, which often hinder maintenance operations, affect the quality of maintenance operations, and make it difficult to improve maintenance efficiency. Software maintenance process, the quality of phase documents, correct cross-reference table, and design phase documents modularity are the critical items to affect maintenance efficiency. From the affected software items identification and isolation, problems fixed, regression testing, and delivery to deployment are major steps of software maintenance (shown as

Fig. 1). Referencing development phase documents are necessary in each maintenance step. Lack of complete, correct, and consistent development documentation requires more time and effort to handle software maintenance.

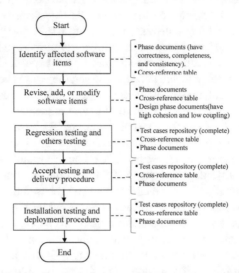

Fig. 1. Relationship between the software maintenance process and phase documents

2.2   Advantages of ChatGPT for Software Maintenance

Many literatures have researched and discussed the use of chatGPT to assist in the development of software systems. Among them, the literature on software debugging and testing is the most common. Many students of programming courses in our department will directly submit the homework with chatGPT. Most of the code generated by chatGPT can meet the requirements of homework questions. However, the input and output formats of some codes do not match the homework. Because students do not want to learn to write programs, they do not know what is wrong. There is no ability to modify the program code, so the program code generated by chatGPT can only be submitted directly as a job. Students use chatGPT as a generator for programming courses. They can complete the assignments without using their brains or hands-on skills. They can't get the ability and technology to write programs. In the end, it can't to meet the talent needs of the job market. If students can look at chatGPT as a teacher or teaching assistant, they will ask chatGPT for code that cannot be executed correctly. ChatGPT will help find out the errors and deficiencies in the code, revise the code by themselves, and practice over and over again to correct the errors and defects in the code. Corrected one by one, and finally completed the program self-wrote with the help of ChatGPT. In addition to having a high sense of accomplishment, also learned the ability and technology to write programs and became the talent that the job market is looking forward to.

ChatGPT has many features and advantages, which are beyond the scope of this article. This paper will focus on how the field of software development and maintenance can improve efficiency, productivity, and product quality with the assistance of chatGPT. Therefore, we pay more attention to the following advantages of chatGPT [11]:

- Have high interaction and communication skills.
- Have the ability to learn and integrate quickly.
- Have strong file analysis capabilities.
- Have the ability to adapt to situations by drawing inferences from one example.
- Ability to review and inspect documents.
- Ability to clearly identify and explain errors or omissions.

However, reports generated by chatGPT may have several defects as there may be incorrect content, may quote false data, information security issues that may involve personal privacy, unable to confirm the time sequence of data generation, unable to process simultaneously large amounts of data or files, inability to effectively analyze visual charts and other deficiencies. Therefore, the reports generated by chatGPT must have a verification mechanism and improvement process, and are not suitable for direct use, otherwise it will inevitably cause unexpected results.

3 Maintenance Partner and ChatGPT-Based IIM

3.1 ChatGPT as a Software Maintenance Partner

To develop a software system with incomplete documentation and poor maintainability, a lot of manpower and time must be invested to barely meet the maintenance requirements put forward by the user unit, causing software maintenance costs to continue to rise significantly. To reduce software maintenance costs and increase maintenance efficiency, it is necessary condition to make up the phase documents and improve the maintainability of software systems. Development phase documents of include requirements specifications, preliminary design documents, detailed design documents, source codes, and test cases for each step. These documents must have the following quality characteristics:

- Each phase documents have correctness, completeness, and consistency characteristics
- Cross-reference table between phase documents need to have correctness and completeness.
- Design phase documents with modular characteristics that have high cohesion and low coupling.

Reverse engineering technology is a key method to make up for insufficient development files and improve design quality [1], but reverse engineering requires to integration of several software tools and also needs to be supported by senior software engineers. In general, maintenance personnel are most lacking experience in using reverse engineering tools and are without senior software engineer support. Therefore, it is almost impossible to apply reverse engineering for software maintenance operations. However, chatGPT has the following advantages: high interaction and communication skills, the ability to learn and integrate quickly, strong file analysis capabilities and the ability to identify and explain errors or omissions of phase documents.

ChatGPT as a maintenance partner is a feasible approach. Think of chatGPT as a veteran consultant with years of experience. With the assistance of chatGPT, incorrect, incomplete, and inconsistent phase documents and cross-reference table can be repaired and made up on time and the modularity of the design phase documents can be improved. Effectively improve each phase documents quality and software maintainability, then improve maintenance efficiency, productivity, and product quality.

3.2   Process of ChatGPT-Based Iterative Improvement Model

The phase documents of software development have a relationship of continuation and reference. In the traditional waterfall development model, developers at each phase belong to different teams. Phase documents have become the key to the extension and interaction of each phase. However, after many years of maintenance operations, phase documents have become incomplete, incorrect, and inconsistent defects and problems, and the design phase documents have lost their reconfigurability. The number of maintenance personnel is limited and less experience, and users have high expectations for maintenance results. Therefore, this paper applies the senior software engineers to train chatGPT to become a senior software engineer with many years of experience and regards chatGPT as an important partner in software maintenance. ChatGPT is responsible for Identifying possible deficiencies and problems in existing documents, then the maintenance personnel will take over to ensure the deficiencies and problems and make revisions. To evaluate chatGPT with mutation testing, the defects and problems that have not yet been found will be assured, and then feedback to chatGPT and through the personnel further training, repeatedly and continuously improve the chatGPT maintenance capability and software maintainability. And effectively improve software maintenance efficiency. This paper proposes an iterative improvement model (IIM) based on chatGPT, which is divided into five steps. The operation process is shown in Fig. 2. The steps are described as follows:

1. Software senior engineer teaches chatGPT software maintenance concepts and technology that include basic, advanced, and feedback training.
2. Maintenance personnel prepares existing software development documents and explains the contents and the reference relationships between phase documents to chatGPT.
3. According to the defects and errors that have not yet been discovered with mutation testing to prepare data and documents are fed back to ChatGPT and further training.
4. According to chatGPT reply information, and then the maintenance staff takes over to confirm the defects and problems and make revisions.

5. If the correctness and completeness of chatGPT are lower than the preset criteria then return to the Training step. Repeatedly and continuously improve the phase documents quality of the software system.

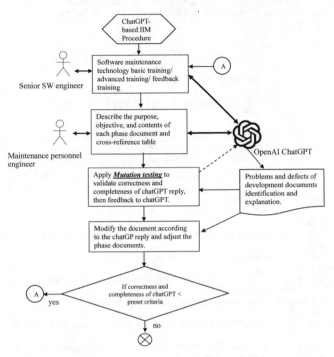

Fig. 2. Flowchart of chatGPT-based IIM

4   ChatGPT Code Inspection and Suitable IIM Flow

4.1   Java Code Errors Detection and Improvement with ChatGPT

To understand the program inspection capability of chatGPT, we use two Java programs with mutation testing to evaluate chatGPT.

- CASE 1: Predesign some errors and defects into program specifications and evaluate the completeness, consistency, and correctness of code inspection with chatGPT. Four types of problems or defects are embedded in the Java source code for evaluation.

Code Specification of Java program as follows:

There are ten students in a class and student scores are 67, 82, 51, 69, 93, 88, 77, 48, 60, and 90 respectively. Write a Java program to output maximum, minimum, and average (one decimal) scores, and pass the number of students (who score great equal to 60).

Java program (preset four types of errors) as follows:

```
class Case1
{
        public static void main(String args[]){
                int[] score={67, 82, 51, 69, 93, 88, 57, 48, 60, 90};      // inconsistent error
                int max=100, min=0, passNo=0;      // initial value error of max (incorrect error)
                double sum=0;
                for(int sc:score){
                        if(sc > max)
                                max=sc;
                        if(sc < min)
                                min=sc;
                        sum+=sc;
                        if(sc > 60)          // decision condition error (incorrect error)
                                passNo++;
                }
                System.out.printf("Max:%d, Min:%d, Average:%f, Passed No.:%d \n", max, min, sum/10, passNo);
                        // output format error (incomplete error)
        }
}
```

ChatGPT replied as follows:

1. Initialization of max and min:
   − Issue: max is initialized to 100 and min to 0. This might lead to incorrect results as actual student scores may not fall within this range.
   − Solution: Initialize max and min to the first value in the score array.
2. Average Calculation:
   − Issue: Average calculation sum/10 is hardcoded to the total number of students (10). This would be incorrect if the number of students changes in the future.
   − Solution: Calculate the average based on the number of elements in the score array.
3. Minimum Calculation:
   − Issue: The logic for finding the minimum score is incorrect. It initializes min to 0, which will always be less than any other score in the array.
   − Solution: Initialize min to a very large number or modify the logic to compare with the first element of the score array.
4. Passing Criteria:
   − Issue: The passing criteria is hardcoded as sc > 60. While this matches the specification, it might be better to define it as a constant or variable to make the code more maintainable.
5. Array Length Hardcoding:
   − Issue: The program hardcodes the length of the score array (10) in the average calculation.
   − Solution: Use score.length to dynamically determine the number of elements in the score array.
6. Typo in Score Array:
   − Issue: There's a typo in the score array. Instead of 77, it's listed as 57.
   − Solution: Correct the value to 77.

ChatGPT identifies all predesign errors and defects and provides the correct, complete, and consistent program.

- CASE 2 Common Coupling (High coupling)

Loose module design: We design a program to call a function that has high coupling and low cohesion. Module coupling design by global variables not with parameters interface and in getInfo() module handles many features (low cohesion).

ChatGPT replied as follows:

To improve cohesion in the program, you can refactor it to separate concerns into more cohesive methods. Here's how you can modify the program:
- Separate Concerns:
  - Break down the main logic into smaller, more focused methods.
  - Each method should have a single responsibility and operate on a well-defined set of data.
- Encapsulate Data:
  - Avoid using static variables for data that can be encapsulated within objects.
  - Pass necessary data as parameters to methods instead of relying on static variables.
- Improve Modularity:
  - Organize related methods into classes to improve modularity and maintainability.
  - Consider creating a Student class to represent student data and perform operations on individual students.

In addition, chatGPT also generate the modified program code.

4.2 Iterative and Improvement Model with Mutation Testing

Mutation testing inserts some defects and errors in source code or documents in advance, and then the defects and error rates found by chatGPT are used to evaluate the effectiveness. Analyze the reply information of chatGPT and collect the content of the defects and errors that have not yet been identified by chatGPT. Understand the causes of chatGPT omission with mutation testing, and analyze the knowledge and skills that chatGPT lacks and feedback training is carried out. Repeatedly and gradually improve until the defects and the error rate identified by chatGPT are greater than the preset criteria. To overcome the challenges faced by chatGPT-based maintenance operations, we recommend the following iterative and continuous improvement style:

- First, corrective maintenance is performed through simple defect reports, fixing only partial documents and code.
- Secondary, From local maintenance to global maintenance.
- Third, From simple maintenance to complex maintenance.
- Fourth, Iterative and incremental training to improve software maintenance knowledge and skills for chatGPT.

The missions of chatGPT-based IIM include verifying the effectiveness of improvements, assuring phase documents completeness, correctness, and consistency, a cross-reference table can be correctly generated, and the modification flexibility of design architecture.

5 ChatGPT Software Maintenance Efficiency Evaluation

In the previous section, we designed two simple Java program cases to measure chat-GPT's basic knowledge and skills in reviewing, inspecting or improving software development documents and modular design. The results found that chatGPT can successfully find our pre-design errors and defects. Timely identify errors and defects between the program specifications and code of Case 1, and even output the code he thinks is correct. Regarding the lack of modular design in Case 2, chatGPT can also explain in detail how to adjust the design of the module, and even output module design code with high cohesion and low coupling force. Based on the tests of the two cases, it can be concluded that chatGPT capable of assisting in software maintenance-related tasks. Of course, the test cases are still far from practical applications. The following is a list of challenges that chatGPT-based maintenance operations may face:

- The files processed by two test cases are simple and one-to-one files. Maintenance operations must encounter one-to-many, many-to-one, and many-to-many files, and their complexity is bound to be much higher than that of the cases.
- Maintenance operations require a variety of phase documents. The free chatGPT version has a certain limit on the file processing capacity and must be processed in batches.
- Software development uses several analysis and design diagrams to enhance phase developer interaction and communication. The free chatGPT version cannot analyze diagrams of software development documents (such as DFD, Sequence Diagram, Class diagram, etc.)
- After the maintenance operation is completed, the files must be adjusted and modified. How to feedback the new version of the development files and revised contents to chatGPT.

To analyze the capability of chatGPT, the major tasks of software maintenance are defined and described as follows:

Task 1: To Analyze and understand the defect reports (maintenance requirements).
Task 2: Identify the affected items of development documents and ensure the document quality that includes correctness, completeness, and consistency.
Task 3: Based on maintenance requirements to defects fix, function extension, or change requirements.
Task 4: Assure the maintenance quality that meets user requests with unit, integration, and regression testing.
Task 5:Revise the related documents to keep software maintainability.

According to five major tasks of software maintenance assigned to the chatGPT, maintenance prersonnel, and chatGPT-based IIM three kinds of maintenance personnel or mechanisms. We analyze the level of competence for each kind of maintenance personnel or mechanism in the current environment. And using weak, strong, need training, need more training, and take more time and effort 5 situations to show the results of analysis. (shown as Table 1).

Table 1. ChatGPT-based software maintenance tasks assistant Table

Personnel/mechanisms Tasks of software maintenance	ChatGPT	Maintenance Personnel	ChatGPT-based IIM
Task 1	Need training	Take time and effort	Strong
Task 2	Need training	Take time and effort	Strong
Task 3	Need more training	Take time and effort	Needs training
Task 4	Need more training	Weak	Needs training
Task 5	Need more training	Weak	Needs training

6 Conclusion

Software maintenance tasks are generally assigned to new or inexperienced IT personnel. In addition, old versions software system lacks correct, complete, and consistent development documents, and poor refactoring feature resulting in maintenance efficiency that cannot meet customer requirements. ChatGPT has powerful analysis and learning capabilities and uses natural language interface to communicate with personnel. in this paper, senior software engineer teach the chatGPT software maintenance concept and code inspection and document review method and personnel provide existing development documentations. Think of chatGPT as a software maintenance team member and a partner with inspection and review capabilities. When manpower is limited, applying the iterative improvement model (IIM) based on chatGPT and combined with mutation testing evaluation can improve chatGPT maintenance capability and technology continuous. The improvement of software maintenance efficiency can timely meet customer maintenance requirements and increase the market competitiveness of enterprises.

References

1. Schach, S.R.: Object-Oriented Software Engineering, vol. 7. McGraw-Hill, New York (2008)
2. Canfora, G., Cimitile, A.: Software maintenance. In: Handbook of Software Engineering and Knowledge Engineering: Volume I: Fundamentals, pp. 91–120 (2001)
3. Ebert, C., Gallardo, G., Hernantes, J., Serrano, N.: DevOps. IEEE Softw. **33**(3), 94–100 (2016)
4. Leite, L., Rocha, C., Kon, F., Milojicic, D., Meirelles, P.: A survey of DevOps concepts and challenges. ACM Comput. Surv. (CSUR) **52**(6), 1–35 (2019)
5. Purohit, K.: Executing DevOps & CI/CD Reduce Manual Dependency. IJSDR **5**(6), 511–515 (2020)
6. Akbar, M.A., Khan, A.A., Liang, P.: Ethical aspects of ChatGPT in software engineering research. IEEE Trans. Artif. Intell. (2023)
7. Jalil, S., Rafi, S., LaToza, T. D., Moran, K., Lam, W.: ChatGPT and software testing education: promises & perils. In 2023 IEEE International Conference on Software Testing, Verification and Validation Workshops (ICSTW), pp. 4130–4137. IEEE. (2023)
8. White, J., Hays, S., Fu, Q., Spencer-Smith, J., Schmidt, D.C.: ChatGPT prompt patterns for improving code quality, refactoring, requirements elicitation, and software design. arXiv preprint arXiv:2303.07839 (2023)

9. Tanzil, M.H., Khan, J.Y., Uddin, G.: ChatGPT incorrectness detection in software reviews. In: Proceedings of the IEEE/ACM 46th International Conference on Software Engineering, pp. 1–12. (2024)
10. Papadakis, M., Kintis, M., Zhang, J., Jia, Y., Le Traon, Y., Harman, M.: Mutation testing advances: an analysis and survey. In Advances in Computers, vol. 112, pp. 275–378 (2019)
11. Beganovic, A., Jaber, M.A., Abd Almisreb, A.: Methods and applications of ChatGPT in software development: a literature review. Southeast Eur. J. Soft Comput. **12**(1), 08–12 (2023)

Legal Case Retrieval by Essential Element Extraction Based on Reading Comprehension Model

Chen-Hua Huang[1,2], Chuan-Hsin Wang[1,2], Yao-Chung Fan[1,2],
and Fang-Yie Leu[1,2(✉)]

[1] Department of Computer Science and Engineering, National Chung Hsing
University, Taichung, Taiwan
`leufy@thu.edu.tw`
[2] Department of Computer Science and Engineering, Tung-Hai University, Taichung,
Taiwan

Abstract. In this study, we propose a Legal Document Retrieval
Pipeline. Given a legal case, we construct a scenario retrieval process
based on various types of Essential Elements for Prosecution (EEP) associated with different criminal charges. We employ a reading comprehension model to extract essential scenario details, meeting the requirements
of individual criminal charges. Subsequently, we extract keywords from
these essential scenarios and utilize the embeddings of these keywords
to compute the cosine similarity between each essential element, thus
identifying the most closely related judgment documents. This approach
dissects the overall direction of judgments into smaller components and
derives similar judgments by matching the details within the judgment
documents. In this study, we use the crimes of forgery and breach of
trust as preliminary case types. We incorporate ChatGPT to assess the
similarity between two judicial documents. We demonstrate that Chat-
GPT's similarity judgments closely align with those of legal experts. The
experiment results demonstrate the effectiveness of our legal document
retrieval pipeline.

1 Introduction

Legal professionals devote significant time daily to searching for similar judgments to better understand precedents, aiding them in presenting reasoned arguments in current cases. Historically, methods like TF-IDF, BM25, and vector
space models dominated legal document retrieval. However, recent advances in
deep learning have introduced Dense methods, though challenges persist due to
the lengthy nature of judgment documents. Techniques such as the Lawformer
model offer solutions, but at the cost of computational resources and time.

Contemporary research often focuses on condensing these documents to
essential content through methods like keyword extraction, selective paragraph
or sentence emphasis, abstractive summarization, or focusing on crime-specific

scenarios. Our paper introduces a novel framework for legal judgment retrieval that hinges on identifying and using "Essential Elements for Prosecution" extracted from crime scenarios. These elements are critical legal requirements that must be present for a conviction, offering a basis for comparing the similarity of cases by analyzing the details within these crime scenarios.

Our approach utilizes a reading comprehension model to frame and answer questions about these essential elements, subsequently employing a keyword extraction model to refine the process. This enables the transformation of keywords into embeddings, facilitating the comparison of documents based on their similarities in essential legal elements.

We also introduce an automated scoring mechanism using ChatGPT to evaluate our retrieval system, which shows promising results over traditional methods, both in efficiency and accuracy. This paper presents two main contributions: a framework for extracting and utilizing legal prosecution elements to enhance judgment retrieval, and an automated process for scoring similarity, significantly improving the precision of document retrieval in legal settings.

2   Related Work

Legal document retrieval has garnered significant interest, evolving with advances in information retrieval techniques [1,13]. Recent years have seen a shift towards deep learning methods to enhance retrieval accuracy by analyzing the content [9,14], structure [4], and citations [3] of legal judgment documents.

Earlier works include [2], where TF-IDF and TextRank were used for keyword extraction, and similarity was assessed using BM25 and VSM models. [11] utilized TextRank-based summarization to manage model input within a 512-token limit, enhancing entailment models' effectiveness with confidence-scored rankings.

[16] proposed a dense retrieval method combining TF-IDF vectorization to identify highly similar judgments, while [15] implemented a BERT model to score relevance between queries and judgments, integrating BM25 for optimal retrieval.

Our study and [7] both focus on segmenting judgment documents, albeit differently. [7] divides judgments into factual scenarios, analysis, and conclusions, using a text-to-text pre-trained model for retrieval. In contrast, our approach segments based on essential elements for prosecution, aiming to improve similarity retrieval.

Additionally, [12] and [6] illustrate diverse retrieval strategies using dense methods, RNNs, and vector combinations for semantic, topical, and entity-based comparisons.

Both [8] and our work emphasize the role of essential elements for prosecution in determining judgment similarity, albeit [8] relied on manual annotation.

In summary, our research pioneers the use of prosecution's essential elements in legal document retrieval, establishing a new direction for the field.

3   Methodology

In this study, we propose a legal document retrieval pipeline, as illustrated in Fig. 1. Given the content T of a judgment document, we first design an Extractor of Essential Elements for Prosecution (EEP Extractor), denoted as E. The primary function of the EEP Extractor is to extract or mark the corresponding essential elements within the judgment document's content. Specifically, $E(T)$ returns (e_i, t_i), $\forall i \in n$, where e_i represents the essential element i, and t_i is the text extracted by E from T corresponding to e_i.

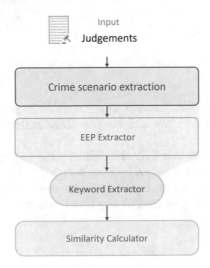

Fig. 1. Pipeline overview

To be a concrete example, we use forgery case analysis to illustrate our idea. According to law in Taiwan, for forgery case, the corresponding legal essential elements e_i are:

- e_1: Must be a document, including private documents, official documents, and variations of documents.
- e_2: Involves forgery or alteration behavior.
- e_3: Is capable of causing harm to the public or others.

The corresponding essential scenario for a forgery case shown in Fig. 2 are:

To obtain the essential element detail from the crime scenario, we employ a reading comprehension model as EEP extractor. Our aim is to assess the similarity between different judgment documents based on the content of the essential element details. To achieve this, we formulate specific questions according to the essential elements required for different types of crimes. The crime scenario, along with the designed questions, serves as the input to the question-answering (QA) model, and the answer generated by the QA model acts as the output

一、卓明祥於民國103年7月11日0時33分，騎乘機車行
經○ ○路時，因攔檢而實施酒測時超標，詎卓明祥竟基
於偽造署押之犯意，於酒測單上偽造「卓明德」之署名1
枚，足生損害於卓明德與交通主管機關處理交通違規事
件之正確性。
1. On July 11, 2014, at 12:33 AM, Zhuo Mingxiang, while
riding a motorcycle on ○ ○ Road, exceeded the alcohol
limit during a routine inspection. However, Zhuo Mingxiang,
with the intention to deceive, forged the signature of 'Zhuo
Mingde' on the breathalyzer form, causing harm to Zhuo
Mingde and undermining the accuracy of the traffic
violation proceedings by the traffic control authorities.

➡ t_1 : 於酒測單上偽造「卓明德」之署名1
枚
forged the signature of 'Zhuo Mingde' on
the breathalyzer form

➡ t_2 : 偽造「卓明德」之署名1 枚
forged the signature of 'Zhuo Mingde'

➡ t_3 : 足生損害於卓明德與交通主管機關
處理交通違規事件之正確性
causing harm to Zhuo Mingde and
undermining the accuracy of the traffic
violation proceedings by the traffic
control authorities.

Fig. 2. Example of Extracted Essential Elements for A Given Forgery Case

of the EEP Extractor. This process enables us to extract the crucial components that constitute the crime and lay the foundation for assessing document
similarity.

Moreover, for all t_i, we use a Keyword Extractor to perform keyword extraction, converting all t_i into a set of keywords $k_{i,j}$. In this step, we transform T
into $(e_i, k_{i,j})$. Furthermore, we introduce a Representation Generator denoted
as $M()$, which converts a given set of $k_{i,j}$ into a representation of e_i, referred to
as r_i.

$$M(\{k_{i,j}\}) \longrightarrow r_i$$

Then, for a given judgment document T, we obtain its document representation in the format (e_i, r_i). Based on the document representation of T, the
similarity between two judgment documents T and T' can be defined by the
following equation:

$$\sum_{\forall i} \sigma(r_i, r_i')$$

Here, σ represents the similarity function, and the equation calculates the
sum of similarities between corresponding representations r_i and r_i' for all essential elements i.

4 Experiment

4.1 Experiments Detail

1. **Evaluation** To evaluate the effectiveness of our framework, we need to score
   the results retrieved by the system. Since we do not utilize an existing dataset
   for the experiment, we require an evaluation metric to assess the quality of
   the retrieval system. However, evaluating the accuracy of a legal judgment

similarity system poses a significant challenge. Manual evaluation requires substantial costs and runs the risk of varying standards and definitions of similarity among evaluators. Therefore, we require a cost-effective and unbiased evaluation metric to address this issue.

OpenAI's large language model ChatGPT, released in November 2022, has demonstrated human-like capabilities in various tasks. This led us to wonder whether ChatGPT is capable of evaluating two legal judgment documents. Therefore, we utilized the ELAM dataset [17], consisting of 5000 crime scenario records. The main task of this dataset is to determine the similarity between two legal judgment documents, with three labels: similar, partially similar, and dissimilar, distributed as 1321/2059/1605 instances respectively. For our evaluation, we selected 250 instances from both the similar and dissimilar labels as our testing data.

As shown in Table 1, the experimental results demonstrated a high F1 score of 73, with Recall and Precision performing similarly. With the impressive performance of ChatGPT, we can conclude that ChatGPT indeed possesses the ability to distinguish whether two legal judgment documents are similar. As a result, we will employ ChatGPT as the benchmark for evaluating the similarity retrieval system for legal judgment documents.

Table 1. Evaluation Metrics on ELAM Dataset

	Precision	Recall	F1
0	72	74	73
1	73	74	72

2. EEP Extractor

   (a) Extractive QA on essential scenario detail extraction: We provided ChatGPT with information about the essential elements for prosecution and crime scenario. Subsequently, we requested ChatGPT to analyze the essential scenario corresponding to each essential element. We expected ChatGPT to generate its analysis results in a specific format. Finally, we used regular expressions to extract the essential scenario details from the responses provided by ChatGPT.

   Prompt:

   Essential elements for the prosecution of forgery: 1. Must be a document, 2. The perpetrator engages in forgery or alteration; 3. Results in damage to the public or others. I will provide you with a judgment scenario and the defendant's information. Please tell me what actions the defendant took, constituting the criminal elements of forgery of documents. The answer should be in traditional Chinese characters and follow the specified format. Format: 1. Must be a document: ... 2. The perpetrator engages in forgery or alteration: ... 3. Results in damage to the public or others: ... . Crime Scenario: "Description of the crime scenario". The defendant is "Defendant's Name".

(b) Generative QA on essential scenario detail extraction: We provide Chat-GPT with individual questions that we have designed based on the essential elements for prosecution, along with the crime scenario provided as context. We then request ChatGPT to answer according to the given question, aiming to generate the essential elements of each criminal scenario.

Prompt:

I will provide you with a crime scenario and a question. Please answer my question based on the content of the scenario. The more detailed your answer, the better. Please provide your response in Traditional Chinese. Criminal scenario: "Content of the crime scenario". The defendant's name is "Defendant's Name". "The question that deigns by essential elements for prosecution".

3. Keyword Extractor

(a) KeyBERT: KeyBERT [5] is a natural language model specifically designed for keyword extraction. It requires a piece of text and the desired number of keywords, denoted as k, as input. The model accurately returns the top k keywords.

In KeyBERT's application, English text requires space-separated words, while Chinese lacks precise spacing between characters. To use KeyBERT with Chinese, a segmentation tool is necessary to break the text into words. Poor segmentation can lead to disjointed segments and inaccurate keyword extraction.

To enhance KeyBERT, we adopted a segmentation optimization strategy, combining a legal dictionary, Jieba Chinese segmentation tool, and Named Entity Recognition (NER) models for accuracy. We crawled the Judicial Yuan Glossary of Common Legal Terms for Courts and Litigati on Procedures and LegisPedia, resulting in 1948 entries. Integrating our custom dictionary into Jieba aimed to improve segmentation accuracy.

(b) ChatGPT: We utilize ChatGPT for the task of keyword extraction, helping us extract keywords from the crime scenario.

Prompt: I will provide you with a description from a court judgment. Please extract the keywords. The more, the better. Description: "Content of the crime scenario". Output in Python list format, provided in Traditional Chinese.

Both ChatGPT and KeyBERT tend to misidentify terms like personal names, place names, and dates as keywords during extraction, despite their lack of relevance to judgment retrieval.

To mitigate this, we utilize part-of-speech (POS) tagging and Named Entity Recognition (NER) categories from the Jieba Chinese text segmentation tool to filter Keyword Extractor results. We exclude entities in categories like PERSON, DATE, GPE, ORDINAL, CARDINAL, and LOC, specifically filtering out part-of-speech tags "nr", "ns", and "ng". Identified entities from the EEP Extractor results are added to Jieba's dictionary to address this.

Before similarity computation, we concatenate essential element keywords into coherent sentences and obtain their embeddings. Pairing up these embeddings based on essential elements, we compute pairwise similarity for comparisons. Subsequently, we compute similarity between essential element pairs to derive scores, aggregating them for the overall judgment document similarity.

(a) TWLegalBERT

We collected 14,000 court judgments covering 20 different types of cases, including fraud, murder, theft, assault, and more. From these judgments, we extracted crime scenarios. Additionally, we used Taiwan's legal constitution, civil code, and criminal code as a corpus to pretrain the BERT model with whole-word masking. We then utilized this model to perform similarity calculations. The model is open on HuggingFace.

(b) CoSENT: We employed the publicly available Chinese Cosine sentence model [10] from Hugging Face for calculating text similarity.

4.2   Baseline

We evaluated the performance of the judgment document system based on the evaluation mentioned above. From the collected dataset, there were 1299 judgements of breach of trust and 16076 judgements of forgery. We randomly selected 50 instances from each category as testing data to assess the system's performance.

1. BM25: We will preprocess all judgment documents into crime scenarios, including both the testing data and the candidate in the database. The crime scenario from the testing data will serve as the query, and we will use BM25 to retrieve similar judgment documents.
2. Dictionary-based BM25: This approach combines a manually selected dictionary of 271 high-frequency tokens from crime scenarios related to forgery and breach of trust with a legal glossary obtained through web scraping, resulting in a total of 2219 entries. The judgment documents, both testing data and those in the database, are preprocessed using this dictionary to create a concatenated set of keywords. The testing data's keywords serve as a query, and BM25 is utilized for retrieving similar judgment documents.
3. Dense retriever: We preprocessed all judgment documents into crime scenarios and obtained embeddings using TWLegalBERT and CoSENT for both testing data and database documents. Calculating cosine similarity between the testing data embeddings and those of database documents facilitated the retrieval of similar judgment documents.

4.3   Experiment Results

As can be seen in Table 2 and Table 3, using TW-LegalBERT for obtaining embeddings did not yield satisfactory performance. We attribute this to a misalignment between the complex crime scenarios and legal terms used in the model's pretraining corpus and the straightforward terms extracted during inference. This discrepancy may hinder the model's ability to identify similar judgment documents. Consequently, our discussion primarily centers on the results obtained using the CoSENT model as the retriever.

Table 2. Precision and NDCG Results on breach of trust

Retriever	EEP extractor	KW extractor	Precision					NDCG				MAP			
			Top1	Top3	Top5	Top10	Top20	Top3	Top5	Top10	Top20	Top3	Top5	Top10	Top20
TW-LegalBERT	Generative QA	ChatGPT	59.09	57.97	55.31	54.33	52.21	58.46	57.04	54.96	53.22	57.00	56.52	54.23	50.41
		KeyBERT	72.09	56.62	57.71	53.25	51.14	59.77	58.01	54.90	52.66	61.11	59.33	55.58	50.34
	Extractive QA	ChatGPT	68.89	60.90	54.91	49.78	50.22	63.43	58.83	52.91	51.35	62.00	58.78	53.56	48.41
		KeyBERT	56.10	54.62	51.14	51.36	49.83	54.47	52.38	52.03	50.35	52.11	51.53	50.20	45.50
	w/o	ChatGPT	64.29	55.64	55.11	51.64	49.51	58.62	55.92	53.01	50.58	58.67	56.06	52.97	47.65
		KeyBERT	77.27	64.44	61.23	57.71	53.61	67.29	64.55	61.17	56.60	68.78	65.43	61.10	54.34
	Dense		72.09	57.97	56.22	52.58	49.40	62.36	58.78	54.94	51.60	63.78	60.33	55.76	50.06
CoSENT	Generative QA	ChatGPT	81.82	69.34	64.81	59.40	57.93	72.22	68.51	62.83	59.96	72.67	69.42	63.95	58.11
		KeyBERT	83.72	66.15	61.54	58.09	55.10	70.54	66.56	61.12	57.47	65.67	64.58	60.79	55.69
	Extractive QA	ChatGPT	68.18	65.44	63.76	59.39	56.42	66.17	65.29	61.35	58.15	71.44	66.19	61.34	54.87
		KeyBERT	84.44	70.90	61.33	58.63	55.91	73.98	66.91	61.13	57.90	70.44	67.29	62.19	53.96
	w/o	ChatGPT	74.42	61.54	58.82	60.18	54.58	63.95	61.22	60.56	57.66	64.33	62.08	60.21	54.02
		KeyBERT	73.33	69.12	63.32	58.66	54.42	69.49	66.28	62.07	57.39	70.22	67.23	62.51	55.96
	Dense		79.55	62.77	61.04	55.58	52.14	66.05	62.83	58.10	54.31	67.78	64.59	60.21	53.52
Dict+BM25			87.80	69.23	56.85	56.93	53.47	73.84	65.14	59.07	55.81	71.11	65.79	59.21	48.47
BM25			85.71	70.80	66.85	62.04	57.80	75.76	71.12	65.02	59.89	73.11	68.51	61.99	49.06

Table 3. Precision and NDCG Results on forgery

Retriever	EEP extractor	KW extractor	Precision					NDCG				MAP			
			Top1	Top3	Top5	Top10	Top20	Top3	Top5	Top10	Top20	Top3	Top5	Top10	Top20
TW-LegalBERT	Generative QA	ChatGPT	59.57	47.83	43.91	41.39	36.44	51.08	47.32	43.93	39.53	50.89	47.47	43.81	38.68
		KeyBERT	44.68	39.29	38.03	38.20	35.96	40.57	39.32	39.07	36.93	40.89	38.95	37.75	35.17
	Extractive QA	ChatGPT	61.70	50.00	41.99	39.83	35.88	53.79	47.62	42.82	38.36	52.22	47.63	42.91	38.19
		KeyBERT	68.09	53.24	48.28	40.22	35.67	57.50	52.37	45.38	39.57	57.44	53.25	46.97	40.50
	w/o	ChatGPT	51.06	43.26	40.61	39.02	35.15	45.91	43.45	41.1	37.24	45.33	42.54	39.66	35.45
		KeyBERT	56.82	48.53	46.52	42.73	38.98	50.37	47.58	44.72	41.05	50.78	47.69	44.27	40.00
	Dense		61.36	65.45	39.91	36.55	33.78	48.67	43.81	39.20	35.48	45.78	42.53	38.54	34.19
CoSENT	Generative QA	ChatGPT	72.34	53.57	50.65	48.92	46.44	58.95	54.17	51.51	48.62	60.00	56.12	52.11	47.29
		KeyBERT	70.21	57.35	50.66	45.05	44.42	60.65	55.33	49.08	45.77	61.00	55.94	49.81	44.64
	Extractive QA	ChatGPT	63.83	48.91	44.83	43.01	39.18	53.43	48.56	44.90	41.36	53.56	49.13	44.80	40.54
		KeyBERT	66.67	47.86	45.26	41.00	40.83	53.43	48.13	43.69	41.62	52.89	48.60	43.93	40.27
	w/o	ChatGPT	65.91	52.21	49.33	46.04	42.43	55.72	52.01	48.45	45.55	55.56	51.85	48.06	43.62
		KeyBERT	74.47	51.41	46.35	43.20	41.06	57.24	51.00	45.86	42.82	58.11	52.65	47.06	42.56
	Dense		67.39	55.40	50.00	46.10	41.67	57.71	53.74	48.94	44.36	56.89	53.39	48.91	43.63
Dict+BM25			65.12	51.67	50.00	46.08	45.71	55.74	52.66	48.25	46.47	51.22	47.97	44.09	38.23
BM25			76.09	57.94	55.83	47.91	45.63	62.02	58.32	52.79	47.97	60.33	56.74	51.14	41.82

1. Breach of Trust

Table 2 shows the results of precision NDCG and MAP score.

Compared to sparsity-based baselines like BM25 and Dictionary + BM25, our approach slightly lags in precision@1 and precision@5 but outperforms in MAP@20, suggesting better ranking of relevant judgment documents.

When employing the CoSENT model as the retriever, our approach generally surpasses the Dense retriever baseline in Precision, NDCG, and Recall.

Notably, when using Extractive QA with ChatGPT as the Keyword Extractor, precision@1 and precision@3 are lower, attributed to a limited average of 36 keywords extracted per breach of trust crime scenario. This limitation may hinder expressing semantic nuances, leading to a loss of scenario details and difficulty finding similar judgment documents.

In conclusion, our proposed framework, using CoSENT as the retriever, achieves the best results. However, attention to the quantity and accuracy of extracted keywords is crucial, as insufficient keywords can impact the framework's ability to accurately locate relevant judgment documents.

2. Forgery

   While our model achieved satisfactory results on Breach of Trust, it performed suboptimally on Forgery, even falling behind the baseline. This may be attributed to the unique characteristics of Forgery or the model's failure to adequately adapt to its specifics. Future work could involve further adjustments to the model architecture to enhance generalization.

5   Conclusion

This paper introduces a framework for legal document retrieval. We extract the essential scenario details within the crime scenarios based on the essential elements for prosecuting criminal offenses. By matching these extracted elements, we retrieve similar judgment documents. We propose a pipeline to determine the similarity between two judgment documents. We overcomes the challenge of requiring extensive datasets to evaluate legal document retrieval systems. Experimental results confirm that the framework proposed in this paper yields the better MAP score.

References

1. Barmakian, D.: Better search engines for law. Law Libr. J. **92**, 399 (2000)
2. Gao, J., et al.: FIRE2019@AILA: legal retrieval based on information retrieval model. In: FIRE (Working Notes), pp. 64–69 (2019)
3. Geist, A.: Using citation analysis techniques for computer-assisted legal research in continental jurisdictions. Available at SSRN 1397674 (2009)
4. Grimmelmann, J.: The structure of search engine law. Iowa L. Rev. **93**, 1 (2007)
5. Grootendorst, M.: KeyBERT: minimal keyword extraction with BERT (2020). https://doi.org/10.5281/zenodo.4461265
6. Hu, W., et al.: BERT_LF: a similar case retrieval method based on legal facts. Wirel. Commun. Mob. Comput. **2022** (2022)
7. Li, H., et al.: Sailer: structure-aware pre-trained language model for legal case retrieval. arXiv preprint arXiv:2304.11370 (2023)
8. Ma, Y., et al.: Lecard: a legal case retrieval dataset for Chinese law system. In: Proceedings of the 44th International ACM SIGIR Conference on Research and Development in Information Retrieval, pp. 2342–2348 (2021)
9. Maxwell, K.T., Schafer, B.: Concept and context in legal information retrieval. In: Legal Knowledge and Information Systems, pp. 63–72. IOS Press (2008)
10. Ming, X.: text2vec: a tool for text to vector (2022). https://github.com/shibing624/text2vec

11. Rossi, J., Kanoulas, E.: Legal information retrieval with generalized language models. In: Proceedings of the 6th Competition on Legal Information Extraction/Entailment. COLIEE (2019)
12. Shao, Y., et al.: BERT-PLI: modeling paragraph-level interactions for legal case retrieval. In: IJCAI, pp. 3501–3507 (2020)
13. Turtle, H.: Text retrieval in the legal world. Artif. Intell. Law **3**, 5–54 (1995)
14. Van Opijnen, M., Santos, C.: On the concept of relevance in legal information retrieval. Artif. Intell. Law **25**, 65–87 (2017)
15. Vuong, Y.T.H., et al.: SM-BERT-CR: a deep learning approach for case law retrieval with supporting model. Artif. Intell. Law 1 28 (2022)
16. Wehnert, S., Sudhi, V., Dureja, S., Kutty, L., Shahania, S., De Luca, E.W.: Legal norm retrieval with variations of the BERT model combined with TF-IDF vectorization. In: Proceedings of the Eighteenth International Conference on Artificial Intelligence and Law, pp. 285–294 (2021)
17. Yu, W., et al.: Explainable legal case matching via inverse optimal transport-based rationale extraction. In: Proceedings of the 45th International ACM SIGIR Conference on Research and Development in Information Retrieval, pp. 657–668 (2022)

Implementation of Switch Slicing in 5G/B5G/6G Networks

Li-Wen Peng[1], Fang-Yie Leu[1(✉)], and Heru Susanto[2]

[1] Computer Science Department, Tunghai University, Taichung City, Taiwan
{G11350011,leufy}@thu.edu.tw
[2] Computational Science, The Indonesian Institute of Sciences, Jakarta, Indonesia

Abstract. The 5th/B5G^{th}/6th generation mobile networks (5G for short) virtualize a physical device into n virtual components to support packet transmission, $n \geq 1$. 5G further classifies a network into three major categories, namely uRLLC (ultra-Reliable and Low Latency Communications), eMBB (enhanced Mobile Broadband), and mMTC (massive Machine Type Communications), because 5G would like to improve the shortcomings of 4G-network transmission, including not transmitted fast enough and no application differentiation among applications. Also, network slicing is a hot research topic since different slices can support different requirements of different applications. However, network slicing currently has not achieved its mature stage. Many efforts are required to define and implement network slicing. Therefore, in this study, we propose an UDF slicing approach by virtualizing a physical switch into many virtual switches with which to transmit data so that each virtual switch can define its own transmission requirements, i.e., latency, data delivery speed, and bandwidth along the best routing path.

Keywords: virtual switch · topology · 5G network slicing · SDN

1 Introduction

Currently, the functionalities of mobile phones have become increasingly diverse. Originally, they could only make phone calls. Presently, smartphones are capable of streaming online videos and providing other functions. In this evolution, the Internet connectives play crucial roles in improving our everyday lives since people need low latency, low cost, and high bandwidth communication to colorful their everyday lives, while these needs are basic requirements of 5G/B5G/6G networks (5G for short).

In an era of rapid technological advancement, we must keep pace with the time and wireless communication technology. With the emergence of 5G networks to accommodate various services, network slicing as an essential technique of 5G has attracted researchers' attentions. Basically, Software-Defined Networking (SDN) and Network Function Virtualization (NFV) as the basic functions of 5G networks play crucial roles in data transmission. SDN reconstructs network architecture with software, while NFV transforms network equipment from physical to virtual. Their combination enables new

L. Barolli (Ed.): IMIS 2024, LNDECT 214, pp. 369–378, 2024.
https://doi.org/10.1007/978-3-031-64766-6_37

lives to network slicing which divides a physical network into multiple virtual end-to-end networks. Each is independent from the others. Any failure in a network does not affect operations of the others.

At present, countries worldwide have also begun researching related technologies of 5G networks. However, the core technologies of 5G are not yet fully mature. Therefore, we seek to gain a better understanding of 5G network slicing and utilize SDN virtualization technology to virtualize network functions for slicing them.

Recently, 5G networks still have trouble to handle sudden increases in traffic. In network slicing, virtual resources may not always meet the demands requested by customers, meaning that its resource scheduling need to be further improved. So, this research would like to control packet flow through network slicing, classify packets, and identify the suitable transmission methods.

In the proposed approach, packets travel through multiple interconnected switches to reach their destination hosts. Utilizing the Ryu SDN Controller, the shortest/best path between the source and destination is determined, and OpenFlow is used to physically establish paths. When packets enter a switch, they are classified at the input port. Three corresponding queues for different classes, i.e., ultra-reliable and Low Latency Communications (uRLLC), enhanced Mobile Broadband (eMBB) and massive Machine Type Communications (mMTC), are created at all output ports, and packets are sent to the appropriate queue based on their classification. At last, we develop an algorithm to determine the transmission priority for ensuring efficient transmission of packets.

The rest of this article is organized as follows. Section 2 introduces the three major scenarios of 5G networks and the software used in this study. Section 3 explains the basic architecture of the proposed system and its operations. The implementation details and current achievements of this study are presented and discussed in Sect. 4. Section 5 concludes this study and addresses our future research.

2 Literature Review

Currently, 5G can be divided into three major application scenarios. Typical examples of uRLLC include wireless control in industrial automation, remote surgeries, smart grid automation, transportation safety, and autonomous driving which request high reliability (error rate below 10^{-5}) and low latency (less than 1 s between UE and base stations). eMBB would like deliver high volume of data smoothly. Transmitting AR/VR programs is a typical example. The characteristic of mMTC is to connect a large number of IoT devices with approximately one million sensors per square kilometer. It involves transmitting low data volumes and has lower requirements for data transmission latency.

2.1 Network Slicing

Network slicing means establishing multiple logical networks on physical networks. Each logical network connection, called a slice, has its own network configuration, such as network bandwidth, transmission paths, etc. Therefore, logically, each network slice can be dedicated to a type of applications or meet the specific network requirements requested by users.

Following the SDN network architecture, the OpenFlow protocol is used to decouple the control plane and data plane of a network component, like switch or router. In this study, the Ryu is utilized to control the operations of User Plane Function (UPF), i.e., a switch. It defines packet forwarding rules for the Flow Table in UPFs. This enabling us to programmatically rearrange the flow of network packets without modifying the hardware [1]. Network slicing is roughly divided into two categories:

(1) Computer-based slicing:

Computer-based slicing slices resources of a computer system into several virtual computers. Using containers such as Docker, IP ports can be configured for communication and packets are exchanged among containers. Network topologies can be constructed among Docker containers, and resource limitations can be imposed during the creation of a container to reserve resources for a slice within a computer system.

(2) Switch-based slicing:

In this study, network slicing is implemented inside a switch. Packets are classified based on Quality of Service (QoS) using Meter Bands in the Meter Table of a switch. Then, different packet groups are created using Group Table to meet various requirements, like multi-cast. This classification actually isolates data flows based on QoS. Different packet forwarding paths are established by using Flow Table, aiming to achieve the goal of switch resource "slicing". Actually, this approach can also be used by different network application scenarios.

2.2 SDN and Mininet

SDN brings the concept of virtualization to networking (from hardware to abstract software) by decoupling the control plane of a router and its data plane. It defines networks with software [2].

Ryu as an SDN controller was developed and designed by Nippon Telegraph and Telephone Public Corporation (NTT), Japan. One of its functionalities is acting as the controller of OpenFlow protocol. So it is often referred to as a controller [3].

Mininet is a platform with which we can create virtual networks by connecting virtual terminals, routers, switches, etc., for enabling the creation of SDN-enabled local area networks on a physical computer. Virtual hosts created by using Mininet can send packets, like real computers, and users can log in to virtual hosts by using Secure Shell.

2.3 Docker and Container

Docker is an open-source software platform which can be used to develop, deliver, and run applications. It provides an open interface, allowing researchers to develop their applications regardless of whether they are run on Windows, Linux, or macOS. What the Docker images, a Docker host, runs include applications and their dependencies. Developers working on Linux or macOS can create Linux-based Docker hosts, and generate images suitable for Linux containers. Developers working on Mac can also yield images suitable for Linux or Windows containers.

A Docker image is a template employed to repeatedly create container instances. Docker images can be easily developed by writing a Docker file, which consists of command-line instructions.

A container is an execution instance of an image, virtualized at the operating system level, encapsulating the code, libraries, and environment configurations required by an application. Each container is isolated from others, ensuring the security issues of a platform.

2.4 OpenFlow

OpenFlow is a network communication protocol operating at the data link layer. It enables the control of network switches' or routers' data forwarding planes, thereby altering the network path when necessary. OpenFlow can trigger remote controllers by controlling packet forwarding behaviors, including adding, modifying, or removing packet forwarding rules and actions. It can also control how network packets travel through the network switches [4].

2.4.1 Flow Table

A switch's OpenFlow Pipeline contains at least one or more Flow tables. Generally, an OpenFlow switch has n Flow tables, $n \geq 1$, each of which contains multiple flow entries to match information conveyed network packets and perform corresponding actions for these packets. These Flow tables are sequentially numbered, and each Flow table has its own identifier. A flow entry has:

(1) Match fields: this field includes packet matching rules for processing a packet passed from the previous Flow Tables. However, the pipeline processing on each network packet starts from Flow table 0.
(2) Priority: Matching priority (numeric), the smaller the number, the lower the priority.
(3) Counters: Counters count the number of packets successfully matched by this Flow Entry.
(4) Instructions: Instructions used for processing those packets successfully matched by this Flow Entry.
(5) Timeouts: Managing the idle time of the Flow Entry.
(6) Cookie: A value determined by the controller is used for modifying the Flow Entry by the controller.

2.4.2 Group Table

A Group Table has multiple Group Entries, each of which points to a Group, allowing OpenFlow protocol to provide more advanced packet forwarding features.

The fields of a Group entry include:

(1) Group identifier: A 32-bit unsigned integer, uniquely identifying the Group Entry.
(2) Group type: Indicating the processing behavior for packets.
(3) Counters: Statistics on packets processed by this Group Entry.
(4) Action buckets: An ordered list of Action Buckets, and each of the Buckets contains a set of actions and their parameters.

2.4.3 Meter Table

The Meter Table is also composed of multiple Meter Entries, each of which defines the meters for each Flow. With this structure, OpenFlow Switch can implement various simple QoS functions, like limiting the transmission bandwidth of an UE. The fields of a Meter Table entry are as follows:

(1) Meter identifier: A 32-bit unsigned integer, serving as a unique identifier for a Meter Entry.
(2) Meter bands: Each Meter Band specifies the bandwidth and the behavior when processing related packets.
(3) Counters: Statistics on those packets processed by this Meter Entry.

3 System Architecture and Implementation

This system we propose consists of six main components: (1) topology setup; (2) packet flow control; (3) packet classification; (4) establishment of queues for a packet type; (5) network resource slicing; (6) program allocation that the network topology is constructed by using Mininet, which is connected to Ryu Manager. The shortest/best paths are built as the packet transmission routes in the Mininet topology. Python scripts are deployed for classifying the packets, assigning corresponding packet encoding by using Differentiated Services Code Point (DSCP), and directing them to the appropriate queues based on the settings of the Flow Table. The ultimate goal is to enable packet transmission along the shortest/best path. Figure 1 depicts the system architecture.

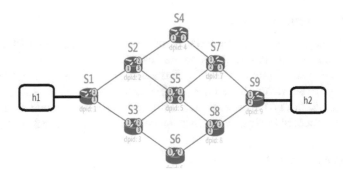

Fig. 1. The system architecture diagram designed for this study.

(1) Ryu: SDN Controller which is responsible for controlling all switches and defining their packet forwarding rules.
(2) Switch: Forwarding packets according to the rules set by the controller.
(3) Docker: Limiting host resources to achieve computer slicing objectives.
(4) Mininet: Constructing a virtual network topology for implementing 5G network slicing.

3.1 Network Topology

We establish a virtual topology and configure the ports connecting hosts and switches as shown in Fig. 1.

3.1.1 The Shortest/Best Path

After the k shortest/best paths are defined (often $k = 2$ or 3), we further weight each shortest/best path for each pair of hosts, identify which switches the path travels through and record the routing information in the corresponding Flow Tables.

A Network Awareness module is utilized to collect real-time changes in network resources, such as topology information and host leave and add. After the shortest/best paths are built, the paths with the maximum remaining bandwidth or the shortest delays as the best paths are chosen.

We measure the delay of the link with a delay detector which sums up the delays of Link Layer Discovery Protocol (LLDP) packets and that of Echo data packets as the round trip delay which will be reduced by the delay time of LLDP and that of Echo data packet both between SDN controller and the switch. The remailing time is the delay along the link from the input port to the output port of the switch that the packet enters. Then, the delay will be stored in the network topology. Following that, the key route weight parameters are set and then the packet forwarding path based on the shortest-delay path can be established if the shortest delay is chosen.

Of course, if remaining bandwidth is considered, the smallest remaining bandwidth along a path will be the remaining bandwidth of the path. After that, the path with the widest remaining-bandwidth path will be selected as the best path.

3.1.2 Ryu Controller and Switch

Ryu provides many development interfaces for users. Through Ryu's API, the required network management functions can be implemented.

Each Switch carries one or multiple distinct Slices, and each Slice adheres to the rules specified in the Meter Table, Group Table, and Flow Table to process data flows. Moreover, each Slice can be customized according to the requirements of users. At last, they are interconnected to physical or virtual machines to simulate real-world scenarios.

3.2 OpenFlow Protocol

With the command **ovs-ofctl-O OpenFlow13 add-flow**, users can add Flow Entries into a Flow Table. We also use this command to connect input Port to output Port in a Switch. After completing the transmission direction setting for an input port and an output port, we use the command **sh ovs-ofctl-O OpenFlow13 dump-flows s1** in Mininet to check to see whether the Flow Entries have been successfully added to the corresponding Flow Table or not. Figure 2 shows an example of Flow-table dump.

The command that adds Meter Entries to a Meter Band is **ovs-ofctl-O OpenFlow13 add-meter** to control the output flow for a Port. With **sh ovs-ofctl-O OpenFlow13 dump-meters**, we can check to see whether the Meter Entries have been successfully added to the Meter Band or not.

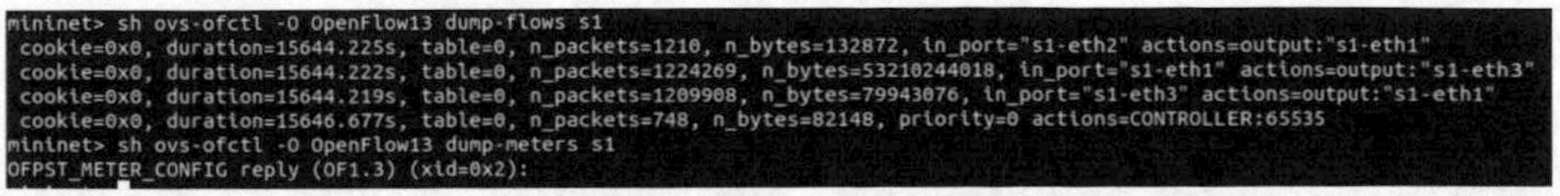

```
mininet> sh ovs-ofctl -O OpenFlow13 dump-flows s1
 cookie=0x0, duration=15644.225s, table=0, n_packets=1210, n_bytes=132872, in_port="s1-eth2" actions=output:"s1-eth1"
 cookie=0x0, duration=15644.222s, table=0, n_packets=1224269, n_bytes=53210244018, in_port="s1-eth1" actions=output:"s1-eth3"
 cookie=0x0, duration=15644.219s, table=0, n_packets=1209908, n_bytes=79943076, in_port="s1-eth3" actions=output:"s1-eth1"
 cookie=0x0, duration=15646.677s, table=0, n_packets=748, n_bytes=82148, priority=0 actions=CONTROLLER:65535
mininet> sh ovs-ofctl -O OpenFlow13 dump-meters s1
OFPST_METER_CONFIG reply (OF1.3) (xid=0x2):
```

Fig. 2. Flow table setup

The command **ovs-ofctl-O OpenFlow13 add-group** can classify packets sent from the same host to prevent packets of the same type from being interleaved and different type from mixing together, i.e., to avoid causing transmission errors along the packet transmission paths.

The command used to add a Queue to the Output Port of a Switch is the **ovs-vsctl set Port [Switch-eth] qos = @newqos**. The upper and lower limits of transmission bandwidth for different Queues are also set. With the **sh ovs-vsctl – get Port [Switch-eth] qos** in Mininet, we can check to see whether the packets are sent to the correct Queue or not.

4   Simulations and Experimental

We first create a network virtualization platform (see Fig. 1). Then Docker is employed to generate Containers as Hosts. At last, packets are transmitted via the shortest/best paths established. Of course, the Flow Entries are setup and queues are created at all output ports of a switch. The simulation software and versions are listed in Table 1.

Table 1.   The simulation environment

Software	version
operating system	Ubuntu–18.04
Docker version	20.10.10
Mininet	2.3.0.dev6
OpenVSwitch	2.9.8
Python	3.6.0

4.1   Topology Map

When a packet P enters Switch 1 (denoted by S1) through port 1, it will be forwarded to an Output port, e.g., port 2 or port 3. However, ports 2 and 3 should not connected to prevent them from forming a loop. Then, P will be sent to Host 2 via be the shortest/best path. Next, we create a file to define the network topology, setting up 9 switches and 2 hosts, and establishing basic links between Hosts and Switches.

4.2 Shortest/Best Path and Settings

After Flow entries are setup, if anyone existing in the table is longer than its set time, this entry will remove. We also develop a weight judgment algorithm. Before a packet P enters the switch, it is judged to see which criteria that P will be considered, i.e., hop count, delay or bandwidth.

Our system also monitors the transmission delay of a packet. If the delay is very longer than those of other packets, the transmission path will be changed to the 2nd shortest/best path and then transmit the corresponding information to the Flow Table to set up the new path. After that, the following packets will travel via this path.

Figure 3 shows the path from Host1 to Host2, beginning at Host1, passing through S1, S2, S5, S8 and S9 in order, and at last reaching Host2.

```
[PATH]10.0.0.1<-->10.0.0.2: [1, 2, 5, 8, 9]
```

Fig. 3. Path from h1 to h2

```
64 bytes from 10.0.0.2: icmp_seq=55 ttl=64 time=0.097 ms
64 bytes from 10.0.0.2: icmp_seq=56 ttl=64 time=0.098 ms
64 bytes from 10.0.0.2: icmp_seq=57 ttl=64 time=0.092 ms
64 bytes from 10.0.0.2: icmp_seq=58 ttl=64 time=0.053 ms
64 bytes from 10.0.0.2: icmp_seq=59 ttl=64 time=0.083 ms
64 bytes from 10.0.0.2: icmp_seq=60 ttl=64 time=26.4 ms
64 bytes from 10.0.0.2: icmp_seq=61 ttl=64 time=0.629 ms
64 bytes from 10.0.0.2: icmp_seq=62 ttl=64 time=0.043 ms
64 bytes from 10.0.0.2: icmp_seq=63 ttl=64 time=0.080 ms
```

Fig. 4. All packet information

Figure 4 shows that packets are continuously transmitted from Host1 to Host2. The field, named icmp_seq, illustrates the number of packets transmitted after initiating a continuous ping command. As shown, when the 60th packet is transmitted, the delay is long due to traffic jam. The shortest path will transfer to the new path. Figure 5 shows that packets pass through S1, S3, S6, S8 and S9, rather than going through those shown in Fig. 3.

```
[PATH]10.0.0.1<-->10.0.0.2: [1, 3, 6, 8, 9]
```

Fig. 5. Shortest path

Figure 6 shows that the system information:

(1) rx-pkts: Number of packets received
(2) rx-bytes: number of bits received

(3) rx-error: Number of receiving errors
(4) tx-pkts: Number of packets sent
(5) tx-bytes: Number of bits sent
(6) tx-error: number of error at sending side (in packets)
(7) port-speed (B/s): the speed at which the port transmits packets

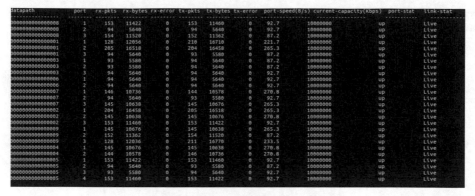

Fig. 6. Resource Monitoring

5 Conclusions and Future Studies

In this study, we first establish a specified network topology and then send packets to designated hosts via the shortest/best path, going through multiple interconnected switches, to reach another specified host. We set up Flow Entries following the shortest/best path to forward the packets to their destinations.

We control how each type of packet is classified upon entering a switch and organize the transmission order. Since uULLC requires real-time performance, its priority is the highest, followed by proportional distribution of mMTC and eMBB traffic. We believe that the transmission of 5G network packets can reduce the impact on transmission congestion and delays.

In the future, we will simplify the network topology for easier observation of packet transmission. We also want to simulate packet transmission between hosts on actual switches and improve the packet classification process. Actual switches have many hosts. We will simulate situations that may occur on actual switches to anticipate and correct potential errors in advance. These constitute our future studies.

References

1. Chen, J., Dezfouli, B.: Predictable bandwidth slicing with Open vSwitch. In: IEEE Global Communications Conference (GLOBECOM), December 2021, Madrid, Spain, pp. 1–6 (2021)
2. Rastogi, A., Bais, A.: Comparative analysis of software defined networking. (SDN) controllers—In terms of traffic handling capabilities. In: 19th International Multi-Topic Conference (INMIC), December 2016, pp. 1–6 (2016)

3. Tivig, P.T., et al.: Slicing 5G core network based on the Ryu SDN controller for everything as a service. In: 26th International Symposium on Wireless Personal Multimedia Communications (WPMC), November 2023, pp. 1–7 (2023)
4. Zhou, X., et al.: Network slicing as a service: enabling enterprises' own software-defined cellular networks. IEEE Commun. Mag. **54**(7), 146–153 (2016)

Base-Station Resource Allocation Based on Frame/Sub-Frame

Yu-Han Chen[1], Fang-Yie Leu[1(✉)], and Heru Susanto[2]

[1] Computer Science Department, Tunghai University, Taichung City, Taiwan
Leufy@thu.edu.tw
[2] Computational Science, The Indonesian Institute of Science, Jakarta, Indonesia

Abstract. In recent years, with the rapid development of network technology, human beings request high quality of wireless services. The demands for self-driving cars, remote medical surgeries and real-time ward monitoring have become urgent to meet the needs of our daily lives. But with a large number of wireless connections connecting to base stations for serving Internet of things (IOTs) and transporting multimedia programs of AR/VR to users, Base stations need faster transmission speeds and more efficient resource usage. Based on the 5G (5th Generation Mobile Networks, 5G) network specifications, scholars and experts have been studying how to fully utilize base station resources and make more efficient use of existing technologies, because due to the narrow bandwidth of wireless links, the network services of base stations have always been one of the biggest bottlenecks in wireless communications. Therefore, this study intends to propose a novel base station resource allocation scheme, which allocates resources to users within a certain period of time and ensures that the network service quality meets the requirements of users. The purpose is to make the operation of the base station comply with Numerology's specifications for Sub-carriers Spacing (SCS) and time frame/sub-frame.

Keywords: 5G · simulator · network · modulation · SRA

1 Introduction

Network communication technology has advanced rapidly in recent years, and the fifth-generation mobile communications (5G) [1] request low latency, high bandwidth and low-cost scheduling. However, there is still a lot of rooms for the improvement of 5G core technology at current stage, Typical examples are the quality of instant messaging [2] and resource scheduling [3], as well as the scheduling of network slices that carry application packets [4]. Network slicing is the technique that virtualizes physical network components into a number of virtual ones with which to serve users in a software-defined network (SDN) environment. However, the requests for the establishment of network slicing in 5G network basically are issued by UE. The corresponding resource scheduling efficiency and effectiveness needs to be further studied. Thus, in this study we propose a new Scheduling and Resource Allocation (SRA) algorithm [5], that effectively support

L. Barolli (Ed.): IMIS 2024, LNDECT 214, pp. 379–385, 2024.
https://doi.org/10.1007/978-3-031-64766-6_38

resource scheduling for network slicing in base stations, so as to reduce telecom operators' operation on costs and enhance network transmission performance, particularly at base stations.

The method adopted is to configure the resource blocks (RB) and the actual configuration on time length of a base station to increase the utilization of Resource element (RE), so that the operations of the base station can meet the transmission requirements of the three types of packets [6, 7]: enhanced Mobile Broadband (eMBB), Massive Machine Type Communications (mMTC) and ultra-reliable and Low Latency Communications (uRLLC). This study will also compare the transmission throughputs, packet loss rates, spectrum efficiency, etc. with those of other algorithms to highlight the advantages of the proposed method.

The rest of this article is organized as follows. Section 2 briefly describe related work of this study. Section 3 presents the proposed system model. Simulations and their results are stated and discussed in Sect. 4, respectively Sect. 5 concludes this study and addresses our future studies.

2   Related Work

Current specifications of the 5G network are released by 3GPP (3rd Generation Partnership Project), providing network services in a low-latency, low-cost, and high-throughput manner to meet the needs of a variety of applications. The main application scenarios of 5G are as follows [6, 7]:

(1) eMBB: which serves users with a high-bandwidth environment. High-resolution live. Broadcast is an example. This scenario often transmit volume of data.
(2) uRLLC: which supports low-latency and high-reliability application environments, such as remote medical surgery and drone online remote control.
(3) mMTC: which offers transmission services in a massive-connection manner, especially in the environments controlled by connecting sensors. Smart cities and smart buildings are two typical examples.

Nasralla in [8] used Hybrid for downlink Scheduling and differentiated the differences between QoS aware and QoS unaware.

Liu *et al.* [9] utilized Q-learning approach to strain downlinks scheduler, and also established an novel SRA algorithm to enhance downlink-bandwidth allocation.

2.1   SRA

In traditional 5G, Channel State Information (CSI) [10] conveys communication quality information for the channel established between BS and UE. CSI includes system control information, Channel Quality Information (CQI) and Precoding Matrix Indicator (PMI). After receiving CQI from UE, BS chooses an appropriate modulation and coding scheme (MCS) for the UE, based on the value of Adaptive Modulation and Coding (AMC) which is a field of a CQI Table. Then, the SRA effectively allocates appropriate radio resources to the UE, letting this UE and BS itself achieve better system performance. Our base station and UE architecture is shown in Fig. 1.

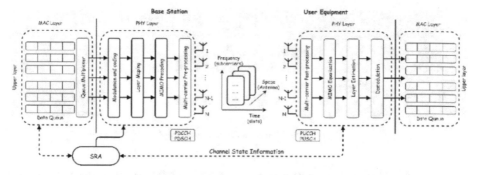

Fig. 1. Scheduling and Resource Allocation SRA [5]

2.2 Packet Scheduling Algorithm Analysis

Traditional SRA algorithms can be classified into QoS aware [4] and QoS unaware [4]. Basically. QoS unware has three typical schemes: Proportional Fairness (PF) [11–13], Round Robin (RR) [9] and Max CQI [3, 10]. QoS aware [12] generally has following schemes: Delay Aware [13], Queue Aware [9], Target Bit-rate Aware [9] and Hybrid [9].

3 System Model

In this study, we propose an QoS unaware approach focusing on resource blocks and time line scheduling. Dynamic adjustment on bandwidth allocated to UEs improves the efficiency of resource usage. For this, we adopt 1 s as the unit, i.e., 1000 ms, and 10 ms frame as the allocation cycle to ensure that sufficient resources are allocated within 1 Frame, aiming to fulfill the UE's transmission needs. We perform a total of three different resource allocation methods under the assumptions that UE requests K Mbps transmission speeds, the BS involved is installed with 20M Hz bandwidth and the modulation methods for all RBs are the same, i.e., m bits/Hz, taking Numerology 1 as an example (Figs. 2 and 3).

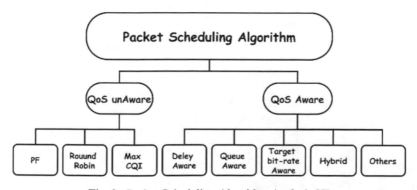

Fig. 2. Packet Scheduling Algorithm Analysis [5]

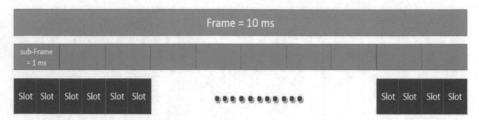

Fig. 3. Frame, sub-frame and slots

Method 1. During data transmission in a frame, an UE is evenly allocated the same number of RBs for each of the 10 sub-Frames, i.e., for each sub-frame, BS fairly allocates n RBs to the UE, where $n = \lceil$ K Mbps/1000 s, /(180 KHz * m bits/Hz)$\rceil$.

Method 2. For the first q sub-frames, BS allocates the bandwidth that UE requests in the underlying frame, i.e., K/100 Mbps, where $q = \lceil$(K/100)/(100 * 180 kHz * m bits/Hz)$\rceil$ where 100 in the denominator represents 100 RBs per base station, If $q >$ (K/100/(180 * 100 * m)), meaning that 100RBs in the first q-1 sub-frames (i.e., 0, 1…q − 2) are all allocated to this UE and in the $q − 1^{\text{th}}$ sub-frame, K/100/(180 * m) − (q − 1) * 100 RBs will be allocated to this UE. This method is suitable for use in uRLLC. That is, the data required by this UE is directly transmitted at the beginning of a frame.

Method 3. Assuming that before the end of this frame, the UE has not been allocated RBs, all resources should be given to the ue in the last q sub-frames to transmit its transmission volume.

Through these three methods, all UEs can complete the transmission requirements of K/100 Mbps in 10 sub-Frames. However, in this study, we employ method 1 only, attempting to see what may happen when traffic is evenly delivered in 10 sub-frames for UEs.

```
resource allocation based on UE requests
m = modulation;
K = ue's requesting bandwidth in Mbps;
ueRB = (K/100)/(180*m) in Kbps
/* the number of RBs in integer that BS needs to allocate to UE in a frame */
n = celi(ucRB); /*no. of RBsrequested by UE*/
q = celi(n/100);
/* q sub-frame :the minimum no. of sub-frames that BS needs to allocate RBs that UE
regquests in the underlying frame to UE*/
while ueRB do
    case 1 : /* BS evenly allocates celi(n/10) RBs to UE for every sub-frame */
        for sub-frame 0 to 9 do
            if celi(ueRB/10)>100 then
                print(error);
                break ;
            allocation (celi(ueRB/10))

    end
    case 2 : /*allocation at the beginning */
        for sub-frame 0 to q-1 do
            if ueRB ≥ 100 then allocation(100)
                ucRB = ucRB - 100
                / * allocate 100 RBs to UEs in underlying sub - frame * /;
            if 0<ueRB <100 then allocation(ueRB);
    end
    case 3: /*allocation at the end of a frame*/
        for sub-frame 10-q to 9 do
            if ueRB ≥ 100 then allocation(100)
                ucRB = ucRB - 100;
            if 0<ueRB <100 then allocation(ueRB);
    end
```

4 Experimental Results

In this section, the 5G-air-simulaation is employed as the simulation tool of this experiment. Two types of packets, VIDEO and VoIP are used to simulate eMBB and uRLLC in 5G, respectively. The number of UE utilized ranges between 1 and 21. The compared schemes include Framing Level Scheduler (FLS), Proportional Fai (PF) and Maximum-Largest Weighted Delay First (M-LWDF). Test metrics are Delays, Throughputs and Packet Loss rates.

From the Figs. 4 and 5, it can be seen that the delays of VoIP and Video are not very longer than those of MLWDF and FLS since the two approaches send real-time packet as early as possible and VoIP packets are real-time ones. PF does not consider QoS, the delays are long.

As shown in Figs. 6 and 7, the throughputs are a little higher than those of MLWDF and FLS, particularly on Video, since traffic of video is heavy, meaning that the proposed approach is feasible and also suitable for real-time and heavy traffic.

As illustrated in Figs. 8 and 9, the packet loss rates are higher than those of MLWDF and FLS since our approach sends data evenly in the 10 sub-frames. Sometimes, some real-time packets are dropped due to time out on the way toward their destinations.

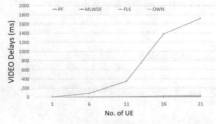

Fig. 4. Video delays

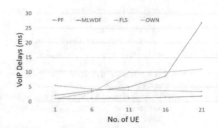

Fig. 5. VoIP delays

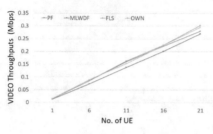

Fig. 6. Video throughputs

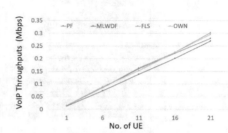

Fig. 7. VoIP throughputs

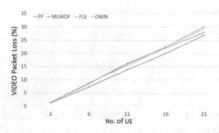

Fig. 8. Video packet loss rates

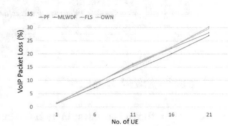

Fig. 9. VoIP packet loss rates

5 Conclusions and Future Studies

For the base station scheduling proposed in this study, we can conclude that by limiting transmission based on the unit of 10-ms frame, it is a feasible to evenly deliver data send by UE to the destination, particularly for data sent by constant bit rates. Actually, the proposed algorithm can also be applied to other applications of 5G systems, like streaming video or voice data, to make sure that the transmission throughputs meet the requirements of an UE-requested slice.

More improvements can be made for the algorithm and UE allocation, which will also have a further impact on the current 5G base station schedule. For example, increase priority, optimizing packet scheduling and increase modulation data. We will also derive the reliability model and behavior model so that users can realize the system's behaviors and reliability before using them. These constitute our future studies.

References

1. Dangi, R., Lalwani, P., Choudhary, G., You, I., Pau, G.: Study and investigation on 5G technology: a systematic review. Sensors **22**(1), 26 (2022). https://doi.org/10.3390/s22010026
2. Choi, J.G., Bahk, S.: Cell-throughput analysis of the proportional fair scheduler in the single-cell environment. IEEE Trans. Veh. Technol. **56**(2), 766–778 (2007)
3. Sadiq, B., Madan, R., Sampath, A.: Downlink scheduling for multiclass traffic in LTE. EURASIP J. Wirel. Commun. Netw. **2009**, 1–18 (2009). 510617. https://doi.org/10.1155/2009/510617
4. Ramli, H.A.M., Sandrasegaran, K., Basukala, R., Patachaianand, R., Afrin, T.S.: Video streaming performance under well-known packet scheduling algorithms. Int. J. Wirel. Mob. Netw. (IJWMN) **3**(1), 25–38 (2011)
5. Ramli, H., Basukala, R., Sandrasegaran, K., Patachaianand, R.: Performance of well known packet scheduling algorithms in the downlink 3GPP LTE system. In: IEEE Malaysia International Conference on Communications (MICC), Kuala Lumpur, Malaysia, December 2009, pp. 815–820 (2009)
6. Popovski, P., Ttrillingsgard, K.F., Simeone, O., Durisi, G.: 5G wireless network slicing for eMBB, uRLLC, and mMTC: a communication-theoretic view. IEEE Access **6**, 55765–55779 (2018)
7. Patel, W., Tripathy, P.: eMBB, URLLC, and mMTC 5G wireless network slicing: a communication-theoretic view. In: International Conference on Artificial Intelligence and Smart Communication (AISC), April 2023, pp. 1291–1296 (2023)
8. Nasralla, M.M.: A hybrid downlink scheduling approach for multi-traffic classes in LTE wireless systems. IEEE Access **8**, 82173–88218 (2020)
9. Liu, J.C., Susanto, H., Huang, C.J., Tsai, K.L., Leu, F.Y.: A Q-learning-based downlink scheduling in 5G systems. Wirel. Netw. 1–22 (2023)
10. ShareTechnote. https://www.sharetechnote.com/html/5G/5G_CSI_Report.html
11. Basukala, R., Ramli, H.M., Sandrasegaran, K.: Performance analysis of EXP/PF and M-LWDF in downlink 3GPP LTE system. In: IEEE First International Conference on Internet, Kathmandu, Nepal, November 2009, pp. 1291–1296 (2009)
12. Ali, A.H., Nazir, M.: Radio resource management with QoS guarantees for LTE-a systems: a review focused on employing the multi-objective optimization techniques. Telecommun. Syst. **67**(2), 349–365 (2018)
13. Yao, Q., Huang, A., Shan, H., Quek, T.Q.S., Wang, W.: Delay-aware wireless powered communication networks—Energy balancing and optimization. IEEE Trans. Wirel. Commun. **15**(8), 5272–5286 (2016)

Discussion on the Labor Shortage Problem in Taiwan's Construction Industry

Kuei-Yuan Wang[1], Ying-Li Lin[1]($\boxtimes$), Chien-Kuo Han[2], and Wen-De Lin[3]

[1] Department of Finance, Asia University, Taichung, Taiwan
{yuarn,yllin}@asia.edu.tw
[2] Department of Food Nutrition and Healthy Biotechnology, Asia University, Taichung, Taiwan
jackhan@asia.edu.tw
[3] Yanqun Construction Co., Ltd., Taichung, Taiwan

Abstract. Taiwan's rapid economic development have caused the dilemma of the shortage of various skilled workers in the construction industry. This study discusses and analyzes the problem of labor shortage in Taiwan's construction industry. Based on the interview contents and SWOT strategy analysis, this study used TOWS Matrix analysis to provide some suggestions to the government and managers.

1 Introduction

For a long time, the market demand of the construction industry has increased in Taiwan. Labor shortage has been a serious problem in various types of work in Taiwan's construction industry. This study will explore the problem of labor shortages, follows are the purposes of this study: (1) Understand the current status and development of the construction industry. (2) Understand the causes of labor shortage in the construction industry. (3) Generate feasible strategies through SWOT analysis and TOWS Matrix analysis.

According to Article 6 of the Construction Industry Law, the construction industry is divided into comprehensive construction enterprises, specialized construction enterprises and civil engineering contractors (National Regulations Database, 2022). Lee [1] pointed out that the construction process of the construction industry requires highly integrated management and control. The construction industry is an industry that requires a high degree of integration of capital, technology, and manpower. Different from other industries, it is often sold first and then produced later in the construction industry. In addition, the construction industry is often expensive, diverse, and immovable. Therefore, these factors make the characteristics of the construction industry different from general traditional industries. The characteristics of the construction industry are summarized as follows: (1) closely integrated with surrounding related industries; (2) linked to government policies and major public construction; (3) products are customized and immovable; (4) obvious regional characteristics; (5) business model combines manufacturing control and integrated management; (6) large groups, small and medium-sized enterprises and family enterprises coexist in the market; (7) high-risk industries; (8)

L. Barolli (Ed.): IMIS 2024, LNDECT 214, pp. 386–392, 2024.
https://doi.org/10.1007/978-3-031-64766-6_39

affected by political, economic and price fluctuations; (9) the sources of construction workers and materials are irregularly; (10) labor-intensive and capital-intensive industries; (11) skilled workers are hired on contract, resulting in the inability to accumulate experience; (12) project-based organizations lack the concept of sustainable management; (13) production process and cost scale change greatly; (14) vulnerable to external environmental factors; (15) focus on project performance and core technology; (16) high degree of resource integration.

Lin [2] pointed out that without various professional and technical workers in each project, even with perfect design and advanced construction methods, the project still cannot be completed. In Taiwan's construction industry, skilled workers have generally been older in recent years, and young people are unwilling to engage in the construction industry, resulting in widespread labor shortages and disconnection problems in Taiwan's construction industry. This resulted in the inability to inherit skills or immature professional technologies, further leading to poor project quality.

Lin [3] pointed out that although the construction industry is high-income, it still lacks of labor force. The reason is related to the government's inaccurate economic construction planning and the young that favors easy work. Lin [3] summarized the reasons for the shortage of workers in Taiwan's construction industry as follows: (1) The US-China trade war has caused a large number of mainland Taiwanese businessmen to return to Taiwan to build factories. The increased demand for factory construction has made the construction market, which was already short of workers, even more serious. (2) The physical workload of technical workers in the construction industry is heavy and the rate of occupational accidents is high, resulting in young people being unwilling to work in the construction industry. (3) Older workers are gradually withering away from the workplace. Each year, a considerable proportion of technically proficient people quit the workplace. Since the number of new employees has not increased, and older skilled workers have withdrawn from the construction industry year by year. The shortage of workers in the construction workplace has become increasingly serious. The productivity of the overall construction industry has declined year by year. (4) The government restricts the introduction of foreign workers. In order to protect employment opportunities for Taiwanese people, Taiwan has always adopted policies that highly protect domestic workers. (5) The education level of young people in Taiwan has generally improved. Taiwanese parents are generally unwilling to let their children engage in the construction industry to learn difficult construction techniques, which also blocks the source of construction industry workers. (6) The social status of workers in the construction industry is low. Traditional society does not value workers highly. This aspect requires the government to vigorously promote policies and promote the professional image of workers to enhance the social status of skilled workers in the construction industry.

It can be seen from the above that Taiwan's construction industry is facing a serious labor shortage problem. This issue deserves attention and discussion, and further solutions to the problem can be found.

2   Methodology

2.1   In-Depth Interviews

This study used in-depth interviews to collect data. Kahn and Cannell [4] described the interview as "a purposeful conversational conversation." Lin [2] pointed out that in-depth interviews are a kind of conversational behavior in which the researcher visited the researched and conducts conversations and inquiries. The advantages of the in-depth interview method include that interviews are more suitable for complex situations, help to collect deeper information, information can be supplemented, and questions can be explained and clarified, etc.. Interviews can be widely used and the identity of the interviewee can be confirmed and the response rate is relatively high. The disadvantages of the in-depth interview method include that interviews are time-consuming and expensive, instrument errors are prone to occur, difficult to control the quality of respondents, the quality of data is also different when using multiple interviewers, the problem of researcher bias, and the problem of interviewer bias, and cannot answer anonymously, etc..

2.2   Samples

This study conducted in-depth interviews between April 1 and April 30, 2022. There were five interviewees in this study, including construction industry managers, steel bar lashing workers, formwork erection workers, concrete workers, plumbers, etc. This study recommended these four key types of work mainly because they have a higher proportion of costs in the entire project, a higher degree of construction risk, a longer proportion of the total project period, and are important construction items. In addition, these types of jobs are more difficult to learn and develop skills.

2.3   Interview Outline

The interview outline for this study were as follows:

a. What do you think is the reason for the labor shortage problem in Taiwan's construction industry?
b. What do you think is the impact of Taiwan's current shortage of workers in the construction industry?
c. What do you think are the "advantages" of Taiwan's construction industry?
d. What do you think are the "disadvantages" of Taiwan's construction industry?
e. What do you think are the "opportunities" of Taiwan's construction industry?
f. What do you think are the "threats" of Taiwan's construction industry?

3   Research Results

3.1   Reasons and Impacts of Labor Shortages in Taiwan's Construction Industry

This study used in-depth interviews to summarize the reasons why interviewees considered that the construction industry faced labor shortage problems as follows: (1) reliance on a large number of labor forces; (2) the universalization of university education changes the values of mass employment; (3) young people are unwilling to engage in construction industry; (4) the age of technical workers in the construction industry is generally higher; (5) physically demanding and high-risk industries; (7) professional and technical training takes a long time; (8) the public does not pay attention to the engineering technician certificate system; (9) The management system is imperfect.

From what the interviewees mentioned, it could be seen that the impacts of labor shortage on Taiwan's construction industry: (1) Quality is not easy to control; (2) Project progress lags behind; (3) Construction costs increases; (4) Inability to systematize; (5) Work safety incidents and occupational disasters occur frequently; (6) Temporary workers, semi-apprentices or illegal workers are hired to make up for the shortage of Taiwan's construction industry; (7) The shortage of workers causes workers to continuously increase their wages, causing construction costs to continue to be increased High; (8) The project cannot perform the contract and is difficult to manage; (9) Breach of contract and loss of reputation, or even poor management and bankruptcy.

3.2 SWOT Analysis

a. Advantages of Taiwan's Construction Industry
From the opinions of the interviewees, it could be seen that the advantages possessed by Taiwan's construction industry include: (1) Abundant material resources, market resources and information; (2) Rich construction methods, technologies, machinery and equipment, and engineering experience; (3) Complete capabilities management system and information management; (4) Sound financial capabilities; (5) Strong integration capabilities.

b. Disadvantages of Taiwan's Construction Industry
From the opinions of the interviewees, it could be seen from the experience of the interviewees that the disadvantages of Taiwan's construction industry include: (1) The outflow of talents and the difficulty in recruiting new personnel; (2) Limited funds, which restrict the capital turnover and expansion of business scale of professional and technical construction operators; (3) Business contacts or the development of new customers must rely on the operator's personal connections; (4) New construction methods and new materials involve patent rights, which affects the construction industry's bidding qualifications and bargaining space; (5) High-risk working environment, it is easy to have a huge impact on the project costs of the construction industry.

c. Opportunities of Taiwan's Construction Industry
From the interviewees' observations, it can be seen that the opportunities of Taiwan's construction industry include: (1) Taiwan's stable economic growth provides opportunities for sustainable development of the construction industry; (2) The construction of major government public projects is promoted through private investment, creating investment opportunities worth nearly NT$1 trillion; (3) The government's opening-up policy has created a broader overseas market and provided more opportunities for Taiwan's construction industry.

d. Threats of Taiwan's Construction Industry
From the interview contents of the interviewees, it can be seen that the threats faced by

Taiwan's construction industry include: (1) The construction industry is greatly affected by economic prosperity, government policies, etc.; (2) After joining the WTO, the domestic market must be open and eliminate non-tariff trade. It will make the domestic construction market more competitive; (3) The rising awareness of environmental protection and labor safety among Taiwanese people has resulted in increasingly more and strict restrictions on environmental labor laws, which has increased construction costs for construction companies.

e. Summary of SWOT Analysis

This study compiled the SWOT analysis results of the construction industry based on the interview content. As shown in Table 1:

Table 1. SWOT Analysis Results

Strength	Weakness
S1: Abundant material resources, market resources and information S2: Rich construction methods, technologies, machinery and equipment, and engineering experience S3: Perfect capability management system and information management S4: Sound financial capabilities S5: Strong integration capabilities	W1: Shortage of professional and technical workers W2: Limited funds W3: Difficulty in obtaining business W4: Patent rights for new construction methods and building materials W5: High-risk industries
Opportunity	Threat
O1: Economic boom and growth O2: Major government public construction projects O3: Broader overseas market	T1: Economic climate fluctuates greatly T2: A highly competitive market T3: Environmental labor law restrictions

3.3   TOWS Matrix Analysis

This study hoped to help the construction industry find effective suggestions and strategies through TOWS matrix analysis. The analysis results were as follows:

a. SO Strategy:

SO1: Strengthen original business areas and expand business scale (S1, S2, S3, S4, S5, O1)

SO2: Establish image and reputation, increase business opportunities (S2, S3, S4, S5, O1)

SO3: Expand business scope and expand business areas (S1, S2, S3, S4, S5, O2)

SO4: Develop overseas markets and increase business opportunities (S2, S3, S4, S5, O3)

b. **ST Strategy:**

ST1: Strengthen the firm's constitution and secure business opportunities (S2, S3, S4, S5, T1)

ST2: Increase in own capital to strengthen competitiveness (S4, T2)

ST3: Obtain ISO international certification and diversify operations to increase competitiveness (S2, S3, S4, S5, T2)

ST4: Use alliances to enhance competitiveness (S3, S4, S5, T2)

ST5: Implement environmental protection and industrial safety related procedures to avoid environmental protection and industrial safety incidents (S3, S5, T3)

c. **WO Strategy:**

WO1: Establish human resources systems and establish cooperative relationships to strengthen competitiveness (W1, O2, O3)

WO2: Expand business scale through alliances (W2, O1, O2, O3)

WO3: Maintain good relationships to secure business sources (W3, O1, O3)

WO4: Improve productivity through joint contracting and introduction of new construction methods (W4, O2, O3)

WO5: Implement international quality certification and industrial safety procedures to adapt to future markets (W5, O2, O3)

d. **WT Strategy:**

WT1: Market segmentation, strategic alliances and diversified operations (W2, W3, T2)

WT2: Strategic alliance, introduction of new construction methods, and cost reduction to increase competitiveness (W4, T2)

WT3: Strengthen environmental protection and labor safety management and organize lectures and training (W5, T3)

4 Conclusions and Suggestions

This study used in-depth interviews to explore the labor shortage in Taiwan's construction industry. This study used SWOT analysis and TOWS matrix analysis to make the following recommendations for Taiwan's construction industry:

a. Strengthen original business areas and expand business scale
b. Establish image and reputation, increase business opportunities
c. Expand business scope and expand business areas
d. Develop overseas markets and increase business opportunities
e. Strengthen the company's constitution and secure business opportunities
f. Increase in own capital to strengthen competitiveness
g. Obtain ISO international certification and diversify operations to increase competitiveness
h. Use alliances and alliances to enhance competitiveness
i. Implement environmental protection and industrial safety related procedures to avoid environmental protection and industrial safety incidents

j. Establish human resources systems and establish cooperative relationships to strengthen competitiveness
k. Expand business scale through alliances
l. Maintain good relationships to secure business sources
m. Improve productivity through joint contracting and introduction of new construction methods
n. Implement international quality certification and industrial safety procedures to adapt to future markets
o. Market segmentation, strategic alliances and diversified operations
p. Strategic alliance, introduction of new construction methods, and cost reduction to increase competitiveness
q. Strengthen environmental protection and labor safety management and organize lectures and training

This study hoped to make the following suggestions to the authorities: (a) It is recommended to strengthen the promotion of industry-university cooperation and the acquisition of professional licenses. Then, students are encouraged to enter the construction industry after graduation and obtain certificates to solve the labor shortage dilemma faced by Taiwan's construction industry and improve the skills of professional technicians. (b) It is hoped that the government will re-examine laws and policies related to the construction industry so that professional and technical workers could receive corresponding benefits and protections, so as to increase the motivation of young people to invest in the construction industry.

Acknowledgments. Authors thank anonymous reviewers for their valuable suggestions.

References

1. Lee, H.Y.: Overview of construction industry development in 2012. In: CNCCU-SINYI Research Center for Real Estate (ed.) Taiwan Real Estate Yearbook 2013, pp. 311–317. CNCCU-SINYI Research Center for Real Estate (2013)
2. Lin, S.T.: Research on the shortage of professional and technical workers in Taiwan's construction industry [Unpublished master dissertation]. National Central University (2014)
3. Lin, C.N.: How to solve the problem of labor shortage in the construction industry (I), Engineers Times, 1266 (2021). http://www.twce.org.tw
4. Kahn, R.L., Cannell, C.F.: The Dynamics of Interviewing: Theory, Technique, and Cases. Wiley, Hoboken (1957)

The Impact of Investor Sentiment on Abnormal Returns and Abnormal Volumes - The Study of ESG Event

Yung-Shun Tsaia, Shyh-Weir Tzang(✉), Chun-Ping Chang, and Ruei-Tsz Chuang

Department of Finance, Asia University, Taichung, Taiwan
swtzang@gmail.com

Abstract. In the securities market, the information of individual investors is relatively incomplete, and it is easy to generate abnormal returns and volume due to emotional fluctuations after the impact of events shock. This paper uses the component stocks of Yuanta Taiwan ESG ETF as observation samples to examine the impact of investor sentiment and ESG events on abnormal returns and volume. The empirical results of this study are as follows: (1) ETF component stock have abnormal returns and abnormal volumes before and after ESG events (2) Investor sentiment is negative correlated with abnormal returns and abnormal volumes. (3) The impact of ESG events will decrease the abnormal returns and abnormal volume of stocks.

Keywords: abnormal returns · abnormal volume · ESG investor sentiment

1 Introduction

From the past to the present, we have found that the market is prone to abnormal phenomena caused by special conditions, which cannot be explained by traditional financial theory. Such as January effect, weekend effect and even weather effect, all have abnormal returns in the stock market. [40, 41] questioned the efficient market. He believed that irrational behavior will appear on investors, and this behavior occurs randomly and is more likely to proceed in the same direction, thus affecting market prices. As for retail investors, they are weak in information, and their investment behavior is easily affected by the quality of the news. [45] The quality of emotions will affect investors' judgment attitude towards things, especially when they are inclined to one side. Sometimes, you may ignore news from the other side or overreact to news about your current emotions. The proportion of retail investors in Taiwan has hit new highs in recent years, and now exceeds 70%. Therefore, in the Taiwan market, retail investors have an increasing influence on the stock market.

Nowadays, ESG thinking is increasingly popular in society, and many related studies also use ESG as a factor to explore. For example, [29] also used a quasi-institutional theory method to prove the positive value effect of ESG factors on companies. Even [34, 42] and [44] all agree and support the need to take ESG into consideration, because most

© The Author(s), under exclusive license to Springer Nature Switzerland AG 2024
L. Barolli (Ed.): IMIS 2024, LNDECT 214, pp. 393–411, 2024.
https://doi.org/10.1007/978-3-031-64766-6_40

ESG has an impact on financial performance, and when ESG is taken into consideration, this behavior is considered a Direct and responsible, it allows decision-makers to find appropriate financial information before investing. In Taiwan, Taiwan Index Company has also compiled the "Taiwan Sustainability Index" with FTSE International Co., Ltd. The sample used in this research is based on the Yuanta Taiwan ESG Sustainability ETF, and this ETF is the first ESG index in Taiwan. ETF, which tracks the above-mentioned "Taiwan Sustainability Index", in which the constituent stocks not only include basic performance indicators, but also pay more attention to the ESG performance of the company. Therefore, this study uses the Yuanta Taiwan ESG Perpetual ETF as a research sample, adds investor sentiment indicators, and uses the event study method to observe whether investor sentiment in the Taiwan market has abnormal return changes in response to ESG events.

Investors usually have excessive expectations for emerging hot industries and are more easily affected by emotions when making investment decisions. However, this aspect has not been discussed in the past literature. In addition, investors' investment decisions must have an impact on the market through actual transactions, so the observation of transaction volume can provide more useful information. Therefore, this study explores the impact of investor sentiment on abnormal stock returns and abnormal trading volumes - ESG event research. Based on the above, this study explores whether changes in investor sentiment caused by ESG events drive abnormal returns and trading volumes. The purpose is as follows: (1) Whether there are abnormal returns and abnormal trading volumes in ETF constituent stocks before and after ESG events. (2) Whether investor sentiment has an impact on abnormal returns and abnormal trading volume. (3) Whether the impact of ESG events will affect abnormal returns and abnormal trading volumes of stocks.

2 Literature Review

2.1 Investor Sentiment and Measurement

From a psychological perspective, they show that people are Making decisions will be unstable due to emotional fluctuations. Therefore, it is assumed that people are normal and cannot be completely rational as assumed in the past. Therefore, many studies are exploring the impact of emotions on investor behavior [1]. [20] broke down the so-called investor sentiment into two categories: direct and indirect. There is an important indicator to judge the above, which is the investor sentiment index. It is constructed by professional institutions using questionnaires to collect surveys from ordinary investors.

Extreme situations often occur in emotions. For example, [16] combined the representational bias and conservatism and other concepts in psychology with investor behavior and derived the results to find that investors will overreact or overreact when faced with differences in their perceptions of earnings announcements. Deficiency phenomenon. [22] found that overconfidence often occurs when investors have private information. If the investment result is successful, it will be attributed to their own ability. On the contrary, failure will be attributed to bad luck. This reason will also cause the market to overreaction or underreaction. The behavioral model established by [31] found that people tend to make investment decisions based on news and current stock market conditions, which is the main reason for underreaction or overreaction. [45] The quality of emotions will affect investors' judgment attitude towards things. Especially when they are inclined to one side, they may ignore the other side of the news or overreact to the news of the current emotions. Therefore, we want to know whether the mood is good or bad and whether there is any excessive phenomenon, which will cause changes in stock prices.

[14] found that in the investment transactions of the general public, the market turnover rate can be regarded as an indicator of their emotional judgment when investing. [4] used the following three types as sentiment indicators in the relationship between sentiment and Taiwan stock returns, first, the issuance ratio of new shares, second, the market turnover rate, and third, the balance ratio of bonds. To test whether emotions can explain market returns. In the end, it was found that the only variable that could significantly explain market returns was market turnover rate, and it was negatively correlated with the market stock price in the next period.

2.2 Investor Sentiment and Stock Price Returns

Many scholars have no consensus on whether returns are good predictors of investor sentiment indicators. In foreign literature, [37] and [25] both found that emotional indicators have a good effect on rewards; in domestic literature, [3, 5, 7] and [2] believes that sentiment indicators have predictive power on stock price returns. However, three other groups of researchers, [18, 21] and [20], believe that this prediction has no good effect.

[23] noisy trading model describes how they affect the equilibrium of stock prices. Because traders often overreact, causing stock prices to deviate from the true value in a short period of time, then arbitrageurs will appear. Unpredictable investor sentiment or risk aversion reduces transactions or even leaves the market, causing prices to continue to deviate. [9] used the principal component analysis method to form a sentiment indicator from many sentiment proxy variables, mainly discussing the impact of sentiment on stock returns. The results found that this sentiment indicator has a significant positive relationship with stock returns in the current period, but if the previous sentiment indicator is used, it has a significant negative relationship with stock returns in the current period.

[23] developed a model between irrational and rational investors and found that the former investors were the main cause of expected deviations from theoretical returns. [19] proposed the hypothesis that extreme optimism causes stock prices to be higher than their true value. The results confirmed that there will be lower returns after a period of high investor sentiment, so the market price will eventually return to its true value regardless of whether it is high or low. [13] studied how cross-sectional stock returns are affected by investor sentiment and found that differences in sentiment levels also affect stock returns based on company characteristics. [32] conducted research on the stock market crash in October 1987 and found that fundamental factors, including changes in future earnings or interest rates, could not explain the dramatic value changes at that time, but investor sentiment did. It has a systematic impact on market prices.

[8] used market capitalization weighting to calculate high-sentiment investment portfolio strategies during the long-short period to explore whether there can be stock price returns. The emotional variables used financing increase and turnover rate. The results found that when the stock price rises, it has a negative impact on the future. [10] studied the relationship between investor sentiment and stock returns in the electronics and financial industries. The result of the study is that sentiment will affect returns, and the impact will be to varying degrees depending on different industries. [17] studied investors' views on the price-to-earnings ratio in the stock market. Taking the United States as an example, he found that the expected return with a low price-to-earnings ratio was better than the expected return with a high price-to-earnings ratio. There were also expected market returns and The price-to-earnings ratio shows a negative relationship. [6] used the change in financing amount as a proxy variable for investor sentiment, and empirically found on the Taiwan stock market that stock price fluctuations have a significant impact on daily changes in financing amount.

2.3 ESG and Share Price Returns

[38] used an integrated decision-making model to explore the relationship between company value and ESG factors, and the results pointed out that ESG factors will bring positive value to the company. [29] also used a quasi-institutional theory method to prove the positive value effect of ESG factors on companies. In terms of news, [15] found that when a company releases good news, most stock prices will continue to rise, whereas when bad news is released, stock prices will fall. [28, 35] and [33] all pointed out that for companies that pollute the environment, the company's market price will suffer losses the day after the incident, and the negative news issued by the media will also Let the stock price return react, causing short-term adverse effects on the company.

Next is the impact of corporate social responsibility on stock prices. [30, 36] and [39] all mentioned that when companies fulfill their social responsibilities, It can not only motivate employees, improve reputation and corporate brand value, but also create corporate differentiation, thereby bringing in new customer groups. These are helpful for risk management and cost reduction, and can also increase the company's revenue to generate positive value and then increase the positive stock price. Reaction [26] pointed out that there is a strong correlation between companies that perform high social responsibilities and stock price returns. [12] research pointed out that when governments of various countries have formulated laws and regulations related to corporate social responsibility, they can establish expectations of corporate social responsibility. If there are social problems that need to be solved, corporate behavior will play an important role. [34, 42] and [44] all agree and support the need to consider ESG, because most ESG has an impact on financial performance, and when ESG is considered, this behavior is considered to be a direct and responsible behavior, allowing decision-makers to find appropriate financial information before investing. [24] found that after controlling for investment forms and risks, companies with high environmental ratings performed better than companies with low environmental ratings.

[11] proposed a model theory after research. The theory is about the relationship between corporate social responsibility and financial performance: although the initial implementation of corporate social responsibility will increase costs and reduce profits,

when the degree of corporate social responsibility increases, it will bring good reputation and image, which will also increase profits. [27] put forward three conclusions. First, companies that pay attention to socially responsible investments are likely to outperform other companies of the same type in the long run; second, companies that properly handle social and environmental issues the company will have good prospects and future. [24] studied those environmental factors are highly correlated with stock price performance, so they proposed stock price performance recommendations that take the environment into consideration. [43] pointed out that financial performance is no longer the only reference for investors, but social participation, corporate governance and environmental protection on ESG issues are the trend in sustainable investment portfolio management.

3 Research Methods

3.1 Research Hypothesis

The samples of this study were analyzed using the event analysis method on the date when the Yuanta Taiwan ESG Perpetual ETF was listed to test the three research hypotheses established in this article:

Hypothesis 1: ETF constituent stocks have abnormal returns and abnormal trading volumes before and after ESG events.
Hypothesis 2: There is a positive correlation between investor sentiment, abnormal returns and abnormal trading volume.
Hypothesis 3: ESG event impact will increase stock abnormal returns and abnormal trading volume.

3.2 Definition of Variables

Abnormal return (AR)
It refers to the difference between the expected return and the actual return of a security at a specific point in time.

$$ARE_{it} = RE_{it} - E(RE_{it}) \tag{1}$$

Among them, ARE_{it}, RE_{it} and $E(RE_{it})$ respectively represent the abnormal return rate, actual return rate and expected return rate of the security in period t.

Average Abnormal Return (AAR)
Since ARE_{it} is not only affected by expected events, but also affected by other events, the events must be reduced or the impact on the rate of return must be eliminated. The abnormal returns of all samples are averaged and the interference events are assumed to be independent of each other, so that when the sample is large enough, the interference events can offset each other or reduce their impact.

$$AARE_{it} = \frac{\sum_{i=1}^{N} ARE_{it}}{N} \tag{2}$$

Cumulative Average Abnormal Return (ACAR)

Because sometimes there is no way to know the exact date when the abnormal return reflects, only the range of time when the event occurs, so the cumulative average abnormal return rate from the beginning of the event to period t needs to be considered.

$$ACARE_t = \frac{1}{N} \sum_{i=1}^{N} \sum_{k=1}^{t} ARE_{ik} \tag{3}$$

3.3 Event Study Model

Risk adjustment model, OLS estimation method, assume that each security's return follows the following pattern:

$$R_{it} = \alpha_i + \beta_i R_{mt} + \varepsilon_{it} \tag{4}$$

Among them, R_{it} is the return rate of security i in the t-th period; βi is called the risk coefficient or regression coefficient, which indicates the sensitivity of its security return rate to the market index; Rmt is the return rate of the market index in the t-th period; and finally εit is the random error term. The expected return of security i at time t is:

$$E(R_{it}) = \hat{\alpha}_i + \hat{\beta}_i R_{mt} \tag{5}$$

Among them, $\hat{\alpha}_i$ and $\hat{\beta}_i$ are the estimates of the least squares method.

Hypothesis Test

In the event study method, if only the abnormal returns detected on individual securities cannot be concluded, the abnormal returns obtained during the event period should be averaged and accumulated, and statistical tests should be used to detect whether the average abnormal returns and cumulative abnormal returns are significantly different from zero. To test abnormal returns, we use the uncorrelated adjustment test method.

$$\hat{\sigma}^2_{AARE_t} = \frac{1}{N^2} \sum_{i=1}^{N} \frac{1}{T-p} \sum_{t=1}^{T} (ARE_{it} - \overline{ARE_i})^2 \tag{6}$$

where T is the length of the estimation period, p is the number of unknown parameters in the model, and N is the total number of securities. Under the null hypothesis, the test statistic of the average abnormal return rate in event period t is:

$$t_{AARE} = \frac{AARE_t}{\sqrt{\hat{\sigma}^2_{AARE_t}}} \sim N(0, 1) \tag{7}$$

Among them, statistics follow normal distribution. Let $Cov(ARE_{ip}, ARE_{ip}) = 0$, the test statistic of the accumulated average abnormal return is:

$$t_{ACARE} = \frac{ACARE_t}{\sqrt{\hat{\sigma}^2_{ACARE_t}}} \sim N(0, 1) \tag{8}$$

Among them, $\hat{\sigma}^2_{ACARE_t} = M\hat{\sigma}^2_{AARE_t}$, the number of periods from which the M event period starts to accumulate to the tth period.

Event Period

Use t to represent the t-th trading day after the event day, -t to represent the t-th trading day before the event day, and then take the listing day of Yuanta Taiwan ESG Perpetual ETF as the event day, and the event day to 120 days after the event day as the event day. Period, and the event day is period 0, and the event date -122 to -2 is the estimation period. This paper discusses the regression of investor sentiment, so the time after the event is 120 days.

3.4 Investor Sentiment Indicator

Margin Balance (BTER)

In the past literature, [4] found that the bond-equity ratio can also be used as a sentiment variable. When the securities lending is lower or the financing is higher, it is found that investors will be optimistic about the subsequent development. When the bond-equity balance ratio the bigger it is, the lower the market sentiment.

$$\text{BTER} = \text{average}[(\frac{\text{n1Margin balance}}{\text{n1Margin balance}} + \cdots + \frac{\text{n70Margin balance}}{\text{n70Margin balance}}) \times 100\%] \quad (9)$$

Rate of Price Spread (CR)

The market is mainly viewed based on the overall market. The calculation method is to divide the number of rising stocks by the number of falling stocks within a certain period of time to calculate the ratio. Then, use this ratio to see the strength of the upward trend in the market to determine the future market development trend.

Net Buy/Sell of Three Institutional Investors (TMC)

$$\text{TMC} = (\text{n1 over buy} - \text{n1 over sell}) + \cdots + (\text{n70 over buy} - \text{n70 over sell}) \quad (10)$$

4 Empirical Result

4.1 Research Sample Description

The source of this article is taken from the Taiwan Economic Journal (TEJ) database, using "Yanta Taiwan ESG Perpetual ETF" as the research scope, examining the Taiwan stock market from 2019/02/27 to 2020/02/25, based on the ETF listing date (2019/08/23) Observe before and after to observe the changes in abnormal returns. Frequency: Daily data, the number of samples is 241. There were a total of 73 ETFs when this ETF was listed, but three of them (respectively 2823 China Life, 2204 China, and 2227 Yuri Auto) lacked data, so they were eliminated and the total number of sample companies in this study was 70.

4.2 Statistical Analysis

The Table 1 shows the narrative statistics of the empirical variables in this study. The average AAR abnormal return rate is −0.0137, and it is normal distribution. The average cumulative abnormal return rate of ACAR is 0.4029, the median is 0.4189, and it is not normal distribution. The average abnormal turnover rate of AAT is −0.0532, and it is not normal distribution. The average cumulative abnormal turnover rate of ACAT is − 5.7828, and it is not normal distribution. The average buying and selling excess of the three major TMC legal persons is −8302.7, and it is not normal distribution. The average CR rise and fall ratio is 0.0492, and it is not normal distribution. The average BTER equity ratio is 0.0533, and it is not normal distribution.

Table 1. Statistical Analysis

	AAR	ACAR	AAT	ACAT	TMC	CR	BTER
Mean	−0.0137	0.4029	−0.0532	−5.7828	−8302.7	0.0492	0.0533
Median	−0.0289	0.4189	−0.0572	−5.1941	−5722	0.0847	0.0504
Maximum	0.7744	3.8016	0.2829	0.0492	273599	1.0000	0.1027
Minimum	−0.7374	−3.7829	−0.2022	−12.8291	−340574	−0.9714	0.0235
Std. Dev.	0.2509	2.1075	0.0616	3.8084	93588.6	0.4527	0.0154
Skewness	0.2625	−0.3918	1.0134	−0.2233	−0.3238	−0.2321	0.6763
Kurtosis	3.3548	2.1115	5.9972	1.7976	3.7861	2.2698	3.1056
Probability	0.1332	0.0009***	0.0000***	0.0003***	0.0055***	0.0233**	0.0001***
Observations	241	241	241	241	241	241	241

1. AAR is the average abnormal return rate, ACAR is the average cumulative abnormal return rate, AAT is the average abnormal turnover rate, and ACAT is the average cumulative abnormal turnover rate. Period: 2019/02/27–2020/02/25, frequency is daily data, and the number of sample days is 241. 2. TMC is the trading index of the three major legal persons, CR is the rise and fall ratio, and BTER is the bond-to-equity ratio. Period: 2019/02/27–2020/02/25, frequency is daily data, and the number of sample days is 241. 3. *** represents a significant level of 1%; ** represents a significant level of 5%; ∗ represents 10% level of significance.

4.3 Hypothesis Test

This section uses hypothesis testing to test whether there are abnormal phenomena and their significance in average abnormal returns, average cumulative abnormal returns, average abnormal turnover, and average cumulative abnormal turnover.

Abnormal Return

It can be seen from the Table 2 that the probability of the average AAR is 0.3957, and the probability that the median AAR is 0.2316. None of the above cannot reject the difference between the average and the median. The probability of the average ACAR is 0.0033, and the probability that the median ACAR is 0.0035, both are significantly different from 0.

Table 2. Abnormal Return (Full Sample AAR = 0)

Test	Statistic	Value	Probability
AAR Mean = 0	t-statistic	−0.8508	0.3957
AAR Median = 0	Wilcoxon	1.1962	0.2316
ACAR Mean = 0	t-statistic	2.9677	0.0033***
ACAR Median = 0	Wilcoxon	2.9214	0.0035***

1. Average abnormal return rate AAR, cumulative average abnormal return rate CAAR, average abnormal turnover rate AAT, cumulative average abnormal turnover rate CAAT. 2. Sample AAR Mean = −0.0137, AAR Sample Median = −0.0289∘ ACAR Sample Mean = 0.4029, ACAR Sample Median = 0.4189. 3. *** represents a significant level of 1%; ** represents a significant level of 5%; * represents 10% level of significance.

Abnormal Turnover

In Table 3, the average abnormal turnover rate has a significance level of 1%. Both the median and the average are significantly different from 0. The average is −0.0532 and the median is −0.0572, which represents a negative abnormal turnover rate. The average cumulative abnormal turnover rate has a significance level of 1%. Both the median and the average are significantly different from 0. The average is −5.7828 and the median is −5.7828, which means there is a negative abnormal turnover rate.

Table 3. Abnormal Turnover

Test	Statistic	Value	Probability
AAT Mean = 0	t-statistic	−13.4014	0.0000***
AAT Median = 0	Wilcoxon	10.5724	0.0000***
ACAT Mean = 0	t-statistic	−23.5721	0.0000***
ACAT Median = 0	Wilcoxon	13.4541	0.0000***

AAT Sample Mean = −0.0532, AAT Sample Median = −0.0572. ACAT Sample Mean = −5.7828, ACAT Sample Median = −5.1941. *** represents a significant level of 1%; ** represents a significant level of 5%; * represents 10% level of significance.

4.4 Regression Analysis

It can be seen from the data in the Table 4 that all variables have no single root, and the time series data are very stable, so they are less likely to be biased when performing regression estimation using the least squares method. Regression analysis will be used below to test whether investor sentiment used in this study has an impact on average abnormal returns and abnormal turnover.

Table 4. Unit Root Test

Variable	t-Statistic	Prob.
AAR	−14.2427	0.000***
ACAR	−14.3612	0.000***
AAT	−10.5515	0.000***
ACAT	−11.9016	0.000***
TMC	−14.6875	0.000***
CR	−11.9774	0.000***
BTER	−13.5087	0.000***

1. AAR is the average abnormal return rate, ACAR is the average cumulative abnormal return rate, AAT is the average abnormal turnover rate, ACAT is the average cumulative abnormal turnover rate, TMC is the Net Buy/Sell of Three Institutional Investors, CR is the Rate of Price Spread, and BTER is the Margin balance. 2. *** represents a significant level of 1%; ** represents a significant level of 5%; * represents 10% level of significance.

Table 5. AAR Regression Analysis

Variable	Coefficient	Std. Error	t-Statistic	Prob.
Period: 1–120				
C	0.0000	0.0892	0.0000	1.0000
TMC	−0.1171	0.1209	−0.9689	0.3346
CR	−0.1717	0.1200	−1.4303	0.1553
BTER	0.0487	0.0905	0.5375	0.5920
R-squared	0.0693			
Period: 122–241				
C	0.0000	0.0881	0.0000	1.0000
TMC	0.1587	0.1171	1.3549	0.1781
CR	−0.3734***	0.1159	−3.2216	0.0017

(*continued*)

Table 5. (*continued*)

Period: 122–241				
BTER	0.0483	0.0898	0.5382	0.5914
R-squared	0.0911			

1. AAR is the average abnormal rate of return, TMC is the Net Buy/Sell of Three Institutional Investors, CR is the Rate of Price Spread, and BTER is the Margin balance. 2. *** represents a variable that reaches a significant level of 1%; ** represents a variable that reaches a significant level of 5%; * represents the variable that reaches the 10% level of significance. 3. Regression:

$$AARt = c0 + c1 * TMCt + c2 * CRt + c3 * BTERt.$$

It can be seen from the data in the Table 5 that the rise and fall ratios have a significant negative correlation after the event. The P-value is 0.0017, the correlation coefficient is −0.3734, and the average rise and fall ratios are 0.0492, which is greater than 0. That is to say, the proportion of rises during this period a large drop indicates over-optimism, so this event has a downward pressure and reversal effect on the average abnormal return.

Table 6. ACAR Regression Analysis

Variable	Coefficient	Std. Error	t-Statistic	Prob.
Period: 1–120				
C	0.0000	0.0909	0.0000	1.0000
TMC	−0.1217	0.1232	−0.9876	0.3254
CR	0.0983	0.1223	0.8036	0.4233
BTER	−0.1477	0.0923	−1.6007	0.1122
R-squared	0.0331			
Period: 122–241				
C	0.0000	0.0682	0.0000	1.0000
TMC	0.1163	0.0907	1.2832	0.2020
CR	−0.0932	0.0897	−1.0381	0.3014

(*continued*)

Table 6. (*continued*)

Period: 122–241				
BTER	−0.6575***	0.0695	−9.4580	0.0000
R-squared	0.4553			

1. ACAR is the average cumulative abnormal return rate, TMC is the Net Buy/Sell of Three Institutional Investors, CR is the Rate of Price Spread, and BTER is the Margin balance. 2. *** represents a significant level of 1%; ** represents a significant level of 5%; * represents 10% level of significance. 3. Regression: $ACAR_t = c0 + c1 * TMC_t + c2 * CR_t + c3 * BTER_t$.

It can be seen from the data in the Table 6 that the bond-equity ratio has a significant negative correlation after the event, with a P value of 0.0000 and a correlation coefficient of -0.6575, which means that investors have a significantly optimistic mood after the event.

Table 7. AAT Regression Analysis

Variable	Coefficient	Std. Error	t-Statistic	Prob.
Period: 1–120				
C	0.0000	0.0921	0.0000	1.0000
TMC	0.0673	0.1248	0.5394	0.5907
CR	−0.0908	0.1240	−0.7328	0.4651
BTER	−0.0519	0.0935	−0.5555	0.5796
R-squared	0.0071			
Period: 122–241				
C	0.0000	0.0924	0.0000	1.0000
TMC	0.0052	0.1227	0.0425	0.9662
CR	−0.0322	0.1215	−0.2653	0.7912
BTER	0.0312	0.0941	0.3311	0.7411
R-squared	0.0019			

1. AAT is the average abnormal turnover rate, TMC is the Net Buy/Sell of Three Institutional Investors, CR is the Rate of Price Spread, and BTER is the Margin balance. 2. *** represents a significant level of 1%; ** represents a significant level of 5%; * represents 10% level of significance. 3. Regression: $AAT_t = c0 + c1 * TMC_t + c2 * CR_t + c3 * BTER_t$.

It can be seen from the data in the Table 7 that the average abnormal turnover rate has no significant correlation with the three major legal persons' buying and selling excess, rise and fall ratios, and securities capital ratio before and after the event. Therefore, it is believed that the average abnormal turnover rate has no correlation with the three.

Table 8. ACAT Regression Analysis

Variable	Coefficient	Std. Error	t-Statistic	Prob.
Period: 1–120				
C	0.0000	0.0755	0.0000	1.0000
TMC	0.4661***	0.1024	4.5536	0.0000
CR	−0.2063**	0.1016	−2.0303	0.0446
BTER	0.4008***	0.0767	5.2271	0.0000
R-squared	0.3326			
Period: 122–241				
C	0.0000	0.0630	0.0000	1.0000
TMC	0.1686**	0.0837	2.0128	0.0465
CR	−0.0691	0.0829	−0.8335	0.4063
BTER	−0.6973***	0.0642	−10.8579	0.0000
R-squared	0.5351			

1. ACAT is the average cumulative abnormal turnover rate, TMC is the Net Buy/Sell of Three Institutional Investors, CR is the Rate of Price Spread, and BTER is the Margin balance. 2. *** represents a significant level of 1%; ** represents a significant level of 5%; * represents 10% level of significance. 3. Regression: $ACATt = c0 + c1 * TMCt + c2 * CRt + c3 * BTERt$.

It can be seen from the data in the Table 8 that the average cumulative abnormal turnover rate has a significant correlation with the three major legal persons' buying and selling excess, the rise and fall ratio and the securities-to-equity ratio beforehand, but only the three major legal persons' trading excess and the securities-to-equity ratio have a significant correlation afterwards. The result of the three major legal persons' trading exceedance, indicating that the significance after the event has been slightly reduced. The result of the securities-equity ratio, indicating that this event has an optimistic impact on investors.

4.5 The Effect of ESG Event

AAR, The Difference Between Before and After

From the data in the Table 9, we can know that before data is not significantly different from 0, and after data is significantly different from 0. There is no obvious impact of before sample, and there is a significant negative impact of after sample, which means that this event will produce a negative average abnormal return.

Table 9. Test AAR = 0

Hypothesis	Statistic	Value	Probability
Period: 1–120, AAR Mean = 0	t-statistic	0.6346	0.5269
Period: 1–120, AAR Median = 0	Wilcoxon	0.2383	0.8116
Period: 122–241, AAR Mean = 0	t-statistic	−2.0846	0.0392**
Period: 122–241, AAR Median = 0	Wilcoxon	2.1684	0.0301**

1. Period: 1–120, AAR Sample Mean = 0.0153, AAR Sample Median = 0.0196. 2. Period: 122–241, AAR Sample Mean = −0.0445, AAR Sample Median = −0.0513. 3. *** represents a significant level of 1%; ** represents a significant level of 5%; * represents 10% level of significance.

ACAR, The Difference Between Before and After

It can be seen from the data in the Table 10 that it is significantly different from 0 for the data before the event, and it is also significantly different from 0 for the data after the event. There is a significant positive impact before the event. Although there is an obvious impact after the event, it is a negative impact, which means that this event will produce a negative average cumulative abnormal return.

Table 10. Test ACAR = 0

Hypothesis	Statistic	Value	Probability
Period: 1–120, ACAR Mean = 0	t-statistic	14.4207	0.0000***
Period: 1–120, ACAR Median = 0	Wilcoxon	9.0102	0.0000***
Period: 122–241, ACAR Mean = 0	t-statistic	−5.2453	0.0000***
Period: 122–241, ACAR Median = 0	Wilcoxon	4.8305	0.0000***

1. Period: 1–120, ACAR Sample Mean = 1.7161, ACAR Sample Median = 2.1117. 2. Period: 122–241, ACAR Sample Mean = −0.9238, ACAR Sample Median = −0.6880. 3. *** represents a significant level of 1%; ** represents a significant level of 5%; * represents 10% level of significance.

AAT, The Difference Between Before and After

From the data in the Table 11, it can be seen that the pre-event is significantly different from 0, and the post-event is also significantly different from 0. Both before and after the event have obvious negative effects, indicating that this event will produce a negative average abnormal turnover.

Table 11. Test AAT = 0

Hypothesis	Statistic	Value	Probability
Period: 1–120, AAT Mean = 0	t-statistic	−8.5980	0.0000***
Period: 1–120, AAT Median = 0	Wilcoxon	6.9138	0.0000***
Period: 122–241, AAT Mean = 0	t-statistic	−10.3938	0.0000***
Period: 122–241, AAT Median = 0	Wilcoxon	7.8723	0.0000***

1. Period: 1–120, AAT Sample Mean = −0.0422, AAT Sample Median = −0.0418. 2. Period: 122–241, AAT Sample Mean = −0.0636, AAT Sample Median = −0.0835. 3. *** represents a significant level of 1%; ** represents a significant level of 5%; * represents 10% level of significance.

ACAT, The Difference Between Before and After

It can be seen from the data in the Table 12 that the ex-ante is significantly different from 0, and the ex-post is also significantly different from 0. There is also a significant negative impact before and after the event, which means that this event will produce a negative average abnormal return, and the trend will be more obvious after the event.

Table 12. Test ACAT = 0

Hypothesis	Statistic	Value	Probability
Period: 1–120, ACAT Mean = 0	t-statistic	−17.5973	0.0000***
Period: 1–120, ACAT Median = 0	Wilcoxon	9.4947	0.0000***
Period: 122–241, ACAT Mean = 0	t-statistic	−45.8309	0.0000***
Period: 122–241, ACAT Median = 0	Wilcoxon	9.5052	0.0000***

1. Period: 1–120, ACAT Sample Mean = −2.4710, ACAT Sample Median = −2.2107. 2. Period: 122–241, ACAT Sample Mean = −9.0994, ACAT Sample Median = −8.8079. 3. *** represents a significant level of 1%; ** represents a significant level of 5%; * represents 10% level of significance.

Test the Difference Between Before and After

The Table 13 shows abnormal returns and abnormal turnover. In the ESG event, the ex-post-ex-ante difference test shows that the P-value of the average abnormal return is 0.0557, the P-value of the average cumulative abnormal return is 0.0000, and the P-value of the average abnormal turnover is 0.0067 and the P-value of the average cumulative abnormal turnover is 0.0000. The above are all significant, and the t-statistic is negative, indicating that this event has a negative impact on abnormal returns and abnormal turnover before and after.

Table 13. Test the difference between before and after the ESG event

After-Before	T-statistic	Average	Result
AAR	−1.9324	−0.0598	After < Before, significant
	(0.0557)*		
ACAR	−9.6314	−2.6400	After < Before, significant
	(0.0000)***		
AAT	−2.7604	−0.0214	After < Before, significant
	(0.0067)***		
ACAT	−105.9921	−6.6284	After < Before, significant
	(0.0000)***		

1. AAR is the average abnormal return rate, ACAR is the average cumulative abnormal return rate, AAT is the average abnormal turnover rate, ACAT is the average cumulative abnormal turnover rate. 2. *** represents a significant level of 1%; ** represents a significant level of 5%; * represents 10% level of significance.

5 Conclusion

This study explores whether ESG events cause investors' emotional changes, thereby driving abnormal returns and abnormal turnover. The results of this study found that investors were optimistic after the event, but over-optimism occurred during the period. Therefore, the event had a downward pressure and reversal on the average abnormal returns.

1. ETF constituent stocks have abnormal returns and abnormal trading volumes before and after ESG events. From the hypothesis testing analysis, it is concluded that ETF constituent stocks have abnormal returns and abnormal trading volumes before and after ESG events.
2. Investor sentiment is negatively correlated with abnormal returns and abnormal trading volume. It is concluded from regression analysis that investor sentiment is negatively correlated with abnormal returns and abnormal trading volume.
3. The impact of ESG events will reduce abnormal stock returns and abnormal trading volume. From the difference test, it was found that the impact of ESG events will reduce abnormal stock returns and abnormal trading volume.

When referring to past historical documents, we found that more people have discussed the impact of ESG on stock prices and company value, and less people have discussed whether investing in ESG brings abnormal returns. Therefore, this study wants to use the Yuanta Taiwan ESG Perpetual ETF to conduct research and discussion. Historical literature in the past pointed out that although investing in ESG companies can bring higher returns, after the establishment of ETFs, it became easier for the public to invest in ESG, and the emergence of optimism led to a rise in stock prices, resulting in negative abnormal returns. Therefore, you must be more careful when investing. Pay attention to the phenomenon of swarming and over-optimism.

When investing in ESG themes, there are few investment methods to choose from in the Taiwan investment market, so there are also fewer samples to study when conducting research. Therefore, I hope to add ETFs of similar nature in other markets for research, and It is hoped that the establishment date can also be staggered to avoid the impact of the market stock price.

References

1. Wang, Y., Chi, X., Zhou, G.: Review and prospect of behavioral finance literature - research on Taiwan market. Econ. Paper Ser. **44**(1), 1–55 (2016)
2. Gu, J.: An empirical study on factors influencing the psychological emotions of investors in Taiwan's stock market. Master's thesis, Department of Finance, Chaoyang University of Science and Technology (2003)
3. Lin, J.: Application of sentiment indicators in futures markets - taking Nikkei 225 index futures as an example. Master's thesis, Department of Finance, Ming Chuan University (2003)
4. Zhou, B., Zhang, Y., Lin, M.: The interactive relationship between investor sentiment and stock returns. Secur. Mark. Dev. Q. **19**(2), 153–190 (2007)
5. Chen, D.: Research on market sentiment and stock returns. Master's thesis, Department of International Trade, National Chengchi University (2001)
6. Xu, N., Guo, M., Zheng, N.: The interactive relationship between investor sentiment and stock price return fluctuations. Taiwan Finan. Q. **6**(3), 107–121 (2005)
7. Xu, M.: Comparison of the impact of market sentiment and fundamentals on short-term stock prices. Master's thesis, Department of International Trade, National Chengchi University (2002)
8. Ye, Z., Zhu, J., Li, C.: A study on the relationship between investor sentiment and stock returns in Taiwan's stock market. Lingdong Gen. Educ. Res. J. **2**(4), 121–142 (2008)
9. Tsai, P., Wang, Y., Zhang, Z.: Research on investor sentiment, company characteristics and Taiwan stock returns. Econ. Res. **45**(2), 273–322 (2009)
10. Tsai, Y.: Analysis of the correlation between investor sentiment and stock returns in the electronics and financial industries. Master's thesis of the Department of Finance, Shih Hsin University (2020)
11. Adam, A.M., Shavit, T.: How can a ratings-based method for assessing corporate social responsibility (CSR) provide an incentive to firms excluded from socially responsible investment indices to invest in CSR? J. Bus. Ethics **82**(4), 899–905 (2007)
12. Aguilera, R.V., Rupp, D.E., Williams, C.A., Ganapathi, J.: Putting the S back in corporate social responsibility: a multilevel theory of social change in organizations. Acad. Manag. Rev. **32**(3), 836–863 (2007)
13. Baker, M., Wurgler, J.: Investor sentiment and the cross section of stock returns. J. Finan. **61**, 1645–1680 (2006)
14. Baker, M., Wurgler, J.: Investor sentiment in the stock market. J. Econ. **21**(2), 129–151 (2007)
15. Ball, R., Brown, P.: An empirical evaluation of accounting income numbers. J. Account. Res. **6**(2), 159–178 (1968)
16. Barberis, N., Shleifer, A., Vishny, R.: A model of investor sentiment. J. Finan. Econ. **49**, 307–343 (1998)
17. Basu, S.: The relationship between earnings' yield, market value and return for NYSE common stocks: further evidence. J. Finan. Econ. **12**(1), 129–156 (1983)
18. Baur, M.N., Quintero, S., Stevens, E.: The 1986–88 stock market: investor sentiment or fundamentals? Manag. Decis. Econ. **17**(3), 319–329 (1996)
19. Brown, G.W., Cliff, M.T.: Investor sentiment and asset valuation. J. Bus. **78**, 405–440 (2005)

20. Brown, G.W., Cliff, M.T.: Investor sentiment and the near-term stock market. J. Empir. Finan. **11**(1), 1–27 (2004)
21. Clarke, R.G., Statman, M.: Bullish or bearish? Finan. Anal. J. **54**, 63–72 (1998)
22. Daniel, K., Hirshleifer, D., Subrahmanyam, A.: Investor psychology and security market under- and overreactions. J. Finan. **53**, 1839–1886 (1998)
23. De Long, J.B., Shleifer, A., Summers, L.H., Waldmann, R.J.: Noise trader risk in financial markets. J. Polit. Econ. **98**, 703–738 (1990)
24. Derwall, J., Guenster, N., Bauer, R., Koedijk, K.: The eco-efficiency premium puzzle. Finan. Anal. J. **61**(2), 51–63 (2005)
25. Fisher, K.I., Statman, M.: Investor sentiment and stock returns. Finan. Anal. J. **56**, 16–23 (2000)
26. Forrest, S., Ling, A., Lanstone, M., Waghom, J.: Enhanced energy ESG framework. Goldman Sachs (2006)
27. Güller, E., Ruttmann, R.: Responsible investment: a systemic approach to investing in market leading companies. Credit Suisse (2010)
28. Hamilton, J.T.: Pollution as news: media and stock market reactions to the toxics release inventory data. J. Environ. Econ. Manag. **28**(1), 98–113 (1995)
29. Hawn, O., Ioannou, I.: Do actions speak louder than words?. The case of corporate social responsibility, July 6 (2012)
30. Heal, G.: Corporate social responsibility: an economic and financial framework. Geneva Papers Risk Insur. Issues Pract. **30**(3), 387–409 (2005)
31. Hong, H., Stein, J.C.: A unified theory of underreaction, momentum trading, and overreaction in asset markets. J. Finan. **54**, 2143–2184 (1999)
32. Kamara, A., Miller, T.W., Siegel, A.F.: The effect of futures trading on the stability of standard and poor 500 returns. J. Futures Mark. **12**(6), 645–658 (1992)
33. Khanna, M., Quimio, W.R.H., Bojilova, D.: Toxics release information: a policy tool for environmental protection. J. Environ. Econ. Manag. **36**(3), 243–266 (1998)
34. Kiernan, M.J.: Investing in a Sustainable World. Why Green is the New Color of Money on Wall Street. Amacom Books, New York (2008)
35. Konar, S., Cohen, M.A.: Information as regulation: the effect of community right to know laws on toxic emissions. J. Environ. Econ. Manag. **32**(1), 109–124 (1997)
36. Kong, N., Salzmann, O., Steger, U., Ionescu-Somers, A.: Moving business/industry towards sustainable consumption, The role of NGOs. Eur. Manag. J. **20**(2), 109–127 (2002)
37. Neal, R., Wheatley, S.M.: Do measures of investor sentiment predict returns? J. Finan. Quant. Anal. **33**, 523–547 (1998)
38. Nielsen, K.P., Noergaard, R.W.: CSR and mainstream investing: a new match?–an analysis of the existing ESG integration methods in theory and practice and the way forward. J. Sustain. Finan. Invest. **1**(3–4), 209–221 (2011)
39. Schaltegger, S., Burritt, R.: In the International Yearbook of Environmental and Resource Economics 2005/2006: A Survey of Current Issues, Corporate Sustainability, pp. 185–222 (2005)
40. Shiller, R.J.: Stock prices and social dynamics. Brookings Papers Econ. Act. **2**, 457–498 (1984)
41. Shiller, R.J.: Irrational Exuberance. Princeton University Press, New York (2000)
42. UNEP FI: Fiduciary responsibility: legal and practical aspects of integrating environmental. Social and Governance Issues into Institutional Investment (2009)
43. Verheyden, T., Eccles, R.G., Feiner, A.: ESG for All? The impact of ESG screening on return, risk, and diversification. J. Appl. Corp. Finan. **28**(2), 47–55 (2016)

44. Woods, C.: Funding climate change: how pension fund fiduciary duty masks a collective (In) action problem. Working Paper, School of Geography and the Environment, Oxford University (2009)
45. Wright, W.F., Bower, G.H.: Mood effects on subjective probability assessment. Org. Behav. Hum. Decis. Processes **52**(2), 276–291 (1992)

Effects of Other Customers' Negative Behavior on Focal Customers' Evaluation of Service Results

Mei-Hua Huang, Chiau-Yuan Lee, and Szu-Hsien Lin[✉]

Department of Accounting and Information Systems, Asia University, Taichung City, Taiwan
aleclin@asia.edu.tw

Abstract. This study explores how negative behavior from other customers affects the focal customer's evaluation of the service provider. The focal customer's feelings toward the negative behavior are examined as a mediating variable. The study investigates how four factors (severity of the negative behavior, customer agreeableness, customer similarity, and the organization's customer compatibility initiatives) influence those feelings, and their subsequent impact on the customer's perception of service quality and overall satisfaction. Data were collected via Questionnaire survey. Data from 284 usable surveys reveal that severity of negative behavior, customer agreeableness, and the organization's customer compatibility management efforts all significantly influence the focal customer's feelings. These feelings, in turn, positively impact customer satisfaction with the organization.

1 Introduction

Service has been a major driver of national economic development and firms' revenue. Service quality and customer satisfaction are the keys of the sustainability of service organizations. However, service is a process and customers play an important role in the service delivery process. Except for the service delivery employee and the focal/target customer (the customer receiving the service), fellow customers (other customers receiving or waiting for the service) also take part in the service devilry process. These service encounters are ubiquitous: when customers contact in a very short distance, have an oral conversation, need to wait for services, or have to share the same services cape, etc. [16, 17] and hence it is unavoidable that other customers' behaviors may have impacts (positive or negative) on a focal customer's feeling. In most service encounters, interactions among customers are common. A pleasurable interaction with other customers is very likely to enhance the customer's quality perception and satisfactory with the service. Oppositely, a negative experience with the fellow customers may upset the focal customer, which can be also deemed as a service failure. Consequently, the customer may withhold the service evaluation.

Nevertheless, can such a service failure caused by other customers' negative behaviors be mitigated or prevented? This is important for both academics and service practitioners to understand customers' feelings for the fellow customers' negative behaviors.

L. Barolli (Ed.): IMIS 2024, LNDECT 214, pp. 412–423, 2024.
https://doi.org/10.1007/978-3-031-64766-6_41

Past research in the area of service marketing and management has focused largely on the impact of service failure caused by service employees on customers and on the organization, however, less on the role of fellow customers.

2 Literature Review and Research Hypotheses

2.1 Negative Behavior of Fellow Customers and Feelings of the Focal Customer

This study examines the impact of customer-to-customer negativity on service experiences. Customers' negative behavior is described as the behavior that deviates from norms or violates regular procedures of service experiences [9]. Studies illustrate a number of popular negative behaviors by customers such as baby crying, smoking at inadequate place, inappropriate dress, loud noises or such as abusing or hurting employee body or organizational property [3]. Research suggests that such behavior can negatively affect other customers' experiences, leading to dissatisfaction and reduced enjoyment [4]. This study proposes that the focal customer's feelings towards this negativity act as a mediating variable, influencing the impact of other factors on service quality perception and satisfaction.

The focal customer's feeling is conceptualized as the level of discomfort or upset caused by other customers' behavior. This concept draws on the service marketing concept of the "zone of tolerance". Similar to service quality, the study argues that there exists a tolerance zone for fellow customer behavior. If behavior falls within this zone, the customer may be unhappy but accepting. However, if it falls below the tolerance limit, they may experience dissatisfaction and even consider switching providers.

Building on existing literature of service marketing, and organizational behavior, the study identifies several potential independent variables that might influence the focal customer's feelings: severity of service failure and severity of customer undue behavior, customer agreeableness, customer similarity, and the customer compatibility management effort of service organizations. These factors are theorized as the independent variables of this study, and their relationships with the mediating variable "focal customers' feelings for fellow customers' negative behavior," as well as the related hypotheses, will be discussed in the subsequent sections.

2.2 Severity of Fellow Customers' Negative Behavior

Severity of service failure broadly described as a service failure happens when a service required by a customer or promised by the service organization is not fulfilled or when the delivered service is under the expectation of the customer [1]. As discussed in the earlier section, fellow customers' negative behavior can be seen as a service failure. That is not only employees' failure but also customers' undue behavior can result in a negative perception of the organization's service quality. Accordingly, the concept of service failure severity can be applied to the area of the severity of fellow customers' negative behavior.

Research suggests the severity of this "misbehavior" affects the focal customer's expectations, service perception, and satisfaction [25]. Severe behavior can weaken the customer-organization relationship and damage the organization's retention efforts and image [13]. Additionally, it can negatively impact employee morale, performance, and customer complaints [3] and even lead to customer defection and increased costs [32]. Based on this, the study hypothesizes:

H1: The severity of fellow customers' negative behavior is positively related to the focal customer's unfavorable feeling towards the behavior.

2.3 Customer Agreeableness

Robbins and Judge [28] define personality traits as a person's enduring characteristics. According to "The Big Five Personality Model" [28], personality traits involve five dimensions: extraversion, agreeableness, conscientiousness, emotional stability, and openness. Among them, agreeableness is especially related to the current study, which has been described as the inclination of a person's compliance to others. People high in agreeableness are cooperative, warm, and trusting, whereas people low in agreeableness are cold, opposite, and difficult. This study refers to agreeableness as the extent to which one customer acts friendly to or gets along well with other customers in the same service encounter.

Customers can influence the processes and results of a service delivery, and customer personality traits play an important role in service experiences [29]. Research on emotional contagion reveals that a person's personality traits can positively or negatively influence other people's emotion state, and vice versa [33]. In the service interactions, high agreeable people will adjust their emotion to adapt to the situation, particularly under negative mood circumstances. Even when they are confronted by unpleasant contacts from service providers or other customers, high agreeable customers still try to maintain good social relationships with others, and will be more forgiving and have a lower negative feeling for other customers' undue behavior. [29]. Thus, the study hypothesizes:

H2: The focal customer's degree of agreeableness is negatively related to the unfavorable feeling towards the fellow customers' negative behavior.

2.4 Customer Similarity

Person similarity delineates the extent to which a person perceives another person as similar to the self [5]. According to empirical research in social science, people often attract others having similar perception or characteristics [22]. Research in psychology shows that similar contexts produce akin influences [8]. In consumer behavior, similarity of customer perception is regarded as a critical influencer of personal preference [31]. In service research, Uhrich [31] finds that distinguishing the similarity of customer patterns may facilitate contact satisfaction among the service organization and its employees and customers.

Although in service encounters the critical role of fellow customers has been established, the mutual influence among customers got relatively little attention from researchers [5]. Studies have found that similarity has a positive impact on customers' proactive attitude [22], person actions, cooperative behavior [7], and purchase intention [12]. Thus, the study hypothesizes.

H3: The perceived similarity of customers is negatively related to the focal customer's unfavorable feeling towards the fellow customers' negative behavior.

2.5 Customer Compatibility Management Effort of Service Organizations

Service encounters often involve diverse customers, leading to potential conflicts and negative perceptions of the service. Research suggests this can impact customer satisfaction and loyalty [30]. However, service organizations can utilize customer management strategies to improve satisfaction and overall perception [10]. This highlights the importance of customer compatibility management. One approach involves managing the physical environment to cater to diverse customer needs and minimize dissatisfaction [17]. Additionally, managing inappropriate customer behavior becomes crucial to ensure a positive experience for all customers.

Customer compatibility management strategies can be categorized into prevention and response. Prevention includes customer segmentation, such as attracting similar customer groups or creating separate service areas for different needs [27]. It also involves establishing codes of conduct and training employees to identify and address potential conflicts effectively [20]. Based on this review, the study hypothesizes:

H4: Service organizations' customer compatibility management effort is negatively related to the focal customer's feeling towards the fellow customers' negative behavior.

2.6 Customer Perception of Service Quality

Parasuraman et al. [26] portray service quality as the difference between the expectation and perception of the service. Woodside et al. [34] refer to service quality as the customer's overall evaluation of a given service's entire processes and results. Later et al. [3] regard it as a customer's overall image of an organization's advantage and disadvantage of its service offerings. Stance on the literature, this study defines service quality as the extent to which the focal customer feels favorable about the process and results of the service received. Numerous studies have shown that service quality is positively related to customer satisfaction and organizations' profit and image.

Customers interact each other in many service encounters. Under the circumstances, the focal customer is likely to have a negative feeling for the fellow customers' inappropriate behavior such as destructive actions, disobeying the rules, causing delay, overuse, overcrowding, or displaying incompatible requests. Importantly, the negative feeling probably has an adverse effect on the focal customer's service quality perception. According to the Stimulus-Organism-Response (S-O-R) model, when the focal customer (the organism) is affected by the fellow customers' negative behavior (the stimulus), the person is likely to have a response to that. The response may include internal affective evaluation (e.g., un/pleasant, dis/satisfied) and external cognitive evaluation (e.g., good/bad quality). Accordingly, we can speculate that, in a service encounter, the focal customer will have an unfavorable evaluation of the encounter's service quality when having unpleasant feeling towards the fellow customers' negative behavior. Thus, the study hypothesizes:

H5: The focal customer's unfavorable feeling towards the fellow customers' negative behavior is negatively related to the perception of the service quality.

2.7 Customer Satisfaction with the Service Organization

The Expectation Disconfirmation Model [24] describes customer satisfaction as the customer's emotional response resulted from the comparison between the customer's expectation of a product or service and the actual performance of the product or service. If the performance equals expectation, it is called confirmation, with which the customer will be satisfied. When the performance is higher than expectation, it is positive disconfirmation, with which the customer will be delighted. Finally, as the performance is lower than expectation, it is negative disconfirmation, and the customer will be dissatisfied.

The S-O-R model maintains that the focal customer may be caused to have a negative response if affected by the fellow customers' inappropriate actions. This negative response probably is not only toward the fellow customers who display the undue actions but also toward the specific service encounter and the service organization as a whole [35]. Martin [15] finds that a few customers' negative behavior (e.g., children nosy or drunk) in a public place can decrease all other customers' satisfaction with the organization. Hence, the study hypothesizes:

H6: The focal customer's satisfaction with the service organization is negatively related to the feeling towards the fellow customers' negative behavior.

2.8 The Relationship Between Service Quality and Satisfaction

Although past studies ever have a controversy over the nature and causality of service and satisfaction, nowadays most researchers agree that service quality is an antecedent of satisfaction. Numerous studies in service marketing and consumer behavior support that service quality perception strongly influence satisfaction [21, 35]. Thus, the study hypothesizes:

H7: The focal customer's satisfaction with the service organization is positively related to the perception of service quality.

3 Research Method

3.1 Conceptual Framework

On the basis of the cognitive appraisal theory, the research model of this study is established (Fig. 1). The theory suggests human response to the environment as the process of appraisal → emotional response → behavioral intention. For the current study, the first step of appraisal includes four independent variables: severity of the fellow customer's negative behavior, customer agreeableness, customer similarity, and the service organization's customer compatibility management effort. The second step of emotional response is the central concept and the mediator of the study, the focal customer's unfavorable feeling for the fellow customer's negative behavior. The final step of behavioral intention involves two important outcome variables, the focal customer's service quality perception and satisfaction evaluation (the two key drivers of such behavioral intentions as re/purchase intention and recommendation intention).

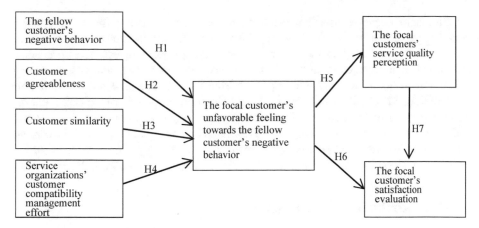

Fig. 1. The Conceptual Framework of the Study

3.2 Measurement

The main body of the questionnaire involves a number of structured scale items pertaining to the research variables of the study. Each of the model constructs was measured with several scale items that were mostly adapted from existing empirical studies. All items were measured with a 5-point Likert type scale ranging from strongly disagree (1) to strongly agree (5). The subjects were asked to indicate their level of agreement with the scale items on the questionnaire.

Severity of the fellow customer's negative behavior is measured with six items mainly adapted from Maxhammer al. [18]. Six measuring items of customer agreeableness are adapted from Mowen and Spears [23] (1999) and McCrae and Costa [19]. Customer similarity also has six items, largely adapted from Jarvis et al. [11]. The variable of the service organization's customer compatibility management effort is proposed by the current study, which consists of two dimensions, prevention of customers' negative behavior and treatment of customers' negative behavior. The prevention dimension is devised on the basis of the concepts of Zeithaml et al. [35] and Bedi and Monica [2]. The mediator the focal customer's unfavorable feeling towards the fellow customer's negative behavior includes six items adapted from Martin, and Pranter [17] and Mansfield et al. [14]. Finally, the focal customer's satisfaction with the company contains five items primarily adapted from Brady and Cronin [6] and Bedi and Monica [2].

3.3 Data Collection

A pre-test was conducted to assess the survey's reliability. Participants were recruited at various high-traffic locations like restaurants and shopping malls. Fifty-two individuals participated, with 41 responses deemed valid. The Cronbach's alpha (α) values of all research variables exceeded 0.6, indicating acceptable reliability for this initial exploratory study. Subsequently, the formal survey employed the same questionnaire.

Similar to the pre-test, participants in the formal survey were required to have experienced negative customer behavior within the past year. Data collection utilized two methods: in-person and mail surveys. For the in-person approach, researchers approached individuals like high school and college students, store employees and customers, and community residents to request their participation and collect completed questionnaires immediately. The mail survey involved obtaining cooperation from various organizations, who then distributed questionnaires to their designated contacts for completion and return. A total of 400 surveys were distributed (130 by mail and 270 in-person), resulting in 380 returned surveys (110 by mail and 270 in-person). After excluding invalid responses, the final sample size was 284, translating to a 74.7% valid return rate.

4 Data Analysis and Discussion

4.1 Reliability and Validity Analyses

The analyses result of Pearson correlations is shown as Table 1. The structure equation modeling (SEM) procedure is employed to evaluate the research model and test all hypotheses of the study. The results for model fitness are as follows: PGFI = .71, NFI = .92, NNFI = .96, IFI = .96, CFI = .96, RMSEA = .053, RMR = .059, SRMR = .059, PNFI = .82, GFI = .85 and AGFI = .8. All these values satisfy the standards suggested by scholars except for GFI and AGFI. In total, most of the fitness indicators of the structural model of the study are in accord with standard values suggested by scholars. Therefore, it is reasonable to speculate that the research model of the study fits the data well.

Results of the hypothesis testing are shown in Table 2. All hypotheses are supported except for hypothesis 3 and 5. Specifically, the direct effect of the severity of the fellow customer's negative behavior on the focal customer's unfavorable feeling towards the negative behavior (H1) is the greatest (r = .89), followed by the impact of the customer's service quality perception (H7, r = .77). Other significant direct effects are found in the impact of the focal customer's feeling towards the fellow customer's negative behavior on the focal customer's satisfaction with the service organization (H6, r = −.39), of the service organization's customer compatibility management effort on the focal customer's unfavorable feeling for the fellow customer's negative behavior (H4, r = −.25), and of customer agreeableness on the focal customer's unfavorable feeling towards the fellow customer's negative behavior (H2, r = −.12). In contrast, no significant relationships are found between customer similarity and the focal customer's unfavorable feeling for the fellow customer's negative behavior (H3, r = −.03), and between the focal customer's unfavorable feeling towards the fellow customer's negative behavior and the focal customer's service quality perception (H5, r = .00).

Table 1. Average, Standard Deviation, and Pearson Correlation

Construct	Average	Std. Dev.	1	2	3	4	5	6
Severity of Negative Behavior	3.651	.797						
Customer Agreeableness	4.007	.642	0.318**	-				
Customer Similarity	2.167	.735	−0.291**	−0.132*	-			
Effort of Customer Compatibility Mgmt	3.033	.960	−0.168**	−0.002	0.018	-		
Unfavorable Feeling towards Negative Behavior	3.662	.702	0.649**	0.191**	−0.258**	−0.225**	-	
Service Quality Perception	3.316	.941	−0.177**	0.082	−0.022	0.634**	−0.111	-
Satisfaction with Organization	3.343	.859	−0.221**	0.034	−0.008	0.657**	0.147*	0.822*

Notes: * denotes P < 0.05; ** denotes p < 0.01

Table 2. Path Coefficients of the Structural Model and Hypotheses Testing

Hypothetic path	Std. path coefficient	t-value	Results
H1(+): Severity of negative behavior → Feeling for negative behaviour	0.89	7.27***	supported
H2(−): Customer agreeableness → Feeling for negative behaviour	−0.12	−1.79†	supported
H3(−): Customer similarity → Feeling for negative behavior	−0.03	−0.41	not supported

(*continued*)

Table 2. (*continued*)

Hypothetic path	Std. path coefficient	t-value	Results
H4(−): Organization's customer computability management effort → Feeling for negative behavior	−0.25	−3.95***	supported
H5(−): Feeling for negative behavior → Service quality perception	−0.00	−0.03	not supported
H6(−): Feeling for negative behavior → Satisfaction evaluation	−0.39	−4.79***	supported
H7(+): Service quality perception → Satisfaction evaluation	0.77	12.08***	supported

Notes: † denotes P < 0.1; * denotes P < 0.05; ** denotes P < 0.01; *** denotes P < 0.001

5 Conclusion

5.1 Research Findings and Discussion

This research examines how fellow customers' negative behavior impacts other focal customers in service encounters. It finds that the more severe the negative behavior, the worse the focal customer feels, similar to how severe service failures by the organization lead to greater dissatisfaction. The study also identifies factors influencing the focal customer's feeling towards the negative behavior. People who are more agreeable (forgiving) are less affected, and the service organization's effort to manage customer compatibility (e.g., seating arrangements) can significantly reduce the negative feelings. This is crucial because it's a controllable factor, unlike customer agreeableness or similarity. Surprisingly, customer similarity (demographics only) did not influence the focal customer's feelings. The study suggests this positive effect, seen in other contexts, might be suppressed by the negativity of the behavior.

Furthermore, the research found that the focal customer's negative feelings towards the fellow customer's behavior did not directly impact their perception of the service quality. This might be because service quality perception is influenced by multiple factors, including the actual service outcome, delivery process, interaction with staff, and physical environment. Unless the negative behavior is extreme, it might not be strong enough to outweigh these other factors. Finally, the study confirms a well-established finding: a strong positive relationship between service quality perception and customer satisfaction.

5.2 Managerial Implications

This study investigates the impact of customer-to-customer interactions on service experiences. It reveals that a single customer's negative behavior can significantly affect other customers, potentially leading to 30% of them abandoning the service. This highlights the importance of managing customer interactions for service organizations. This research identifies two key findings with corresponding managerial implications. Firstly, witnessing negative behavior from other customers can lead to dissatisfaction and reduced repurchase intention among observing customers. This negative impact is comparable to the effect of service failures itself. Therefore, it's crucial to address potential conflicts effectively. Secondly, the study shows that agreeable customers are more forgiving of other customers' negativity and have higher overall satisfaction. This underscores the value of recognizing and rewarding agreeable behavior to encourage loyalty and positive customer interactions.

Based on these findings, the study suggests several strategies for service organizations to manage customer interactions effectively, including (1) empowering employees: train frontline staff to identify and address potential customer conflicts, equipping them with the skills to resolve issues and minimize negative impacts, (2) customer segmentation: consider attracting homogenous customer groups and separating diverse groups to reduce potential friction, (3) customer code of conduct: implement guidelines for appropriate behavior to deter negative interactions, and (4) physical environment design: create comfortable and user-friendly spaces that facilitate positive customer interactions.

Furthermore, the study emphasizes the link between positive consumption experiences and improved service quality perception. Service companies can focus on two approaches. They are (1) preventing negative experiences: implement customer compatibility management programs as described above, which can help mitigate the impact of negative customer interactions, and (2) boosting positive experiences: enhance the physical environment and service processes to create a more enjoyable experience for all customers. Overall, by minimizing negative experiences and amplifying positive experiences, service organizations can improve customer satisfaction and perception of service quality, ultimately leading to greater business success.

5.3   Research Limitation and Future Research

This research explores the impact of fellow customers' negative behavior on focal customers in service encounters. It argues that such behavior not only affects the focal customer's external responses (e.g., complaints) but also their internal feelings (e.g., dissatisfaction). The study proposes a psychological concept, "unfavorable feeling," to explain the link between negative behavior and its consequences. While this feeling significantly impacts the focal customer's satisfaction with the service organization, it surprisingly doesn't affect their perception of service quality. Hence, the role of this variable needs to be examined further.

The study identifies three factors influencing "unfavorable feeling": the focal customer's agreeableness, their similarity to the negative customer (demographics only), and the service organization's customer compatibility management efforts. It suggests exploring additional factors like psychographic and lifestyle similarities in future research. Finally, the study acknowledges limitations due to its diverse sample across industries. It suggests that future research might benefit from categorizing service types and exploring unique negative customer behaviors and their consequences within each category.

References

1. Ahmad, S.: Perspectives service failures and customer defection: a closer look at online shopping experiences. Manag. Serv. Qual. **12**, 19–29 (2002)
2. Bedi, M.: An integrated framework for service quality, customer satisfaction and behavioral responses in Indian banking industry - a comparison of public and private sector banks. J. Serv. Res. **10**, 157 (2010)
3. Bitner, M.J., Booms, B.H., Mohr, L.A.: Critical service encounters: the employee's viewpoint. J. Mark. **58**, 95–106 (1994)
4. Bougie, R., Pieters, R., Zcelenberg, M.: Angry customers don't come back, they get back: the experience and behavioral implications of anger and dissatisfaction services. J. Acad. Mark. Sci. **31**, 377–393 (2003)
5. Brack, A.D., Benkenstein, M.: The effects of overall similarity regarding the customer-to-customer-relationship in a service context. J. Retail. Consum. Serv. **19**, 501–509 (2012)
6. Brady, M.K., Cronin, J.J.: Customer orientation effects on customer service perceptions and outcome behaviors. J. Serv. Res. **3**, 241–251 (2001)
7. Emswiller, T., Deaus, K., Willits, J.E.: Similarity, sex, and requests for small favors. J. Appl. Soc. Psychol. **1**, 284–291 (1971)
8. Finkel, E.J., Baumeister, R.F.: Attraction and rejection. In: Baumeister, R.F. (ed.) Social Psychology, pp. 419–460. The State of the Science, Oxford University Press, New York, (2010)
9. Fullerton, R.A., Punj, G.: Repercussions of promoting an ideology of consumption: consumer misbehaviour. J. Bus. Res. **57**, 1239–1249 (2004)
10. He, Y., Chen, Q., Alden, D.L.: Consumption in the public eye: the influence of social presence on service experience. J. Bus. Res. **65**, 302–310 (2012)
11. Jarvis, C.B., Mackenzie, S.B., Podsakoff, P.M.: A critical review of construct indicators and measurement model misspecification in marketing and consumer research. J. Consum. Res. **30**, 199–218 (2003)
12. Jiang, L., Hoegg, J., Dahl, D.W., Chattopadhyay, A.: The persuasive role of incidental similarity on attitudes and purchase intentions in a sales context. J. Consum. Res. **36**, 778–791 (2010)
13. Lau, G.T., Ng, S.: Individual and situational factors influencing negative word of mouth behavior. Can. J. Adm. Sci. **18**, 163–178 (2001)
14. Mansfield, J., Winter, J., Waner, K.: A study of team performance in business communication: can the FIRO-B help. Acad. Bus. Discip. J. **4**, 1–19 (2012)
15. Martin, C.L.: Consumer-to-consumer relationships: satisfaction with other consumers' public behavior. J. Consum. Aff. **30**, 146–169 (1996)
16. Martin, C.L.: Retrospective: compatibility management: customer-to-customer relationships in service environments. J. Serv. Mark. **30**, 11–15 (2016)
17. Martin, C.L., Pranter, C.A.: Compatibility management: customer-to-customer relationships in service environments. J. Serv. Mark. **3**, 6–15 (1989)
18. Maxham, G., Netemeyer, J., Richard, G.: A longitudinal study of customers' evaluations of multiple service failures and recovery efforts. J. Mark. **66**, 57–71 (2002)
19. McCrace, R.R., Costa, P.T.: Discriminant validity of NEO-PIR facet scales. Educ. Psychol. Meas. **52**, 229–237 (1992)
20. McQuilken, L.: The influence of failure severity and employee effort on service recovery in a service guarantee context. Australas. Mark. J. **18**, 214–221 (2010)
21. Mohammad, A.A.S., Alhamadani, S.Y.M.: Service quality perspectives and customer satisfaction in commercial banks. Middle East. Finan. Econ. **14**, 60–72 (2011)

22. Montoya, R.M., Horton, R.S., Kirchner, J.: Is actual similarity necessary for attraction? A meta-analysis of actual and perceived similarity. J. Soc. Pers. Relat. **25**, 889–922 (2008)
23. Mowen, J.C., Spears, N.: Understanding compulsive buying among college students: a hierarchical approach. J. Consum. Psychol. **8**, 407–430 (1999)
24. Oliver, R.L.: A cognitive model of the antecedents and consequences of satisfaction decisions. J. Mark. Res. **17**, 460–469 (1980)
25. Oliver, R.L., Swan, J.E.: Consumer perceptions of interpersonal equity and satisfaction in transactions: a field survey approach. J. Mark. **53**, 21–35 (1989)
26. Parasuraman, A., Zeithamal, V.A., Berry, L.L.: A conceptual model of service quality and its implication for future research. J. Mark. **49**, 41–50 (1985)
27. Pranter, C.A., Martin, C.L.: Compatibility management: roles in service performances. J. Serv. Mark. **5**, 43–53 (1991)
28. Robbins, S.P., Judge, T.A.: Organizational Behavior. Pearson Education, Limited, London (2017)
29. Tan, H.H., Foo, M.D., Kwek, M.H.: The effects of customer personality traits on the display of positive emotion. Acad. Manag. J. **47**, 287–296 (2004)
30. Tombs, A.G., McColl-Kennedy, J.R.: The social-servicescape: a conceptual model. Mark. Theory **3**, 447–475 (2003)
31. Uhrich, S.: Explaining non-linear customer density effects on shoppers' emotions and behavioral intentions in a retail context: the mediating role of perceived control. J. Retail. Consum. Serv. **18**, 405–413 (2011)
32. Verhoef, P.C., Lemon, K.N., Parasuraman, A., Roggeveen, A., Tsiros, M., Schlesinger, L.A.: Customer experience creation: determinants, dynamics and management strategies. J. Retail. **85**, 31–41 (2009)
33. Wild, B., Erb, M., Bartels, M.: Are emotions contagious? Evoked emotions while viewing emotionally expressive faces: quality, quantity, time course and gender differences. Psychiatry Res. **102**, 109–124 (2001)
34. Woodside, A.G., Frey, L.L., Daly, R.T.: Linking service quality customer satisfaction. J. Health Care Mark. **9**, 5–17 (1989)
35. Zeithaml, V.A., Bitner, M.J., Gremler, D.D.: Services Marketing. McGraw-Hill, New York (2024)

The Investment Performance of Taiwan and Hong Kong: A Comparative Analysis

Mei-Hua Liao[1], Yiu Chan[1], and Ya-Lan Chan[2](✉)

[1] Department of Finance, Asia University, Taichung, Taiwan, R.O.C.
liao_meihua@asia.edu.tw
[2] Department of Business Administration, Asia University, Taichung, Taiwan, R.O.C.
yalan@asia.edu.tw

Abstract. This study aims to examine the performance of index exchange-traded funds (ETFs) in the Hong Kong and Taiwan markets. ETFs have gained increasing popularity among investors in both markets, warranting a thorough investigation into their price and performance fluctuations. Disparities exist in the performance of ETFs across different industries in bullish and bearish market environments, with varying degrees of impact from market trends. For instance, the primary industry in the Taiwan market is electronics, while the Hong Kong market is focused on finance. By analyzing the performance of ETFs in the Hong Kong and Taiwan markets, while considering different risk-return profiles, market environments, and industry influences, this study's findings will offer practical investment recommendations to investors and contribute to policy-making and academic research.

1 Introduction

Mutual funds have become synonymous with risk diversification and low entry barriers. In Hong Kong, formal employees are required to contribute at least 10% of their monthly salary to the Mandatory Provident Fund Schemes (MPF), which also invest in mutual funds. In Taiwan, as of the end of December 2023, domestic investment trust companies have issued 1,032 mutual funds, with assets exceeding NT$670 billion. This illustrates that mutual funds are also a significant investment option. Particularly noteworthy within mutual funds are Exchange Traded Funds (ETFs), whose price fluctuations and performance variations warrant in-depth research.

Mutual funds and ETFs are both composed of a basket of securities, but their main difference lies in the method of selection. Mutual funds are chosen by a team of fund managers, while ETFs are selected based on the constituents of an index. When ETFs track their underlying index or asset portfolio, discrepancies may arise between their actual performance and that of the underlying index or asset portfolio. The reasons for these differences may include management fees, transaction costs, dividend taxes and trading taxes, market liquidity, and other factors. Exploring these factors in depth can provide insights into the performance of ETFs in different economic environments.

In bull and bear market environments, different industry ETFs exhibit varying performances, and changes in market trends have different impacts on these ETFs. These

L. Barolli (Ed.): IMIS 2024, LNDECT 214, pp. 424–432, 2024.
https://doi.org/10.1007/978-3-031-64766-6_42

impacts may stem from macroeconomic factors and policy changes, among other influences. The most significant industry in the Taiwan market is the electronics sector, while in the Hong Kong market, it is the financial sector. The performance of the electronics and financial industries is influenced to varying degrees by industry-specific structures and macroeconomic factors. These two industries face various levels of risk and return. Therefore, by comparing the performance of ETFs in these two industries, we can gain a deeper understanding of their risk-return characteristics.

This study aims to explore how the performance of ETFs is influenced by the turnover rate and exchange rate volatility, and different industries in both the Hong Kong and Taiwan markets. A thorough investigation into these influences is beneficial for investors in risk management and asset allocation. Additionally, it can contribute to policy formulation by future governing bodies and further development in academic research within the relevant field.

2 Literature Review

2.1 Exchange Traded Funds

Huang, Ma, Li and Li (2017) conducted a study using data from the Taiwan Economic Journal Database from January 2001 to December 2015. They selected 101 mutual funds and employed one-way analysis of variance and paired t-tests. The study indicated that a periodic fixed-value investment strategy outperformed a periodic fixed-amount investment strategy during most periods. Additionally, it specifically pointed out that the periodic fixed-value investment strategy is more suitable for medium- to long-term investment behavior, particularly in small-cap and technology sector funds. Liu, Chen, and Liu (2008) also suggested that for long-term investments, periodic fixed-amount investing yields higher returns and lower risks compared to lump-sum investing. For short-term investments, however, the lump-sum method may potentially outperform periodic fixed-amount investing only when the risk-free interest rate is low.

Hsu (2010) utilized static and dynamic models to examine the performance of distinct types of funds in the TEJ database from January 1995 to December 2007. The top three performers in each fund category were selected as the investment targets for the portfolio. It was found that the dynamic model was most effective in identifying the most accurate investment returns and controlling investment risks.

Lin and Chi (2013) argue that the disposition effect and commitment effect in mutual funds negatively impact performance. Even after controlling dynamic investment styles, the disposition effect and commitment effect still have negative effects. Caglayan, Celiker, and Sonaer (2018) analyzed the demand for hedge funds and non-hedge funds over time based on the book-to-market ratio. They found that once the price-to-earnings ratio becomes public information, hedge funds tend to shift their preferences from growth stocks to value stocks. Additionally, hedge funds are better at identifying overpriced growth stocks compared to non-hedge funds.

Fu and Shu (2015) applied the anti-disposition effect model. They found that investors tend to sell stocks with larger book losses before year-end but sell stocks with smaller book gains less. Additionally, they concluded that the higher the shareholding of mutual funds, the weaker the above reaction (anti-disposition effect). Yuan and Zeng (2014) utilized performance indicators such as Sharpe ratio, Treynor ratio, Jensen's alpha, and timing and selection abilities. They found that index funds with superior performance exhibit momentum effects, demonstrating persistence.

2.2 ETF Perform

Li and You (2015) utilized monthly data from 2009 to 2013 for ETFs listed and traded for at least five years in Taiwan, Hong Kong, and mainland China. They found that the local index ETFs in Hong Kong and mainland China did not perform as well in tracking as those in Taiwan. However, foreign index ETFs in Hong Kong outperformed those in Taiwan. Several factors were identified as significant influences on ETF tracking error, including total expense ratio, exchange rate, asset size, trading volume, index replication strategy, and regional tracking performance. Ho, Tsao, and Hsu (2021) also investigated the investment performance of domestic high-dividend ETFs listed on the Taiwan Stock Exchange. They observed tracking error using net asset value and market price and found that excess returns were not significantly present in high-dividend ETFs.

Lin and Chen (2022) conducted a similar analysis and found that funds with strong short-term performance also perform well in the long term, while those with poor short-term performance also perform poorly in the long term.

Sherrill, Shirley, and Stark (2020) employed actively managed domestic equity mutual funds within the United States from 2004 to 2015. They discovered that mutual funds reduced their cash holdings by utilizing benchmark ETFs. The reduction in cash holdings also resulted in a decrease in tracking error, thereby making the fund's returns more closely align with the benchmark.

Peng, Lin, and Zhang (2015) conducted a study on Taiwan ETFs issued for over 4 years, using data from 2009 to 2013 as the research period. They analyzed the performance and risk values of Taiwan ETFs during financial storms using data from the Citi XQ Global Winner system database. The research indicated that ETFs are an excellent investment tool for hedging and maintaining stable profits. Even during financial storms such as the European debt crisis, ETFs maintained a certain level of stability. Lin and Meng (2004) reached a similar conclusion when analyzing Taiwan's first ETF (TTT). Through mean variance analysis and portfolio technical analysis, they observed that ETFs are highly popular investment products in the Taiwanese market, providing investors with better options.

Chen, Wang, and Hong (2017) utilized Corporate Governance Index (CGI) and Corporate Social Responsibility Index data from 2010 to 2012, covering 167 industry classifications from the Taiwan Economic Journal Database. They concluded that corporate governance has a significant impact on both corporate social responsibility and corporate operations. They found that the more directors and the larger the board, the better the performance in terms of corporate social responsibility. Similarly, Chuang, Liu, and Lee (2023) made a similar discovery, indicating that corporate social responsibility positively affects operational performance, leading to improved overall operational efficiency.

Chen, Hong, and Liu (2014) pointed out that several types of funds exhibit distinct characteristics. By employing systematic risk and total risk, they found that there is no significant difference in risk between equity portfolio funds and general equity mutual funds. Additionally, they discovered that the performance of small-scale funds is better than that of large-scale funds. Furthermore, they found evidence of momentum and smart money effects in general mutual funds, with the smart money effect being less pronounced than the momentum effect. Moreover, both momentum and smart money effects are not exclusively present in portfolio funds.

Ma, Huang, and Wu (2011) explored the correlation between different investment styles and investment performance of Taiwan equity mutual funds. Their study period ranged from 1999 to 2009, during which they selected 50 open-end general equity funds established for over ten years as their research subjects. The research findings indicated that a concentrated investment strategy is likely to generate better investment performance. Lee, Lee, and Lee (2011) examined fund performance and found that higher investment concentration leads to better performance.

Lin (2004) found that the best-performing funds tend to increase risk, while the worst-performing funds tend to decrease risk. Funds with average performance exhibit a more complex and dynamic behavior, aiming to achieve better performance and rankings. Funds with inferior performance actively reduce risk to avoid the risk of unemployment. Chen (2012) calculated fund flows, performance, risk levels, turnover rates, and expenses. They discovered that after winning awards, the performance of mutual funds begins to decline. Similarly, Li and Liao (2012) made a similar finding, indicating that the stock selection ability of most award-winning funds is not significantly better than that of non-award-winning funds.

Kao, Chen, Tang, and Cao (2005) studied the impact of different indicators on fund performance. They found that among absolute performance indicators, without considering the fund's timing ability, Jensen's alpha is the most effective measure of mutual fund investment performance. However, when considering the fund's timing ability, the Carhart four-factor model alpha is the most accurate measure of fund performance.

Lu, Zhang, and Shen (2017) pointed out that the active proportion has a positive impact on performance, whether it is in the current period or in the future. This indicates that the higher the active proportion of the fund, the higher the additional returns generated. Converse, Yeyati, and Williams (2020) obtained monthly fund data from EPFR Global from January 1997 to August 2017, including mutual funds and ETFs. They found that compared to mutual fund flows, ETF flows are less connected to local economic conditions and are more sensitive to global financial conditions.

Wang, Lin, and Yeh (2014) studied the risk-taking behavior of Taiwan mutual fund managers during bull and bear markets. They found that during bull markets, managers exhibited poorer fund performance in the first half of the year but increased risk-taking in the second half. Conversely, during bear markets, underperforming managers reduced portfolio risk in the second half of the year.

Lee, Lee, and Lee (2011) found that during bull markets and when the overall fund investment rate is high, investment concentration is higher, and performance is better. Conversely, higher portfolio turnover rates are associated with lower investment concentration and poorer performance. This suggests that higher investment concentration leads to better investment opportunities.

The following research hypotheses will be employed to compare performance of the electronics and financial industries in Hong Kong and Taiwan.

H1: The mean performances of ETFs in Hong Kong and Taiwan are equal.

H2: The mean performances of ETFs in Hong Kong and Taiwan are not equal.

Building upon these studies, our research aims to analyze the performance of different industries within the ETF markets of Taiwan and Hong Kong. Drawing from the literature, we will design testing methodologies, select samples, and observe causal relationships to provide a more concrete understanding of information transmission.

3 Methodology

This study aims to compare the performance of ETFs between Taiwan and Hong Kong markets, to explore their correlation. The research period spans from January 2009 to December 2015. Financial data, stock prices, and stock returns required for empirical analysis will be obtained from the Taiwan Economic Journal (TEJ) Databank. The data sources include announcements from stock exchanges and the over-the-counter market, public information disclosed on the Public Information Observation System, TEJ News Database, and the Fair-Trade Commission.

To investigate the performance of ETFs between Taiwan and Hong Kong markets, we use the following variables and model to solve.

(1) Rate of Return (R)

Rit is the ETFs' return over month t. Because benchmark ETFs are highly liquid, well-diversified assets, they provide mutual funds with the ability to reduce their cash holdings through an allocation to a benchmark ETF.

(2) Turnover Rate (Turnover)

We calculate the turnover ratio (ATR) defined as the ratio between the monthly trading volume of the ETFs and the assets under management (AUM) in that category. The higher the ratio, the more intensely traded the ETF.

(3) Exchange Rate Volatility (ExRate)

When ETFs track foreign indices, ETF managers need to convert their domestic currency into foreign currency to purchase the components of the foreign index. However, when calculating the net asset value (NAV) return of the ETF, it is usually based on the domestic currency. This means that when the foreign currency appreciates against the domestic currency, the ETF will experience gains from currency conversion, resulting in higher NAV returns compared to the index returns.

$$R_{i,t} = \alpha + \beta_1 \text{ Turnover}_{i,t} + \beta_2 \text{ ExRate}_{i,t} + \varepsilon_{i,t} \tag{1}$$

Our Hong Kong ETF descriptive statistics appear in Table 1. The highest monthly return in the ETFs of Hong Kong is 4.14%, and the highest monthly turnover rate is 12.25%, with a total of 13 ETFs. In Hong Kong's 13 ETFs, only 1 ETFs invest in the financial industry.

As shown in Table 2, the highest monthly return rate in Taiwanese ETFs is 38.95%, and the highest monthly turnover rate is 184.71%, with a total of 57 ETFs. Among Taiwan's 57 ETFs, 6 ETFs invest in the electronics industry, and 2 ETFs invest in the financial industry.

Table 1. Hong Kong ETF Descriptive Statistics

	R %	Turnover %	ExRate %
Max	4.14	12.25	0.1290
Min	−4.05	0	0.1283
Mean	0.27	0.51	0.1289
Median	0.09	0.03	0.1289
SD	1.06	1.10	0.00018

Table 2. Taiwan ETF Descriptive Statistics

	R %	Turnover %	ExRate %
Max	38.95	184.71	0.147
Min	−18.43	−289.12	−0.17
Mean	1.35	3.32	−0.00024
Median	1.42	0.1	0
SD	5.57	21.93	0.01312

Table 3 shows that, the highest monthly return in the sample of Taiwanese electronics ETF is 25.67%, and the highest monthly turnover rate is 81.77%. And the highest monthly return in the sample of Taiwanese financial ETF is 38.95%, and the highest monthly turnover rate is 11.87%.

Table 3. Electronics and Financial ETF Descriptive Statistics of Taiwan

	Electronics ETF		Financial ETF	
	R %	Turnover %	R %	Turnover %
Max	25.67	81.77	38.95	11.87
Min	−17.97	−14.67	−16.93	−119.66
Mean	1.51	3.62	−1.34	−1.69
Median	1.73	0.14	1.38	0.04
SD	6.73	9.28	6.23	11.76

The results in Table 3 are consistent with the investment principle of risk-return tradeoff. The average monthly return of Taiwanese electronics ETF is 1.51%, higher than −1.34% of Taiwanese financial ETF, and higher than 0.27% of Hong Kong ETF. The standard deviation of Taiwan electronic ETF returns is 6.73%, which is higher than the standard deviation of financial ETF returns of 6.23% and higher than the standard deviation of Hong Kong ETF returns of 1.06%.

In the regression for the results in Table 4, we use the framework of Eq. (1) with rates of return calculated over the month as the dependent variable. The results show that the monthly return in the ETFs of Hong Kong is associated with an incrassation in the monthly turnover rate. However, the monthly return of Taiwanese ETFs is associated with an incrassation in the monthly changing exchange rate. Among Taiwan ETFs, the impact of monthly exchange rate changes on ETF returns can be seen in electronic ETFs, but not in financial ETFs.

Table 4. Regression Results of Monthly Returns for Hong Kong and Taiwan ETFs

	HK ALL ETF	Taiwan: All ETF	Electronics ETF	Financial ETF
C	0.78	1.37	1.66	1.33
Turnover	0.11***	0.00	−0.04	−0.01
ExRate	−4.46	41.18***	48.98*	28.68

Note: Superscripts *, **, and *** denote significance of the t-test for the difference in means between the two subsamples at 10%, 5%, and 1% levels, respectively.

4 Results

The empirical results of this study show that:

(1) The performance of Hong Kong ETFs is positively related to the turnover rate.
(2) The performance of Taiwan ETF is positively related to exchange rate changes.
(3) Exchange rate changes have an impact on the performance of electronic ETFs in Taiwan, but no impact is seen on financial ETFs.

The Taiwan market is dominated by the export industry, especially the electronics industry for 3C products, so the performance of ETFs is deeply affected by exchange rate changes. The performance of ETFs issued in Hong Kong, known as the Asia-Pacific financial center, is not affected by the exchange rate, but is affected by the turnover rate. The results of this study can be used as a reference for investors and practitioners to make long-term investment decisions and can even be used as a reference for the governments of Hong Kong and Taiwan when formulating investment-related policies.

Acknowledgments. Constructive comments of editors and anonymous referees are gratefully acknowledged. Our research is partly supported by the National Science Council of Taiwan (NSC 109-2813-C-468-099-H and NSC 110-2813-C-468-086-H).

References

Caglayan, M.O., Celiker, U., Sonaer, G.: Hedge fund vs. non-hedge fund institutional demand and the book-to-market effect. J. Bank. Finan. **92**, 51–66 (2018)

Chen, C.Y.R., Wang, J.C., Hung, S.W.: The impact of corporate governance on the corporate social responsibility and firm values. Sun Yat-Sen Manag. Rev. 25(1), 135–176 (2017)

Chen, H.Y.: The information content of fund awards. Rev. Secur. Futures Mark. 24(1), 161–194 (2012)

Chen, R.S., Hung, P.H., Liu, Y.S.: Are funds of funds superior to traditional mutual funds? A perspective on the risk-and-return and fund characteristics. J. Manag. Syst. 2(2), 363–392 (2014)

Chuang, Y.N., Liu, J.T., Lee, C.C.: The effect of corporate social responsibility, investment activities and financial activities on operating performance. Fu Jen Manag. Rev. 30(2), 1–29 (2023)

Converse, N., Yayat, E.L., Williams, T.: How ETFs amplify the global financial cycle in emerging markets. Rev. Finan. Stud. 36(9), 3423–3462 (2020)

Fu, H.P., Hsieh, S.F.: The Chinese lunar new year and reverse disposition effect. Sun Yat-sen Manag. Rev. 23(2), 521–561 (2015)

Hsu, C.C., Lin, C.J., Li, C.H., Chao, C.F.: Performance persistence of equity mutual funds in Taiwan securities market. Ling Tung J. 37, 1–21 (2015)

Hsu, K.H.: Using static and dynamic approaches for the mutual funds portfolio selection. J. Manag. 27(1), 75–96 (2010)

Huang, M.G., Ma, K., Lee, D.Y., Lee, C.W.: Performance advantage analysis and optimal application situations exploration for mutual funds with using value averaging: evidence from Taiwanese stock-type funds. Int. J. Commer. Strat. 1.9(4), 233–256 (2017)

Kao, L.F., Chen, A.L., Tang, H.W., Tsao, M.L.: The performance measures for mutual funds: a simulation approach. Sun Yat-Sen Manag. Rev. 13(3), 667–694 (2005)

Lee, H.Y., Lee, H.W., Lee, L.L.: The relationship between investment concentration and performance of mutual fund. Manag. Sci. Res. 7(2), 49–62 (2011)

Lee, H.Y., Liao, W.L.: The analysis on stock picking, market timing ability and average style for outstanding mutual funds. Taiwan Bank. Finan. Q. 13(3), 69–92 (2012)

Lee, T.S., Yu, T.H.: An investigation of ETF tracking errors in Taiwan, Hong Kong, and Mainland China. Cross-Serai Bank. Finan. 3(1), 1–22 (2015)

Lin, A., Meng, F.J.C.: Is Taiwan's first exchange traded fund efficient? J. Finan. Stud. 12(3), 107–138 (2004)

Lin, C.C., Chen, J.X.: The study of using data mining on mutual fund performance and herd behaviors. Manag. Inf. Comput. 11(1), 216–228 (2022)

Lin, M.C.: Mutual fund tournament in the Taiwan market: risk-taking, turnover, and investors' rewards. J. Finan. Stud. 12(2), 87–142 (2004)

Lin, Y.E., Chih, H.H.: The disposition effect, escalation of commitment and momentum strategy in mutual funds. J. Manag. 30(2), 147–168 (2013)

Liu, Y.C., Chen, H.J., Liu, W.J.: A comparison of dollar-cost averaging with lump-sum investing for mutual funds. J. Manag. Syst. 15(4), 563–590 (2008)

Lu, C., Chang, C.F., Shen, Y.P.: Does active share explain Taiwanese mutual fund performance? Rev. Secur. Futures Mark. 29(3), 1–38 (2017)

Ma, K., Huang, M.G., Wu, I.H.: The relationship between investment style and investment performance for Taiwanese stock. Mutual Funds Taiwan Bank. Finan. Q. 12(2), 41–71 (2011)

Peng, K.C., Lin, T.J., Chang, S.Y.: Impact of Taiwan ETF by European sovereign debt crisis. J. Chin. Econ. Res. 13(1), 105–126 (2015)

Sherrill, D.E., Shirley, S.E., Stark, J.R.: ETF use among actively managed mutual fund portfolios. J. Finan. Mark. 51, 100529 (2020)

Tsao, H.E., Hsu, M., Kuang, H.: Investment performance and tracking ability of Taiwan high dividend ETFs. NPTU Manag. Rev. 4, 51–76 (2021)

Wang, C.E., Lin, M.C., Yeh, C.H.: Compensation incentives, employment risk, and intended risk-taking in the Taiwan mutual fund industry. J. Finan. Stud. **22**(2), 97–127 (2014)

Yuan, S.F., Tseng, C.W.: The relationship between NAV and fund performance and persistence: evidence on equity index fund in Taiwan market. Manag. Inf. Comput. **3**(2), 345–456 (2014)

Research on Optimization Strategies of Pension Investment Portfolio — Taking Civil Servants as an Example

Ying-Li Lin[1], Tzu-Ting Chao[2], Kuei-Yuan Wang[1(✉)], and Hui-Ling Yang[1]

[1] Department of Finance, Asia University, No. 500, Lioufeng Road, Wufeng, Taichung 41354,
Taiwan
yllin@asia.edu.tw, gueei5217@gmail.com,
112135006@live.asia.edu.tw
[2] Department of Business Administration, Asia University, No. 500, Lioufeng Road, Wufeng,
Taichung 41354, Taiwan

Abstract. Taiwan has officially entered an aging society and faces the important challenge of aging population. To cope with this situation, retirees need to plan their finances carefully to ensure a stable retirement life. First, establish a comprehensive retirement financial plan that includes considerations for fixed and non-fixed expenses. Secondly, understand your own pension, savings, and investments, and then rationally adjust your investment portfolio to suit your personal risk tolerance. Next, consider health insurance and long-term care plans to cover possible medical expenses in the future. Additionally, aggressively paying down debt, building emergency reserves, engaging in proper tax planning, and updating estate documents are all important steps. At the same time, continue to learn financial knowledge and adjust your lifestyle, including housing, travel, and leisure plans, based on your retirement income. According to the strategies, retirees can manage their finances more effectively and enjoy a stable and secure retirement life.

1 Introduction

With the advancement of medical technology and the intensification of population aging, many countries, including Taiwan, are facing major changes in their demographic structure. In particular, the post-war baby boom generation has gradually entered retirement age, and the birth rate of newborns has continued to decline, causing the proportion of the elderly population over 65 years old to increase year by year. According to National Development Council, Taiwan will enter an elderly society by 2025. The population over 65 years old will account for more than 20% of the total population.

Many countries around the world have established various pension systems to ensure the economic security of the elderly in accordance with an aged society. However, these retirement systems also encounter funding challenges, as estimated pension income may not be sufficient to cover future expenses. Therefore, in recent years, various countries have discussed the reform of the retirement system, which includes changing from a

L. Barolli (Ed.): IMIS 2024, LNDECT 214, pp. 433–440, 2024.
https://doi.org/10.1007/978-3-031-64766-6_43

defined benefit system to a defined contribution system, extending the retirement age, increasing the contribution rate, reducing benefits, and adjusting the asset allocation of pension funds. Against this background, annuity reforms haves become a hot topic, triggering the attention and actions of many civil servants, police officers and military personnel, who have taken to the streets to defend their rights and interests. As annuity reforms continue to advance, the pensions of these groups have also been affected. Under such circumstances, retirees need to pay more attention to their financial planning to adapt to these changes and ensure future financial security.

This study aims to provide in-depth analysis and optimization strategies of pension investment portfolio of civil servants. The main purposes include understanding the status of pensions, clarifying the financial needs and risk tolerance of retirees, and exploring how different investment strategies affect the benefits and risks of pensions. In addition, this study will provide specific recommendations for improving pension investment portfolios, consider the impact of market dynamics and policy changes on investment strategies, and work to improve retirees' understanding and ability of financial planning.

2 Literature Review

2.1 Definition and Scope of Financial Management Cognition

Danes and Tahira proposed in 1986 that the scope of financial knowledge should include the use of credit cards, insurance planning, loan management, frequency of recording expenses, and other elements of financial management. Chen and Volpe (1998) divided financial knowledge into several areas, including basic financial knowledge, savings and loans, insurance and investment. Jacob, Hudson, and Bush (2000) emphasized that financial knowledge should include the management of checking accounts, the value from credit cards, budgeting, loan repayment, and insurance purchasing. Finally, Hogarth and Hilgert (2002) suggested that financial knowledge should cover general financial knowledge, credit use, mortgages and savings.

2.2 Definition and Purpose of Financial Investment

Li (2023) proposed that financial planning basically refers to managing personal finances, including the remaining income after deducting necessary expenses and special expenses from monthly income, and using various financial management tools to achieve goals. In Taiwan, there are four typical retirement expenses, including living expenses, leisure and entertainment, medical care, and nursing care. Du to higher education, better quality of life, and less younger workers, people no longer insist on retiring at 65. Although the government provides a pension system, in the current environment of low interest rates, high house prices and inflation, many people are spending more than they earn. In order to maintain the quality of life before retirement and avoid becoming a poor elderly person, people must save on their own and make financial plans in advance. Scholars particularly emphasize three elements in retirement financial planning, financial goals, information sources, and investment tools. The goals include analyzing personal financial situation and investment risk preferences, setting an ideal income replacement rate, and considering education funds, home purchase plans, and retirement plans. Besides, monthly cash requirements during retirement and the risks of inflation and medical care, as well as longevity. There are several sources of information to refer to planning retirement finances. After collecting all the information you need, choose appropriate investment tools to pick for your profolio.

2.3   Retirement Life Expectations

The retirement career for small and medium-sized business owners is to achieve retirement goals and aspirations, and to ensure a meaningful and fulfilling retirement life according to Chiu (2023). This includes three main aspects, first, the maintenance of physical and mental health, through the development of health plans to maintain good physical and mental status and to improve the quality of retirement life; second, the assurance of financial stability, including the development of financial plans to meet the needs of retirement life, and planning entertainment, social and health activities; finally, arrange end-of-life matters after retirement, especially for small and medium-sized business owners, involving advance preparations for matters such as inheritance and succession, to ensure a peaceful and regret-free farewell.

2.4   Retirement Financial Planning

Hsu (2020) found that most junior high school teachers have insufficient understanding of the retirement system and expected retirement benefits, which affects their effective retirement financial planning. Therefore, it is recommended that junior high school teachers should actively comprehend retirement-related regulations and trial pension calculations, for example, through the school personnel office and the Ministry of Education website. At the same time, teachers should also cultivate interests in investment and financial management, use multiple channels such as financial magazines or online medias to enhance financial management knowledge, and choose appropriate investment tools based on personal risk tolerance. In addition, to achieve a win-win situation, the education authorities and the financial-related institutions should actively promote the retirement system and provide investment and financial management advice tailored to the characteristics of teachers, such as developing suitable websites and applications, and also providing pension calculations and financial planning information to help teachers plan for retirement, and at the same time, those teachers will be new customers to financial institutions as well.

Chiu (2022) also pointed out that whether you are a non-retired person or a retired person, financial management is a necessary behavior in life and requires patience. It is very important to understand your own risk tolerance for financial products. Some people have a negative or conservative attitude towards financial management and believe that it is the safest to deposit money in a current account or time deposit. However, this view ignores the impact that financial changes such as inflation may have on the security of funds. On the other hand, some customers who pursue high-risk, high-yield speculative investments may face greater risks of loss. Therefore, it is crucial to understand the risks of investing in commodities. In terms of retirement finance, it is necessary to regularly absorb investment and financial information and review the investment profile, because the financial environment is changing rapidly. At the same time, research shows that both retired and non-retired people believe that it is very important to have financial tools with a fixed source of income after retirement to supplement living expenses in addition to government social insurance.

3 Research Methods

3.1 In-Depth Interviews

This study uses in-depth interviews as a research method, which is very suitable for exploring the retirement investment situation of Taiwanese civil servants. In-depth interviews are a qualitative research method that allow researchers to obtain rich, detailed data through direct conversations with participants. Here are some steps and considerations for conducting in-depth interviews. First, develop an interview outline, which is a series of open-ended questions designed to guide interviewees into sharing their experiences, attitudes, beliefs, and behaviors.

In this study, these questions will center around topics such as retirement planning, investment strategies, risk tolerance, and more. Select appropriate participants, selecting public servants of different ages and retirement statuses (retired and non-retired) to obtain a diverse perspective. Then choose to conduct the interview in a distraction-free environment to ensure that the interviewee can express his or her views freely, while conducting appropriate follow-up questions based on the interviewee's answers. Interviews should be audio-recorded and written down. Finally, the data is analyzed to identify patterns, themes, and insights. To ensure the privacy of participants and the confidentiality of data, consent will be obtained from the interviewees and their names will be given by numbers. Gaining a deeper understanding of civil service retirement investments through in-depth interviews will help uncover details and deeper insights not evident in quantitative data (Table 1).

3.2 Interviewees

The interview subjects were selected from civil servants with different backgrounds, including different ages, genders, job levels and work departments, to ensure the representativeness and diversity of the research results. Among them were retired and non-retired civil servants to compare and understand the needs and challenges at different stages of retirement. At the same time, consider the financial status of the participants, their savings, investment habits, family income, etc., as these factors may affect their retirement investment decisions. Civil servants with different levels of preparedness for retirement were also selected, including those who have begun actively planning for retirement and those who are more passive about it or have not yet begun planning.

4 Empirical Results

4.1 Cognition of Retirement Financial Management

In terms of their understanding of retirement financial management, respondents have different views on the current retirement system, satisfaction and adequacy of pensions. Some respondents believe that current pensions are sufficient to support their retirement

Table 1. Interview outline

Retirement Financial Management Awareness

1	Are you satisfied with the current military, civil, educational or labor retirement system and pensions? Is it enough to support your retirement life? At what age do you expect to retire? When should you start preparing for retirement? How much pension is expected to be prepared?
2	Do you think retirement investment and financial management are important? Please explain why
3	Please describe your views on retirement financial management and what channels do you use to obtain relevant knowledge?
4	Please tell me what factors you think are most likely to reduce the quality of retirement life in the future (investment failure, major medical expenses, inflation…), and please explain how you will deal with them
5	What are the key factors you consider when formulating your retirement investment and financial plan? ⟨Accumulation of funds, preparation for property purchase for future generations, improvement of living standards, prevention of personal risks…⟩

Retirement life expectations

1	Please describe your living environment after retirement. Do you expect to live with your children, live by yourself, or live in a nursing home?
2	Please describe your expectations for your health after retirement (lifestyle, exercise habits or dietary changes…)? Are regular health checks expected? Are you fully prepared for medical care for the elderly?
3	Please describe your leisure activities after retirement (dancing, hiking, making tea…)? Do you expect to arrange travel regularly (travel area, duration, etc.)?
4	Please estimate how much monthly living, medical, travel and other related expenses you will need after retirement

Retirement Financial Planning

1	Do you think there is a correlation between retirement life expectations and retirement financial planning? Why?
2	What factors do you consider when formulating an investment and financial plan? ⟨Accumulation of funds, preparation for property purchase for future generations, improvement of living standards, prevention of personal risks…⟩
3	What are your investment tools for retirement planning (time deposits, funds, stocks, bonds, annuity-type commercial insurance, real estate…)? What are the main factors you consider when choosing financial tools (investment return rate, risk level…)? If investment and financial management can tolerate low risk (conservative type), medium risk (stable type), or high risk (active type), please explain the reasons
4	In order to meet the expenses of retirement life, are you likely to change financial investment tools after retirement? What adjustments might be made? And please explain the reasons for choosing these financial investment tools?

(continued)

Table 1. (*continued*)

5	In order to achieve your retirement life expectations, have you made any special planning arrangements in advance in your retirement financial planning for health care, housing and care, leisure and entertainment, economic security and other needs?
6	Based on your current retirement financial planning, do you think you can achieve the above-mentioned retirement life expectations? Why?

Data source: Lu (2016), Li (2023)

life, while others are dissatisfied and believe that pensions are not sufficient to fulfill the requirements in the future. It really reflects several evaluations of the current retirement system and uncertainty about future economic conditions. Respondents generally believe that retirement investment and financial management are very important. The reasons include ensuring the financial needs for retirement and improving the quality of life. Hence, retirees are in urgent need for a stable retirement life, an awareness of financial knowledge, and an appropriate investment tool.

4.2 Retirement Life Expectations

In terms of retirement life expectations, respondents have different expectations and plans for their living environment, health status and leisure activities after retirement. Some respondents wanted to live with family, while others planned to live alone or in a nursing home. This reflects different preferences and plans for retirement lifestyle. Most of the respondents emphasized the importance of health, planning to have regular health check-ups and maintaining good living habits. In addition, they have clear ideas about leisure activities after retirement, including travel, sports, etc., showing their emphasis on the quality of retirement life.

4.3 Retirement Financial Planning

In terms of retirement financial planning, respondents generally believe that there is a significant correlation between retirement life expectations and retirement financial planning. They realize that only through sound financial planning can they achieve their retirement life expectations and goals. Factors that interviewees consider when formulating investment and financial plans include accumulation of funds, preparation for property purchase for future generations, and improvement of living standards. They choose a variety of retirement financial investment tools, including time deposits, funds, stocks, bonds, annuity-type commercial insurance and real estate. This shows respondents' perceptions and preferences for different investment instruments, as well as their considerations in risk management and asset allocation.

Based on the above results, it can be seen that the respondents have extensive and diverse considerations in retirement financial planning. They generally realize the importance of retirement financial management and make plans based on their own financial situation, life needs and risk tolerance. Respondents have different expectations and plans for their post-retirement life, which reflects their pursuit of future quality of life and their emphasis on health and leisure activities.

The results of this survey provide valuable insights into the attitudes and behaviors of Taiwanese people regarding retirement planning. It reveals the different needs and challenges people face in retirement and how they manage their finances to meet these challenges. Through these data, we can gain a deeper understanding of the actual needs and considerations of Taiwan's public servants in retirement planning, thereby providing a more targeted reference for relevant policy formulation and financial services.

5 Conclusion

People in Taiwan generally recognize the importance of retirement financial planning and understand that relying solely on pensions provided by the government is not enough to maintain the desired standard of living in retirement. Therefore, many people actively engage in personal financial planning, including investing in various financial products, such as stocks, funds, bonds, insurance, and real estate. Respondents have a clear understanding of the risks in investing and financial management, and have adopted a variety of strategies to deal with these risks, including diversifying investments, choosing low-risk investment vehicles, and regularly rebalancing their investment portfolios.

As regards expectations and preparations for retirement life, the respondents have clear and diverse expectations for retirement life, including a comfortable living environment, ample leisure and entertainment time, stable health conditions and sufficient travel opportunities. In order to achieve these goals, many people have begun or plan to engage in various forms of financial planning. Demonstrating significant diversity in financial planning, they choose different investment tools and strategies based on their own financial situation, life needs and risk tolerance. In terms of the relationship between retirement planning and age, younger respondents are more inclined to adopt active investment strategies, while older people pay more attention to ensuring financial security and stable income. The respondents have a certain understanding of Taiwan's current pension system, but many expressed dissatisfaction with the current system and believe that the amount of pension is not enough to meet their retirement needs. In terms of education and financial planning, it was found that the degree of education is positively correlated with the level of financial management cognition. Respondents with higher levels of education generally have a deeper understanding of financial management and are able to formulate and implement financial plans more effectively.

Therefore, the research results show that Taiwanese people have extensive and diverse considerations in retirement financial planning. They generally realize the importance of retirement financial management and make plans based on their own financial situation, life needs and risk tolerance. These findings provide important insights into understanding the actual needs and behaviors of Taiwanese people in retirement planning and provide valuable reference for related policy formulation and financial services. In addition, the respondents also showed a high emphasis on risk management in their retirement financial planning. They tend to adopt more prudent and diversified investment strategies to reduce risks and ensure financial security. These findings provide important insights into understanding the actual needs and these behaviors of Taiwanese in retirement planning and provide valuable reference for related policy formulation and financial services.

Acknowledgments. This research was supported by National Science and Technology Council of the Republic of China under contract NSTC 112-2420-H-468-001-.

References

Chen, H., Volpe, R.P.: An analysis of personal financial literacy among college student. Financ. Serv. Rev. **7**(2), 107–128 (1998)

Cuiu Hong, C.-Y.: Discussion on the key factors of Taiwanese retirement financial planning [Unpublished master's thesis]. Soochow University (2022)

Chiu, N.-L.: Research on the influence of SMEs owners' retirement career planning on retirement financial behavior -taking SMEs owners in the central region as an example [Unpublished master's thesis]. Chaoyang University of Technology (2023)

Danes, S.M., Tahira, M.: Management knowledge of college students. J. Student Financ. Aid **17**(1), 4–16 (1986)

Hogarth, J.M., Hilgert, M.A.: Financial knowledge, experience and learning preferences: preliminary results from a new survey on financial literacy. Consum. Interest Ann. **48**(4), 1–7 (2002)

Hsu, Y.-L.: The study of financial belief and financial planning for retirement of junior high school teachers in Tainan city after the pension reform [Unpublished master's thesis]. National Taiwan Normal University (2020)

Jacob, K.S, Hudson, Bush, M.: Tools for Survival:An Analysis of Financial Literacy Programs for Lower-Income Families. Woodstock Institute, Chicago (2000)

Li, C.-Y.: A study on financial cognition, financial behavior and retirement financial planning of elderly teaching and administrative staff in private colleges [Unpublished master's thesis]. Nan Kai University of Technology (2023)

Lu, Y.-C.: A study of Yilan X generation civil servants' and teachers' expectations on post retirement life and planning for retirement financial management [Unpublished master's thesis]. Fo Guang University (2016)

An Exploration of Financial Planning and Wealth Management – A Case Study of Fresh Graduates

Ying-Li Lin[1], Tzu-Ting Chao[2]([envelope]), and Yu-Ai Chang[1]

[1] Department of Finance, Asia University, No. 500, Lioufeng Road, Wufeng, Taichung 41354, Taiwan
yllin@asia.edu.tw, 112535011@live.asia.edu.tw

[2] Department of Business Administration, Asia University, No. 500, Lioufeng Road, Wufeng, Taichung 41354, Taiwan
coffeelovescheese@gmail.com

Abstract. Wealth gap between the rich and the poor continues to widen. Due to the reason of their household incomes or household wealth, more fresh graduates need to repay student loans and more fresh graduates join the BNPL "Buy Now, Pay Later" trend. It is very important to help fresh graduates achieve financial literacy, recognize financial tools, strengthen financial management skills, and catch up to more affluent peers. The evaluation indicators of this study on how fresh graduates plan and manage wealth are as follows: budgeting, investments, banking, saving, mortgages, insurance, and retirement planning. At the same time, fresh graduates, who should allocate their incomes, manage their money, deal with financial issues, make financial decisions, and practice financial behavior efficiently, can accomplish long-term wealth and financial freedom.

1 Introduction

Income and wealth inequality brings the wealth gap between richer and poorer households. It is crucial for individuals to have good financial literacy. A person who effectively manages wealth, controls expenses, and bridges the gap between income and expenditures can better cope with economic pressures such as unevenly rising prices and sudden expenses. Relatively speaking, people who lack financial skills and knowledge may be negatively affected. They may overspend, buy non-essential items, make unplanned purchases, or end up in financial stress. The situation is especially serious for fresh graduates who may be burdened with student loans when they graduate. In addition to student loans, BNPL, a short-term loan "Buy Now, Pay Later" allows fresh graduates to buy goods or services in installments in recent years. In conclusion, positive financial behavior is a crucial aspect of fresh graduates to have adequate financial management skills, to manage spending effectively, to avoid impulse spending, and to invest in asset-based investments. As for budgeting, investments, banking, saving, mortgages, insurance, and retirement planning, practicing above behaviors helps fresh graduates achieve financial stability in the long run.

L. Barolli (Ed.): IMIS 2024, LNDECT 214, pp. 441–449, 2024.
https://doi.org/10.1007/978-3-031-64766-6_44

Financial knowledge can help people become more resilient to financial challenges for bridging the wealth gap. The ability to manage personal finances is extremely important for the economic stability of individuals and families. People with good financial literacy skills who can control their spending, increase their income, and plan effectively for the future are often able to avoid financial problems, while those who lack these skills may find themselves in trouble, especially those who have left their jobs or are new to the workforce. For this reason, Taiwan is facing a common situation, such as progressive rising prices and increased living burdens, and unobvious salary increases. In such an environment, people who effectively manage resources, save expenses, and find other sources of income can avoid financial problems, while those who lack financial management skills and overspend often fall into financial difficulties and may be burdened with debt pressure, as a result, especially fresh graduates. The gap between the rich and the poor is growing, which means that some people can use their resources and knowledge to accumulate wealth, while others are unable to relieve financial stress. Therefore, how to effectively manage money and acquire financial knowledge and skills have become an important part of ensuring personal financial stability.

Therefore, this study believes that it is very important to have sufficient financial management skills and knowledge and cognition related to financial management when students are young. However, not everyone has sufficient financial management knowledge and literacy, and having sufficient financial management literacy does not necessarily lead to correct financial management behavior. Here, this study collates relevant financial literacy and financial management literature and proposes preliminary financial planning and management steps, so that newcomers to society can refer to and implement the direction.

2 Literature Review

2.1 Financial Literacy

The Financial Supervisory Commission (hereinafter referred to as the Financial Supervisory Commission) (2008) defines financial literacy as "Being able to make well-founded judgments and effective decisions in the management and application of money, and must have the ability to read, analyze, manage and the ability to communicate personal financial situations." Only by having correct financial knowledge can you make more effective decisions. Therefore, having financial literacy is an important survival skill that everyone should have. Financial decisions are mainly aimed at future personal financial welfare and are time-sensitive and long-term effect. Morton (2005) proposed that the core concept of financial literacy is the application of money management skills and knowledge. It includes the ability to understand, analyze and communicate on financial issues, including maintaining a balance of income and expenditure, understanding the contents of purchase and sale contracts, and understanding future wants and plans for the life you want to live and the life you can enjoy after retirement. Widdowson and Hailwood (2007) believe that financial literacy can be divided into narrow and broad definitions. The narrow sense refers to aspects such as budgeting, savings, investment and insurance, that is, basic money management skills; while the broad sense of financial literacy is a broad concept, Broadly refers to an individual's understanding of economics, an appreciation of how economic conditions and conditions affect personal and family finances, and the knowledge and skills needed to properly manage finances and make appropriate financial decisions after being aware of possible outcomes.

Financial literacy and knowledge are important personal capitals that change one's money management in response to the changes in the times. They must have practical knowledge of financial institutions, systems and services, have a certain degree of analytical ability, and be able to plan out plans that meet personal needs and benefits, as well as responsible financial management decisions (Li, 2010). Robb and Woodyard (2011) measured emergency reserves, personal credit reports, overdraft loans, regular credit card payments, the availability of special retirement accounts, and risk management.

Lai (2017) studied the definition of financial literacy using Lu (2010) definition of financial literacy, "It refers to the ability to use knowledge and skills to effectively manage one's financial resources for a lifetime of financial security." It includes knowledge, analysis, communication, skills, grasping changes in the overall economic environment and financial system, acquisition of knowledge, and methods of acquiring new information, allowing individuals to use it in areas such as work, investment, capital income, savings, editing budgets, expenditures, making good decisions and being able to plan for your retirement life and future in advance, so that you can live without worries.

2.2 Financial Management Behavior

Financial management behavior, as the name suggests, refers to all money-related activities including consumption, insurance, savings, investment, financing, and budget investment. Yang (2003) believes that personal financial management behavior is to measure one's financial status according to accounting rules, formulate short-, medium-, and long-term financial plans, and manage risks through cash flow management to achieve financial goals. Wang (2005) believes that financial management behavior is the effective use of personal and financial resources to maximize the effectiveness of money to meet life needs. The Taiwan Financial Supervisory Commission (2008) defines it as whether our citizens take effective or correct decisions and behaviors in financial management. That is to say, you can reasonably plan your lifetime work income, investment income and expenditures to achieve financial goals at different stages of life, and use risk management to reduce the impact of unexpected expenditures when accidents occur, so that individuals or families can achieve better results in the future, not worrying about financial problems. Huang (2010) believes that financial management behavior includes investment, savings, consumption, insurance, budgeting and borrowing, etc., and also includes the selection of different investment targets, such as real estate, funds, stocks, bonds, etc. In addition, financial management behavior also includes the pursuit of rewards, risk sharing and time value and other implications. Lai (2017) studied the definition of financial management behavior based on Lu (2010) definition of financial management behavior, "taking effective or correct decisions and behaviors in money management." accounting, budgeting, income, investment, insurance, savings, spending and borrowing are used to achieve financial goals at different stages and meet life needs through effective planning. Based on the above definition of financial management behavior, this study believes that effective planning and specific implementation in the four major aspects of money management and maintenance, credit and loan management, financial management and investment planning, insurance and pension planning are called financial behavior.

3   Research Methods

3.1   In-Depth Interviews

The interview questions in this study refer to the four major aspects of financial knowledge indicators proposed by Xiao, Chen, and Liao (2016): money management and savings, credit and loan management, financial management and investment planning, and insurance and pension planning and design interviews Questions, as shown in the table below (Table 1):

Table 1. Interview outline

Money Management and Savings	
1	Do you think there is the difference between "want" and "need"?
2	Do you have the habit of keeping accounts to manage money?
3	How do you plan to use your salary?
4	Have you ever considered using relevant financial institutions to manage your money? Do you know how to use it?
5	What is your concept of saving?
Credit and Loan Management	
1	Do you know what the current recurring interest rates on credit cards are and what the current housing loan interest rates are?
2	Do you know the importance of credit?
3	Do you know how "card slaves" came into being?
4	What is your view on debt? How do you manage your debt to avoid financial distress?
5	Do you know when a "credit report" needs to be filed?
Financial management and investment planning	
1	What is the difference between physical and financial assets?
2	What do you think of the statement "the higher the return on investment, the higher the risk"?
3	If you wanted to invest in stocks, would you choose to do your own research or seek professional help?
4	What are your thoughts on regular fixed-amount investments? What investment method will you choose?
5	Do you have a basic understanding of bonds, stocks, funds or other investment vehicles? Please explain briefly

(continued)

Table 1. (*continued*)

Insurance and Pension Planning	
1	Do you think that insurance is necessary?
2	What are your dreams for the future? Have you made any plans?
3	What is your plan for retirement savings? What tools will you choose to reach your retirement goals?
4	How much do you know about the importance of starting saving early for retirement planning?
5	Do you have any risk management plan (such as insurance)?

Data source: Lu (2016), Li (2023)

3.2 Interviewees

This study plans to interview 10 freshmen who have just entered society within three years, 5 men (codenamed F1–F5) and 5 women (codenamed M1–M5), to evaluate whether their financial literacy has been specifically implemented into financial management behavior. Are they satisfied with the current financial management? We will also provide some financial management and planning steps through the interview to help them achieve ideal goals ahead of schedule.

4 Empirical Results

This research will use interviews to collect data, understand the respondent's background through the respondent's basic information section, including gender, age, working experience, education, occupation category, personal annual income, etc. Then, in the outline of interview questions, corresponding question sections were designed in four aspects: money management and savings, credit and loan management, financial management and investment planning, and insurance and pension planning, so as to understand the interviewees' views on different financial fields. Cognition and coping strategies.

First of all, in terms of money management and savings, question 2 about the presence or absence of accounting habits, only a 10% completion rate! From the interviews, it was learned that respondents such as M4, M5, F1, F2, F3, F4, and F5 were limited in terms of school loans, rent, filial piety fees, etc. that must be paid every month, and there is not much left after what should be paid and given after the salary is paid. In addition, when renting a house, daily necessities and food are also in the era of rising prices, it's really an expense. Occasionally, colleagues have to cope with dinner parties. Therefore, the achievement rate of accounting for financial management is only as low as 10%. Furthermore, in terms of financial planning for salary planning, interviewees M1, M2, and M3 all made fixed deposits or fund deductions, and had financial investment plans; M4, F2, and F5 made fixed deposits because they planned to buy a house in the future. In terms of planning, the interviewees deeply feel that house prices remain high and require a large amount of money. Even if they can find a bank loan, they still have to prepare the down payment themselves. Therefore, if they hope to buy their own property in the future, they must have a fixed deposit. The interest rate is not high, and it is also a way to force yourself to save. Therefore, the achievement rate of salary planning is about 60%. In addition, when interviewee M5 wanted to open his own cafe, F1 wanted to open a restaurant, F3 wanted to open a bookstore, F4 wanted to travel abroad, etc., they had a more casual attitude, so they did not actively plan their own salary. In terms of money management and savings, although the understanding of financial knowledge is 100%, there is an inconsistent gap of 60% or even 10% in financial management behavior. It can be seen that knowing is one thing, but actually doing it is another. Another story. Furthermore, among the interviewees, four of the five male interviewees were active in money management and savings, while only two of the five female interviewees who had plans to buy a house were active in financial management. From the interviews, we learned that most men are given the responsibility by society to buy a house, start a family and start a business. Although times have changed and it is common for couples to start a family together, most male interviewees still hope to own house assets, whether they own houses or rent them out. As a symbol of ability, real estate can also be regarded as an investment tool. For female interviewee F2, since she is the eldest sister in a family of five, she has always hoped to own a large house that can accommodate more than ten or twenty people so that everyone can live comfortably and be more emotionally united, so she has house buying ideas. Respondent F5, because of his work relationship, hopes that when he gets old in the future, regardless of whether he is married or not, he will have a comfortable home and even have house assets as income for his retirement life, so that he does not have to run around for money. While having good medical care. In terms of money management and savings (Questions 2–5), we can understand the respondents' adoption of accounting, deposits, investments, etc. Especially in question 3, the interviewees mentioned that the excess amount was deposited in bank or invested in stocks or funds, which reflects their attempts and thinking about diversified financial management. However, in question 4, it is necessary to further assess the respondents' familiarity and willingness to use various financial institutions for wealth management.

Secondly, in terms of credit and loan management, only M1, M2, and M3, accounting for 30%, are clear about credit card and home loan interest rates. However, M4, F2, and F5 who have plans to purchase a house in the future do not even know about home loan interest rates? The interviewee explained in the interview that he was unclear about the housing loan interest rate because it was currently far from the down payment reserve, and he wanted to save an amount first and then learn more about it. In terms of knowledge related to credit cards and credit reports, 100% of the respondents have a high level of knowledge. During the interview, most of the interviewees were not card holders. They learned about the hard work of card slaves from colleagues or friends. Under the calculation of compound interest, the debt will increase. Furthermore, interviewees M4, M5, F1, F2, F3, F4, and F5 all have student loans that have not been repaid, so they know the importance of dealing with banks. If they want to get a loan to buy a house in the future, they may not be able to apply for a loan if they don't have good credit. Therefore, regarding credit and loan management aspects, they pay extra attention. In this aspect of financial management, students will pay their student loans on time because they want to maintain good credit. However, in the interviews, some interviewees complained that the heavy pressure of student loans always caused a large part of their salary to be lost, and even delayed the realization of their dreams, often showing the dilemma of being cash-strapped. In terms of credit and loan management (questions 6–10), you can explore the respondents' knowledge and coping strategies on credit cards, liabilities, credit reports, etc. Especially in question 8, exploring the causes and countermeasures of the "card slaves" phenomenon will help understand the respondents' views and practices on avoiding excessive consumption and debt. In this modern society, the importance of personal financial management capabilities has become increasingly prominent, among which credit and loan management are one of the key factors affecting a person's economic health. This study examines how to effectively manage credit and debt and how to avoid financial distress. By analyzing the financial knowledge and behavior of respondents, we can gain a deeper understanding of their knowledge, attitudes, and actions in the financial field. Regarding credit and loan management, the questionnaire covers issues such as credit card interest rates, home loan interest rates, the importance of credit, and liability management. Respondents have a certain understanding of credit card revolving interest rates and home loan interest rates, and believe that maintaining a good credit record is important for future financial activities. In addition, there is also a discussion on the occurrence of the "card slave" phenomenon and how to avoid falling into financial difficulties. Most of the interviewees stated that their view on debt is that their income is less than their expenses, so they need to control expenses and avoid unnecessary borrowing.

Third, in terms of financial management and investment planning, only respondents M1, M2, and M3 know the differences and views on different investment tools such as stocks, regular fixed-amount investments, bonds, funds, etc., accounting for only 30%. They do research because they invest. In other words, they invest appropriately because they have research and understanding. Other interviewees were struggling with student loans, rent, utility bills, family expenses, and daily living expenses, and could not save any money at all. In terms of financial management and investment planning (Questions 11–15), by comparing the differences between physical assets and financial assets in Question 11, you can understand the respondents' understanding of different investment methods. At the same time, exploring the relationship between investment return and risk in question 12 will help assess the respondent's perception of risk tolerance and investment orientation. In addition, questions 14 and 15 focus on the characteristics and differences of different investment tools to help respondents better choose an investment method that suits them. In terms of financial management and investment planning, the questionnaire covers issues such as investment knowledge, risk management and retirement savings planning. Respondents have a certain understanding of the differences between physical assets and financial assets, the relationship between investment returns and risks, and stock investment methods. Most of the respondents said they would consider conducting their own research or seeking professional help to invest in stocks, and had a positive attitude towards the regular fixed-amount investment method. In terms of insurance and pension planning, interviewees generally believe that insurance is important for future life and have begun to make some retirement savings plans, mainly focusing on active savings and investment.

Finally, in terms of insurance and pension planning, the respondents all have a certain level of understanding of financial management knowledge related to insurance and pension planning. Regarding the financial behavior of future risk management plans, 100% of the respondents have basic medical insurance. It can be seen that the intervie-wees all believe that insurance is necessary and actually implement it. However, only 60% of the financial management behaviors implemented in future dreams are actively implemented and 20% are passively implemented. The so-called active implementation refers to the specific implementation of financial management and investment activities, either investing in funds or time deposits; passive implementation refers to the accumu-lation of investment only if there is room for it, without active investment and financial management activities. In terms of insurance and pension planning (questions 16–20), the focus is on the respondents' understanding and planning of the necessity of insurance and retirement preparation plans. In questions 17 and 18, asking respondents whether they have clear goals and plans for their future life and how to plan for retirement savings can help assess their emphasis on and ability to implement long-term financial planning.

5 Conclusion

This study aims to explore the cognition and behavior of new people in society regarding money management, credit and loan management, financial management and investment planning, and insurance and pension planning. Through the interview question outline for the interviewees, we can understand their views, habits and awareness on different financial issues, and discover their possible needs and challenges in financial planning. Overall, this study started from the basic background of the interviewees and deeply explored their practices and thinking patterns in aspects such as money management, credit loans, financial investments, and insurance and pensions. By analyzing the content of the interviews, we can better understand the financial management knowledge and risk management capabilities of the interviewees, and provide an important reference for further improving public financial education and personal financial management capa-bilities. Based on the answers to the interview questionnaire, the respondents' cognitive ambiguities and behavioral differences in money management, credit loans, financial investments, and insurance and pension planning can be concluded. It is recommended to provide more financial education courses for newcomers to enhance their understand-ing of financial planning and risk management and help them establish sound financial management concepts and action plans. In addition, we should also strengthen the pop-ularization of the significance of pension planning and insurance to promote personal financial health and long-term development. Based on the above results, this study puts forward the following suggestions to improve the financial management capabilities of new people in society:

1. Strengthen financial education: The government and educational institutions can strengthen financial education for newcomers to society and improve their under-standing of money management, investment planning, etc.
2. Encourage the establishment of accounting habits: help newcomers to society estab-lish accounting and budgeting habits, which will help them better manage personal finances.
3. Promote insurance awareness: Increase the attention of new people in society to risk management, encourage the establishment of complete insurance plans, and ensure the safety of future life.

4. Provide investment guidance: Provide investment guidance services to newcomers to society, help choose appropriate investment tools, and achieve financial value-added goals.
5. Strengthen retirement preparation: Encourage newcomers to society to start saving and investing early to fully prepare for future retirement life.

Acknowledgments. This research was supported by College Innovation and Entrepreneurship Education Program of the Ministry of Education, Republic of China.

References

Morton H.: Financial Literacy, National Conference of State Legislatures, Washington D.C., pp. 1–13 (2005)

Widdowson, D., Hailwood, K.: Financial literacy and its role in promoting a sound financial system. Reserve Bank New Zealand Bull. **70**(2) (2007)

Robb, C.A., Woodyard, A.S.: Financial knowledge and best practice behavior. J. Financ. Couns. Plan. **22**(1), 60–70 (2011)

Lu, C.-P.: The correlational study of financial literacy and financial behavior of junior high school students in Taipei County [Unpublished master's thesis]. National Taiwan Normal University (2010)

Lai, C.-F.: A study of college students' financial literacy and personal financial well-being-a survey in a university of technology [Unpublished master's thesis]. Shu-Te University (2017)

Wang, C.-T.: A study of personal money management perspectives and their influence on financial planning: a case study of the Taipei elementary school teachers [Unpublished master's thesis]. University of Taipei (2005)

Huang, L.-T.: The survey of financial literacy among Taiwan college students and financial education promoting strategy [Unpublished master's thesis]. Shih Hsin University (2010)

Yang, C.-L.: The study of financial planning service model [Unpublished master's thesis]. National Dong Hwa University (2003)

Hsiao, Y.-J., Chen, J.-T., Liao, C.-F.: The relationship between financial literacy and retirement planning. J. Manage. Bus. Res. **33**(2), 311–335 (2016)

The Impact of Google Search Volume Index (SVI) on Stock Returns of Taiwan AI Supply Chain Stocks

Ying-Li Lin[1], Tzu-Ting Chao[2(✉)], and Chia-Fang Hsieh[1]

[1] Department of Finance, Asia University, No. 500, Lioufeng Road, Wufeng, Taichung 41354, Taiwan
yllin@asia.edu.tw, 112535006@live.asia.edu.tw
[2] Department of Business Administration, Asia University, No. 500, Lioufeng Road, Wufeng, Taichung 41354, Taiwan
coffeelovescheese@gmail.com

Abstract. This study aims to explore the impact of Google Search Volume Index (SVI) on the stock returns of Taiwan's AI supply chain stocks. Google Search trend (Google treads) data, through SVI data generated by keyword searches, are used as proxy variables for natural person (retail) investors' attention to the Taiwan stock market. AI supply chain stocks are used as research samples, and at the same time, according to supply Chain type, the sample is divided into upstream supply chain chip manufacturing companies, key component manufacturing companies and midstream supply chain server component manufacturing companies, and the impact of SVI on stock returns and stock trading volume is analyzed. The empirical results show that the SVI index has a significant impact on company stocks returns, stock trading volume, and stock volume abnormal rate of change.

1 Introduction

In the stock market, stock participants are mainly natural persons (retail investors) and institutional investors (legal persons), and there is often information asymmetry between the two; each legal person institution has a mutually circulated information platform and sophisticated complex analysis of information, while natural persons (retail investors) mostly obtain information from newspaper news, online information, social media, and other channels. Many research results have found that when natural person (retail) investors fail to obtain symmetric investment intelligence, they may implement irrational investment decisions. The total number of users using Internet search in Taiwan today is as high as 21.68 million, accounting for 90.7% of the entire Taiwan population. As far as web browsing is concerned, data shows that the Google Chrome browser platform is still the search portal for most Internet users. Google Search Volume Index, referred to as SVI (Search Volume Index), is mainly used to estimate the number of searches that Internet users use Google to search for specific keywords, and to find relative values in a standardized way. Da, Engelberg and Gao (2011) first proposed that the search volume index (SVI) is a "direct" standard that can measure investor attention, and believed that changes in investor attention are related to retail investor behavior. In recent years, many empirical studies have also used SVI as a proxy variable to observe investor attention to measure the relationship between investor attention and stock volatility. For example: Vlastakis and Markellos (2012) studied the New York Stock Exchange (NYSE) For 30 listed stocks, and they also found that SVI has a significant correlation with market volatility and trading volume, and the hypothesis was verified: as risk aversion increases, investors will need more information. Huang, Lai, et al. (2014) proposed using SVI weekly data to explore the impact of retail investors' online search behavior on stock trading activities. The empirical results also show that online search activities have a positive impact on stock returns, and can also measure different groups of investors' attention. Li, Du, and Wang (2017) used SVI monthly data to explore the relationship with stock returns and trading volume. The study found that whether in the long or short term, SVI showed a significant positive correlation with stock returns or trading volume.

The ChatGPT generative artificial intelligence software, which was launched at the end of November 2022, was developed by Open AI, a company invested by Microsoft. This rapidly rising global craze not only allows people to see the large number of business opportunities for AI chat robots, but also brings about the generation behind it. The new artificial intelligence (AI) technology has attracted technology giants such as Microsoft and Google to join the competition. After ChatGPT sparked global craze and usage, the generative AI craze led by Midjourney is spreading everywhere, and various AI applications are rapidly developing in various fields, driving the demand for high-end artificial intelligence (AI) chips; Taiwan is one of the few production bases with the advantage of manufacturing high-end chips and a complete supply chain for servers. Huge business opportunities are undoubtedly the focus of each company's current and future development. Therefore, the progress of AI applications and related concept stocks have also quickly attracted the attention of many investors.

The rapid development of AI has thrown many shock bombs into the stock market in 2023. The future growth trend of the AI-related supply chain industry will not change in the short term. Therefore, this study will focus on the listed companies in Taiwan's AI supply chain to further discuss the relationship between stock price return, stock trading volume and abnormal trading volume change rate and SVI index.

2 Literature Review

Barber and Odean (2008) proposed the attention theory. Retail investors prefer stocks that can attract high attention, such as those with large amounts of news or high trading volume. This study shows that when investors pay more attention to stocks, they will increase the willingness to buy, change investment decisions, and make the stock price rise, which is consistent with the results of Merton's (1987) empirical investor cognition hypothesis. Barber and Odean (2008) used three variables to proxy investors' attention, namely: (1) the degree of volatility of the stock's trading volume in the past year, (2) whether the stock's stock price had abnormal fluctuations on the previous day, (3) whether the stock has been reported by newspapers and media, assuming that the three variables change, can effectively express the impact of investors' attention. The research results show that when there is a lot of information about a company's stock, investors will tend to invest in the stock, thereby pushing up the stock's temporary rise. Whether in large or small stocks, the price pressure caused by investors' attention can be observed. Most of this pressure comes from natural persons (retail investors) rather than institutional investors.

In the past, some literature used the SVI index as a proxy variable to measure investor attention. Takeda and Wakao (2014) believe that the numeric codes of stocks are not suitable to be used as keywords for SVI because the numeric codes of US stocks are easily confused with AD numbers, so that the results contain information irrelevant to corporate decision-making. Joscph et al. (2011) used the U.S. S&P500 stocks from 2005 to 2008 as a research sample and believed that using the S&P500 stock English code as the SVI keyword search method is better than using the stock's numeric code. Others, such as Bank, Larch, and Peter (2011), Vlastakis and Markellos (2012), Adachi et al. (2017), Li et al. (2017) all use simplified company names or abbreviations of commonly used company names as keywords to search for SVI. Bank, Larch, and Peter (2011) believe that investors' information needs are not only related to stocks. Using company abbreviations as keywords can make SVI data collection more extensive including company products or market information, so their research uses German stocks. Using the company abbreviation in the market as a search keyword, it was found that the SVI index searched with the company abbreviation has a positive relationship with stock trading activities, indicating that using the Internet to search for stock-related information can reduce information asymmetry, and at the same time, it will significantly increase the number of retail investors' investment willingness. Da, Engelberg and Gao (2011) used the search volume of Russell 30003 stocks on the Google search engine from 2004 to 2008 to measure investors' attention to the stock. The results showed that SVI significantly affects the stock return performance of the next day. The result of implicit SVI can represent the degree of attention of retail investors to a stock. When retail investors pay more attention to a stock, the abnormal return of the stock will significantly increase the stock price in the next two weeks, which can bring about short-term abnormalities returns, but reverses within a year. Vlastakis and Markellos (2012) used the 30 largest stocks on the New York Stock Exchange (NYSE) and the Nasdaq Stock Exchange (NASDAQ) as samples, and used SVI to represent the proxy variable of investors' information needs. The results show that Information demand has a significant positive correlation with stock price fluctuations and trading volume. Shen, Chen & Yu (2018) studied the relationship between SVI and the average return rate of the Taiwan Stock Exchange Market Capitalization Weighted Stock Index (TAIEX) and learned that (1) the total SVI of small-cap companies has an impact on the impact on the average return rate of TAIEX is higher than that of mid-cap and large-cap companies. (2) The positive impact of the rising trend of total SVI on TAIEX's average returns is also stronger than the overall total SVI, while the declining trend of total SVI has no impact on TAIEX's average returns. This supports Odean's attention hypothesis, as an increase in investor attention, as measured by Google SVI, is a sign of their willingness to buy, leading to higher stock prices.

3 Research Methods

3.1 Research Sample and Period

This study refers to the practice of Shen, Chen & Yu (2018), using company abbreviation as search keywords to explore the impact of SVI on stock returns. This study uses the period from January 1st, 2023 to December 31st, 2023, and the data frequency is weekly data; regarding the selection of research samples, this study uses the 202305 [Digital Special Issue] of Tianxia Magazine titled "What companies are included in the industrial topic AI supply chain?" In the chapter "Three Tables to Understand Finance", the representative AI supply chain companies listed include upstream 3, midstream 4, downstream 5 and other domestic and foreign companies in the AI supply chain. The screening criteria are as follows:

(1) Only companies whose "registration place" is Taiwan's stock exchange are included, and the rest will be deleted.
(2) The upper, middle and lower reaches of the AI supply chain are divided into groups, and groups with less than 2 companies will be deleted.

After screening based on the above two criteria, the companies belonging to the upstream and midstream supply chains are classified into three groups. The first group is the upstream supply chain chip manufacturing companies, a total of 8 companies, and the second group is the key components of the upstream supply chain. There are 19 component manufacturing companies in total. The third group is midstream supply chain server component manufacturing companies, with a total of 5 companies. A total of 32 upstream and midstream companies are included in the research sample.

3.2 Variable Definition

3.2.1 Dependent Variables

1. Stock return (R)

Stock return is used to measure the performance of a stock's investment income over a period of time. The stock return data is obtained from the "TEJ Stock Price Database" of Taiwan Economic News.

2. Stock trading volume (q)

Stock trading volume refers to the total number of stocks traded within a certain period of time. Stock trading volume data is obtained from the "TEJ Stock Price Database" of Taiwan Economic News.

3. Abnormal change rate of stock trading volume (Q)

The abnormal change rate of stock trading volume refers to the degree of abnormal fluctuations in stock trading volume within a period of time. This study refers to the calibration model in the research literature of Li Yonglong, Du Yuzhen, and Wang

Weixuan (2017). The formula for the abnormal change rate of stock trading volume is as follows: (1) shown:

$$Q_{i,t} = \frac{q_{i,t} - q_{i,avg}}{q_{i,a_vg}} \tag{1}$$

$Q_{i,t}$ = the abnormal change rate of stock trading volume at time t.
$q_{i,t}$ = the trading volume of the stock at time t.
$q_{i,avg} = \frac{\Sigma_{t=1}^{L} q_{i,t}}{L}$ = the average trading volume of the stock during L period i.
L = the total number of transactions in the entire sample period.

The abnormal change rate of the average trading volume of the $(\overline{Q}_{k,t})$ k-th investment portfolio in the t-th period is calculated based on different investment portfolios, as shown in formula (2):

$$\overline{Q}_{k,t} = \frac{\Sigma_{i=1}^{n_k} Q_{i,t}}{n_k} \tag{2}$$

$Q_{i,t}$ = the abnormal change rate of trading volume of stock t at time t.
n_k = the number of companies in the k-th portfolio.

3.2.2 Independent Variables

SVI (search volume index) is the abbreviation of Google search volume index. It is used to find out the relative index of the number of searches for a specific keyword in the Google search engine over a period of time. This study uses "company abbreviation" as the SVI search keyword. For words, the SVI value range provided by Google Trends website is 0–100, and the data file with the extension name can be exported to CSV format. For example, if a keyword is searched 100 times in the first week of a month, 1,000 times in the second week, and 4 times in the third week, the standard SVI index value is 1,000, which is the maximum search volume at all time points; after dividing the maximum number of searches/100, we get 1000/1 00 = 10, which means that after the search volume at each time point is normalized, it will be reduced to 1/10. Therefore, the SVI index in the first week is 100/10 = 10, the SVI index in the second week is 1,000/10 = 10 0, and the SVI index in the third week is 4/10 = 0.4. Because the result is less than 1, the SVI index will take the value 0.

3.2.3 Control Variables

In the capital asset pricing model (CAPM), the only explanatory variable is the market risk premium. Fama and French (1993) empirical research found that the market risk premium in CAPM is not the only explanatory factor. Their research further verified that there are other factors that affect asset returns, namely market factor (MR), size factor (SMB) and Net value to market value ratio (HML). Therefore, when conducting regression analysis, this study refers to the factors in the three-factor model of Fama and French (1993) as control variables. In addition, Carhart (1997) used the three-factor model of Fama and French (1993) as an empirical basis, adding a one-year return momentum factor to explore the persistence of fund performance, forming a new four-factor model to try to explain the persistence of fund performance, and dividing the investment portfolio into those with better early returns and those with worse returns. It was found that if the investment portfolio has a better abnormal return rate, the performance of the investment portfolio in the next year will also be the best. The results show that the momentum factor and abnormal returns do have a positive and significant impact, so this study also increase Momentum Factor (MOM).

3.3 Empirical Model

This study mainly explores the relationship between changes in the SVI index and stock returns. The explained variable is the stock return rate of the sample company in this study, and the explanatory variable is the SVI index. An empirical model is set, as shown in formula (3):

$$R_{i,t} = b_0 + b_1 SVI_{i,t} + + b_2 MR_t + b_3 SMB_t + b_4 HML_t + b_5 MOM_t + \varepsilon_{i,t} \qquad (3)$$

$SVI_{i,t}$ = the SVI value of company i on day t
MR_t = the market return on day t
SMB_t = the company size premium on day t
HML_t = the net value-to-market premium on day t
MOM_t = the momentum factor on day t

4   Empirical Results

Table 1. Overall sample empirical results

	R	q	Q
C	1.20***	21587.31***	−0.60***
	(5.01)	(2.74)	(−4.09)
SVI	0.02***	992.17***	0.02***
	(4.17)	(7.12)	(7.12)
MR	0.13	2467.22	0.05
	(1.56)	(0.90)	(0.90)
SMB	−0.13	−40931.59	−0.08
	(−1.58)	(-1.53)	(−1.53)
HML	0.31**	7540.75**	0.14**
	(2.51)	(1.88)	(1.88)
MOM	0.02	0.60	0.00
	(0.30)	(0.00)	(0.00)
A_{dj} R^2	0.05	0.06	0.06

Note: The value in the parentheses of the table is the t-statistic. ***, **, * Denote coefficient estimates that are reliably significant at the 1%, 5%, 10% levels, respectively

It can be seen from Table 1 that the SVI index has a very significant impact on the company's stock return (R), stock trading volume (q) and stock trading volume abnormal change rate (Q), and shows a positive correlation. When the SVI index is higher, it means that stock market investors pay more attention to the stock. The higher the stock returns they receive, the more active the stock market transactions are.

All sample companies are divided into three sample groups according to supply chain types. The first group is upstream supply chain chip manufacturing companies, the second group is upstream supply chain key component manufacturing companies, and the third group is midstream supply chain server component manufacturing companies. Regression analyses are conducted on different sample groups.

The analysis results after grouping the samples are as listed in Tables 2, 3 and 4. Sample groups (1)–(3) each represent three sample groups.

Table 2. The impact of SVI index on stock returns of grouped companies

	(1)	(2)	(3)
C	1.23	1.17***	0.90**
	(1.32)	(4.23)	(1.77)
SVI	0.02	0.02***	0.04***
	(1.08)	(3.04)	(3.38)
MR	0.13	0.12	0.18
	(0.65)	(1.21)	(0.91)
SMB	−0.19	−0.13	−0.04
	(−0.88)	(v1.36)	(−0.18)
HML	0.11	0.28**	0.49**
	(0.34)	(1.86)	(1.68)
MOM	−0.02	0.01	0.08
	(−0.13)	(0.06)	(0.59)
$A_{dj}\ R^2$	−0.01	0.05	0.13

Note: The value in the parentheses of the table is the t-statistic. ***, **, * Denote coefficient estimates that are reliably significant at the 1%, 5%, 10% levels, respectively

The results in Table 2 show that the first group is the stock returns of upstream supply chain chip manufacturing companies and has no significant relationship with the SVI index, while the other two groups show significant correlation.

The results in Table 3 show that the SVI index of the three groups has a significant impact on the stock trading volume of the sample companies in the group.

The results in Table 4 show that the SVI index of the three groups has a significant impact on the abnormal change rate of stock trading volume of the sample companies in the group.

Table 3. The impact of SVI index on trading volume of stocks of grouped companies

	(1)	(2)	(3)
C	25943.97 (1.63)	4712.46 (0.79)	8415.70 (0.40)
SVI	669.84** (2.57)	856.41*** (8.13)	2945.94*** (6.92)
MR	5807.05 (1.61)	729.78 (0.24)	1529.92 (0.18)
SMB	3552.95 (0.97)	−3103.34 (−1.50)	−10652.27 (−1.27)
HML	−559.72 (−0.10)	6066.77** (1.98)	8973.79 (0/70)
MOM	−5570.74** (−1.85)	−71.73 (−0.04)	8204.58 (1.29)
$A_{dj} R^2$	0.03	0.12	0.22

Note: The value in the parentheses of the table is the t-statistic. ***, **, * Denote coefficient estimates that are reliably significant at the 1%, 5%, 10% levels, respectively

Table 4. The impact of SVI index on the abnormal change rate of stock trading volume of grouped companies

	(1)	(2)	(3)
C	−0.52** (−1.75)	−0.92*** (−8.18)	−0.84** (−2.13)
SVI	0.01** (2.57)	0.02*** (8.13)	0.06*** (6.92)
MR	0.11 (1.61)	0.01 (0.34)	0.03 (0.18)
SMB	0.07 (0.97)	−0.06 (−1.50)	−0.20 (−1.27)
HML	−0.01 (−0.10)	0.11** (1.98)	0.17 (0.70)
MOM	−0.10** (−1.85)	−0.00 (−0.04)	0.15 (1.29)
$A_{dj} R^2$	0.03	0.12	0.22

Note: The value in the parentheses of the table is the t-statistic. ***, **, * Denote coefficient estimates that are reliably significant at the 1%, 5%, 10% levels, respectively

5 Conclusion

This study uses the Google Search Volume Index (SVI) as a proxy variable for investors' attention to the stock market to explore the impact on the stock returns, trading volume, and abnormal trading volume change rates of upstream and midstream companies in the AI supply chain. The research results show that, the SVI index has a very significant impact on the weekly stock return rate of AI supply chain companies. We further used the SVI index to verify that the results were the same, indicating that the sample data is robust and referenceable. In addition, by dividing AI supply chain companies into two upstream groups and one midstream group according to their types, it can also be found that, for the first upstream group (which is the upstream supply chain chip manufacturing company), the SVI index has no significant impact, the analysis results of other groups all show significant impact. If all the factors in the three-factor model of Fama and French (1993) are then included, it is found that only the net value-to-market premium (HML) is significantly related to stock return rate, trading volume and abnormal change rate of trading volume. The remaining two factors, namely company size premium (SMB) and momentum facto (MOM), have no significant impact. Follow-up research suggests that the sample can be expanded to the stock markets of other countries to analyze whether there are differences between the stock markets of different countries in regard to the results of this study, and to analyze whether media exposure has any impact on AI supply chain companies.

Acknowledgments. This research was supported by National Science and Technology Council of the Republic of China under contract NSTC 113-2622-8-008-004-SB2.

References

Barber, B.M., Odean, T.: The behavior of individual investors. In: Handbook of the Economics of Finance, vol. 2, pp. 1533–1570 (2008)

Bank, M., Larch, M., Peter, G.: Google search volume and its influence on liquidity and returns of German stocks. Fin. Markets. Portfolio Mgmt. **25**(3), 239–264 (2011)

Carhart, M.M.: On persistence in mutual fund performance. J. Financ. **52**(1), 57–82 (1997)

Da, Z., Engelberg, J., Gao, P.: In search of attention. J. Financ. **66**(5), 1461–1499 (2011)

Fama, E.F., French, K.R.: Common risk factors in the returns on stocks and bonds. J. Financ. Econ. **33**(1), 3–56 (1993)

Joseph, K., Wintoki, M.B., Zhang, Z.: Forecasting abnormal stock returns and trading volume using investor sentiment: Evidence from Online Search. Int. J. Forecast. **27**(4), 1116–1127 (2011)

Merton, R.C.: A simple model of capital market equilibrium with incomplete information. J. Financ. **42**(3), 483–510 (1987)

Shen, P.-H., Chen, S.-H., Yu, T.: Google trends and cognitive finance: lessons gained from the Taiwan stock market. In: Bucciarelli, E., Chen, S.-H., Corchado, J.M. (eds.) DCAI 2018. AISC, vol. 805, pp. 114–124. Springer, Cham (2019). https://doi.org/10.1007/978-3-319-99698-1_13

Takeda, F., Wakao, T.: Google search intensity and its relationship with returns and trading volume of Japanese stocks. Pac. Basin Financ. J. **27**(1), 1–18 (2014)

Huang, T.-L., Lai, K.-L., Chen, M.-L., Kuo, H.-J.: Information demand, web search behavior, and speculative trading activity. Sun Yat-sen Manage. Rev. **22**(1), 157–183 (2014)

Vlastakis, N., Markellos, R.N.: Information demand and stock market volatility. J. Bank. Financ. **36**(6), 1808–1821 (2012)

Li, Y.-L., Yu-Chen, T., Wang, W.-H.: Google search volume index and its effects on stock return and trading volume of Taiwan stock market. J. Manage. Syst. **24**(4), 565–590 (2017)

Impact of the Bullwhip Effect on Supplier Management: A Case Study of the Machine Tool Components Industry

Ya-Lan Chan[1], Jheng-Fong Ke[1], Sue-Ming Hsu[2], and Mei-Hua Liao[3(✉)]

[1] Department of Business Administration, Asia University, Taichung, Taiwan, ROC
yalan@asia.edu.tw, 111531065@live.asia.edu.tw
[2] Department of Business Administration, Tunghai University, Taichung, Taiwan, ROC
sueming@thu.edu.tw
[3] Department of Finance, Asia University, Taichung, Taiwan, ROC
liao_meihua@asia.edu.tw

Abstract. Due to the impact of the pandemic in recent years, manufacturers have been severely out of stock, causing consumers to place excessive orders. When the pandemic slowed down, some of the original orders from manufacturers were abandoned, thus causing a bullwhip effect. When supply chain risk crises lead to changes in market uncertainty, the bullwhip effect will become more prominent. Business operations need to face unpredictable changes in sales forecasts and orders every day.

This study uses qualitative research methods to explore the response strategies adopted by the machine tool parts industry when facing the bullwhip effect. In order to gain a deeper understanding of how related industries respond to rapidly changing consumer demands, this study conducted interviews with five senior operators. From the interviews, important information can be obtained about the impact of the introduction of information systems on inventory management and the impact on business operations and employees. It is hoped that through the interview results, more useful suggestions and help can be provided to the practical community to cope with the challenges brought by the bullwhip effect.

1 Introduction

Global industry competition is fierce. In an era of rising consumerism, product life cycles are shortened, and major international manufacturers are requiring OEMs to reduce costs and deliver on time. Most foundries in Taiwan require raw material procurement and outsourced manufacturing processes, making supplier management and control more complicated. To meet delivery requirements, part of the upstream supply chain has experienced pressure to stock up on materials to coordinate with downstream production. Downstream suppliers also must ship goods on time to meet customer delivery deadlines, leading to a situation of stockpiling.

From the second quarter of 2021 to the second quarter of 2022, orders for machine tool-related industries began to backlog, exceeding the capacity of current production. Due to the imbalance between upstream and downstream supply and demand, when order information is transmitted from the most downstream user to the upstream supply chain, manufacturers at each level gradually amplify demand, eventually resulting in significant distortions and inaccurate demand forecasts. This demand amplification phenomenon in the supply chain is akin to a swinging whip, hence it is termed the "bullwhip effect." When the bullwhip effect causes supply shortages, suppliers can limit supply based on past customer sales records to prevent sudden large demand in the short term from excessively affecting the order quantity of upstream suppliers and further exacerbating gaming behavior.

Effective supply chain management involves seamless information flow between enterprises and upstream/downstream manufacturers, ensuring timely synchronization of sales and production to alleviate inventory pressures. In the e-commerce era, production-oriented manufacturing plants require precise order management systems to prevent excessive inventory buildup, which can hamper cash turnover rates. Traditional supply chain management procedures often follow a one-way vertical model catering solely to the needs of upstream and downstream manufacturers within a single industry. Electronic and automated operations can reduce costs, but to enhance efficiency, supply chain management mechanisms must collaborate with manufacturers and suppliers in both directions, maintaining operational flexibility and adaptability to environmental changes. Timely cost adjustments via contingency operating models, accurate market trend predictions, and shared risk control throughout the entire supply chain are crucial (Chen Qilin, 2015).

The bullwhip effect occurs when a change in demand downstream of the first supply chain gradually propagates upstream. Upstream manufacturers further from the core are affected in their ordering and inventory management due to this change, leading to increased volatility. Consequently, a disparity arises in the understanding of actual demand orders between upstream manufacturers and final consumers.

This study aims to explore how various suppliers manage inventory under pressure, assess methods of information sharing, and integrate upstream and downstream management to improve the supplier management model. By fostering positive interaction and cooperation between upstream and downstream manufacturers, maximum value can be created, enabling overall inventory control and cost management amidst fierce market competition (Lin Shanju, 2015).

Against this backdrop, addressing the additional costs induced by the bullwhip effect and mitigating cost accumulation within the chain effects have emerged as pivotal concerns as the supply chain expands post-order placement. This study aims to investigate the prevailing circumstances and challenges in supply chain management, assess the role of information technology in aiding supply chain management, examine the implementation hurdles encountered by enterprises when adopting such systems, and propose relevant strategies. The primary motivation lies in understanding how supply chains across various industries should be managed and devising strategies to enhance enterprises' competitive edge.

Through this research, we aim to elucidate the current operational landscape of the OEM industry's supply chain and provide insights and guidance for enterprises contemplating the adoption of supply chain management practices. The objectives of this study are as follows:

462   Y.-L. Chan et al.

1. To provide an overview of the current state of supply chain management in the machine tool component industry.
2. To identify the challenges and potential coping strategies associated with the digitization of supply chains.

2 Literature Review

2.1 Supply Chain Management

The scope of the supply chain is defined by the American Supply Chain Council as encompassing all activities from the lowest supplier to the highest consumer. According to the Department of Commerce of the Ministry of Economic Affairs, it includes all process links from raw materials for product manufacturing to the final sale to consumers, encompassing raw materials, manufacturing equipment, manufacturing production, product inventory, product sales, and services, among others. Professor Martin Christopher, a British logistics expert, defined the supply chain in "Logistics and Supply Chain Management" as: "The supply chain refers to the upstream and downstream processes and activities involved in providing products or services to the final consumer, forming a network of downstream enterprise organizations."

The "2021 Taiwan CEO Prospective Survey" published by KPMG pointed out that supply chain risks have emerged as the top risk. The widespread pandemic has disrupted numerous supply and demand cycles. As the pandemic gradually recedes to a controllable level and industrial demand returns to normal, corporate profits should also recover. However, the economic rebound is not optimistic, and many supply bottlenecks persist. The primary reason for the continued deterioration is the inability of logistics, transportation, and goods supply to keep pace with rising demand. Consequently, inflation has surged rapidly in various countries, leading to soaring product prices. The four major supply chain crises in 2021 are outlined as follows:

1. Cargo ships congested at ports
   In response to factors such as climate and environmental changes, as well as the demand for larger shipping facilities, countries often face delays in port shipping schedules due to celestial phenomena and marine life interference. The raging COVID-19 pandemic has further underscored the inadequacy of port infrastructure, posing a significant crisis to the entire shipping industry and highlighting the urgency of port infrastructure upgrades. Severe congestion of cargo ships at ports, combined with shortages of dock workers and drivers, has severely impeded front-end product production and freight transportation processes, resulting in sluggish operations. Despite high and constrained end-market demand, insufficient infrastructure investment has led to the breakdown of the entire logistics chain.
2. Labor Shortage
   Amidst the pandemic, people are using it as an opportunity to demand better labor conditions through strikes and union activities. There continues to be a shortage of direct manpower required for on-site execution. This labor shortage has significantly exacerbated delays across the entire supply chain, resulting in a series of cascading reactions, ultimately leading to severe delays in the delivery of products and consumer services.

3. Production Failure

   In the era of globalization, supply chain management exhibits highly specialized division of labor. The impacts of low-inventory strategies are interconnected at various levels. When one disrupted link arises, it directly affects the subsequent operational developments, triggering a chain of reactions. Any breakage, stagnation, or overload in the chain accelerates chaos both upstream and downstream.

4. Energy Supply

   Global governments and enterprises are committed to "carbon reduction" in line with global goals addressing climate change and environmental sustainability. However, the supply of new energy is failing to keep up, while traditional energy outputs are inadequate, resulting in energy shortages. Supply chain problems are likely to worsen, inevitably affecting production and trade costs. Ultimately, international consensus on "carbon tariffs" and related regulations on "carbon trading" will determine the trajectory of the supply chain.

2.2 Bullwhip Effect

Lee et al. (1993) organized the generation of the bullwhip effect into the following reasons:

1. Demand Signal: In the supply chain, the inaccuracy and latency of demand signals are among the primary reasons for the bullwhip effect. As demand signals propagate through the supply chain, they may become distorted due to inaccurate forecasts, information delays, or order transmission delays. This leads upstream suppliers to produce and replenish based on inflated demand signals, thereby amplifying demand fluctuations.

2. Order Batching: Many members in supply chains tend to engage in batch ordering to obtain price discounts or save on transportation costs. However, batch ordering results in intermittent increases in order quantities, making demand signals unstable. Upstream suppliers produce based on the demand volume from batch ordering, further exacerbating volatility within the supply chain.

3. Price Fluctuation: Price fluctuations can also lead to the bullwhip effect. When product prices fluctuate drastically, consumers may make bulk purchases during low prices and reduce demand during high prices. Such price fluctuations make demand signals unstable within the supply chain, triggering inappropriate reactions from suppliers and further amplifying volatility.

4. Rationing and Shortage Game: When inventory shortages occur within the supply chain, suppliers may adopt rationing strategies to ensure inventory is adequately allocated to customers. Such rationing strategies make customers perceive supply shortages, leading them to increase order quantities to ensure sufficient inventory. This reaction results in dramatic fluctuations in order quantities, further intensifying the bullwhip effect within the supply chain.

3 Research Methodology

This study employed a qualitative research approach, with data collection conducted through in-depth interviews. Utilizing a semi-structured format, participants were encouraged to freely express their subjective life experiences, opinions, and thoughts regarding the interview questions. Case study is a qualitative research method that involves in-depth examination and analysis of specific events or issues to understand their behaviors or problems, determine their objectives, identify the root causes of issues and potential solutions, or present concrete individual cases. It can also provide possible directions for future research (Hsu, 2012).

The interviews in this study primarily focused on relevant industry suppliers and their respective case companies. Drawing on Lin's (2015) "bullwhip effect," the discussion explored the current situation of the case company's bullwhip effect in four directions. Following Chen's (2015) proposition of six advantages brought about by the application of digital commerce, the discussion outlined the following topics:

1. The temporal and feasibility analysis of the case company's initial intention to introduce a supplier management system.
2. Feedback from suppliers regarding the concept and practices of the case company's implementation of a supplier management system.
3. Challenges and issues faced by the case company in the market, as well as how to establish long-term strategic partnerships with suppliers.

4 Interview Results

This study conducted in-depth interviews with five operators or senior executives from tooling component-related enterprises. Through the interviewees' years of experience in the tooling component industry, the study explored the impact of the bullwhip effect on supplier management and its corresponding responses.

The bullwhip effect refers to the phenomenon in the supply chain where demand fluctuations are amplified in the upstream stages. When this occurs, inaccurate demand forecasting in the market leads to significant fluctuations in inventory levels for companies along the supply chain. Potential reasons for inventory problems resulting from data analysis include:

1. Chain Reaction of External Factors: External environmental changes affect demand variations within the supply chain. When changes occur in other segments of the market, these changes may gradually amplify and affect a company's inventory management.
2. Impact of External Environmental Trends: Changes in external environmental trends affect market demand, and companies fail to predict and respond to these trends promptly, resulting in inventory issues.
3. Inadequate Inventory Management and Judgment of the Management Team: Improper internal inventory management strategies may include excessively high or low inventory levels, inaccurate forecasting methods, and unreasonable order quantities and production plans. Additionally, inadequate judgment by the management team in predicting market demand, formulating inventory strategies, and coordinating the supply chain may lead to inventory problems.
4. Inadequate Risk Management and Response: Companies lack effective risk control measures and response strategies when facing risks in the supply chain. When demand changes exceed expectations, companies fail to adjust inventory and production plans promptly, resulting in inventory problems.

5. Chain Shortage Effect: This refers to a situation where one segment of the supply chain experiences a shortage, affecting the production and inventory of downstream companies. If upstream suppliers fail to deliver required materials on time, downstream companies face the risk of inventory shortages and supply disruptions.

The bullwhip effect may have varying impacts on companies' business operations and their relationships with suppliers and customers:

Company A: Close Communication and Production Adjustment According to Customer Demand
The impact of the bullwhip effect on Company A necessitates close communication and contact with customers to adjust production plans based on their demands. This means that Company A needs to respond more flexibly to fluctuations in market demand and adjust production levels promptly to avoid situations of excessive inventory or supply shortages.

Company B: Minimal Impact
The bullwhip effect has relatively little impact on Company B, and it may not significantly affect its business operations. This could be because Company B's supply chain management is more stable, enabling accurate prediction of market demand and reasonable inventory management, thereby reducing the impact of the bullwhip effect on its business.

Company C: Decreased Shipping Volume or Extended Shipping Time
The impact of the bullwhip effect on Company C may result in decreased shipping volume or extended shipping time. This implies that Company C needs to adjust production and supply chain plans in response to demand fluctuations to accommodate changes in inventory and customer demands. This could lead to reduced shipping volume or longer lead times to fulfill customer orders.

Company D: Establishing Partner Interactions and Real-Time Information Sharing
The impact of the bullwhip effect on Company D requires establishing partner interactions with suppliers and timely sharing of market and demand information. This can help Company D and its suppliers better coordinate and adjust the supply chain to cope with demand fluctuations. Through real-time information sharing, Company D may be able to adjust production plans and inventory management more quickly to adapt to changes in market demand.

Company E: More Time for Accessing and Managing Material Supply
The impact on Company E requires real-time information sharing. This means that the company needs to establish stricter communication channels between suppliers and customers to understand changes in market demand and provide updates on supply chain status in a timely manner. Through real-time information sharing, the company can better respond to market fluctuations, adjust production and supply plans promptly, and meet customer demands. This real-time information sharing enhances agility and collaboration within the supply chain.

5   Conclusion and Future Research

The interview results generally suggest that implementing digital systems can address the issues caused by the bullwhip effect. Digital systems offer enhanced visibility across the supply chain, making data and information from various stages more accessible and shareable. This visibility aids in better understanding key data such as demand variations, inventory levels, and transportation statuses within the supply chain. With more accurate information, supply chain managers can better predict demand fluctuations and adjust supply plans more promptly, thus reducing the bullwhip effect. The automation and real-time capabilities of digital systems accelerate the response time of the supply chain. When demand changes occur, digital systems can swiftly identify these changes and trigger corresponding actions, such as adjusting inventory automatically, rescheduling production, and expediting deliveries. This helps reduce response time, enabling the supply chain to flexibly cope with fluctuations caused by the bullwhip effect.

The interviewed companies have employed various technological means to address issues related to the bullwhip effect and have utilized information technology, the Internet of Things (IoT), and other technologies to enhance the efficiency and stability of their supply chains. These technological means may include using information systems and software to track and manage supply chain activities, improving operational efficiency and accuracy. Additionally, IoT technology may be applied to monitor and control logistics transportation processes to provide better visibility and traceability, thereby enhancing supply chain performance. Through the application of information technology, companies may utilize various supply chain management systems such as Enterprise Resource Planning (ERP) systems, Supply Chain Management (SCM) systems, or logistics management systems to automate and optimize processes within the supply chain. These systems can assist companies in more effectively predicting demand, managing inventory, coordinating production and logistics, ensuring timely delivery of products, and minimizing inventory costs. Furthermore, the application of IoT technology enables companies to monitor various aspects of logistics and the supply chain in real-time. This data can be connected to central systems via IoT for real-time monitoring and analysis by management personnel to promptly adjust supply chain activities and provide early warnings for any potential issues.

In summary, the adoption of information technology and IoT technologies enhances the efficiency and stability of the supply chain to address issues related to the bullwhip effect. The application of these technologies enables your company to monitor and manage supply chain activities in real-time, improving operational efficiency, accuracy, and traceability, thereby providing better supply chain services.

Acknowledgments. This research is partly supported by the National Science Council of Taiwan NSTC 112-2635-E-468-003-.

References

Chang, K.Y.: Raw material shortage, across-the-board price increases: what's happening to the global supply chain? Manager Today (2022). https://www.managertoday.com.tw/articles/view/64397. Accessed 18 Nov 2022

Chang, K.Y.: Massive extension of payment terms to suppliers: management insights behind the scenes - excessive inventory is all due to the "bullwhip effect." Manager Today (2022). https://www.managertoday.com.tw/articles/view/63159?utm_source=copyshare. Accessed 31 Mar 2023

Chen, C.L.: The impact of industrial supply chain configuration on supply chain management strategies and company performance. Master's thesis, Department of Information Management, Tamkang University (2012)

Chen, C.L.: Operational strategies of E-supply chain - a case study of company a's cameras. Master's thesis, Department of Business Administration, National Taipei University (2015)

Handfield, R.B., Nichols, E.L.: Introduction to Supply Chain Management. Prentice-Hall, Upper Saddle River (1999)

Harrington, L.H.: Supply chain integration from the inside. Transp. Distrib. **38**(3), 35–38 (1997)

Houlihan, J.B.: Supply chain management. In: Proceedings of the 19th International Technical Conference of the British Production and Inventory Control Society, pp. 101–110 (1984)

Hsu, Z.A.: Optimization of supply chain management. Doctoral dissertation, Department of Industrial Engineering and Management, National Taipei University of Technology (2012)

Huang, C.C.: The impact of advanced planning and scheduling on the stainless steel supply chain. Master's thesis, Executive Master of Business Administration Program, College of Management, National Cheng Kung University (2014)

Huang, S.C.: The influence of the bullwhip effect on the chain - a case study of the golf industry. Master's thesis, Executive Master of Business Administration Program, College of Management, National Sun Yat-sen University (2022)

Hung, J.R.: Analysis of the application model of supplier management inventory. Master's thesis, College of Technology Management, National Tsing Hua University (2019)

Liu, J.C.: Exploring the impact of the bullwhip effect on the global supply chain of the PC industry: the case of the novel coronavirus. Master's thesis, Name of University (2022)

Lin, S.J.: The impact of the bullwhip effect on inventory management in electronic contract manufacturing plants. Master's thesis, MBA Program, Feng Chia University (2015)

Lee, H.L., Billington, C.: Material management in decentralized supply chain. Oper. Res. **41**(5), 835–847 (1993)

Leenders, M.R., Fearou, H.E.: Purchasing & Supply Management, 11th edn. R. D. Irwin (1997)

Metz, P.J.: Demystifying Supply Chain Management. Supply Chain Management Review, Winter Issue (1998)

Stevens, G.: Integrating the supply chain. Int. J. Phys. Distrib. Mater. Manag.Manag. **19**(8), 3–8 (1989)

Analysis on Company Sustainable Development Goals Disclosure: A Case on Indonesian Commercial Bank Listed on BUKU 4 and 3

Szu-Hsien Lin[1], Mirzha Alamsyah Muda[2], and Mei Hua Huang[1](✉)

[1] Department of Accounting and Information Systems, Asia University, Taichung, Taiwan
`meihuang@asia.edu.tw`
[2] State Polytechnic Malang, Malang, Indonesia

Abstract. This study delves into the incorporation of Sustainable Development Goals (SDGs) into sustainability reports by Indonesian commercial banks listed in BUKU 4 and BUKU 3, investigating the interrelation between this incorporation and materiality assessment. It also examines the focal points of SDGs through a materiality analysis framework and evaluates the quality of sustainability reports regarding SDGs. The research reveals a pronounced emphasis on certain SDGs, particularly those aligning with the core mission of the banks, while also highlighting the need for fair working conditions and ethical considerations. However, a weak link is observed between environmental material topics and SDGs in the banking sector. Caution is advised in generalizing results, with future research opportunities exploring materiality-SDG linkage across diverse sectors.

1 Introduction

In the wake of the global emphasis on sustainable practices, the spotlight has shifted towards Corporate Sustainable Development Goals (SDGs) disclosure, urging organizations to align their operations with the broader goal of sustainability as numerous organizations have implemented disclosure practices to communicate sustainable information to stakeholders, utilizing the triple bottom line approach. Sustainability information has evolved alongside the concept of sustainable development. Initially focused solely on environmental protection, it now encompasses the United Nations' Sustainable Development Goals (SDGs), making for-profit organizations important contributors to environmental and social policies. Indeed, several for-profit entities have embraced contemporary methods [9] and strategies for integrating SDGs into their reporting processes [13].

The financial and banking sector actively discloses non-financial reports, particularly focusing on SDGs, as they are pivotal for global economic advancement and societal well-being. Despite indirectly supporting environmental conservation through lending to eco-friendly enterprises, the financial sector remains distinct within the service industry. The United Nations Global Compact offers sector-specific guidance, such as the SDG Industry Matrix for financial services, highlighting shared value opportunities and tailored initiatives to advance each SDG [14].

L. Barolli (Ed.): IMIS 2024, LNDECT 214, pp. 468–477, 2024.
https://doi.org/10.1007/978-3-031-64766-6_47

In Indonesia, there are two regulations governing the role of banks in advancing responsible banking practices. Initially, there was a banking mandate stipulating the implementation of social and environmental responsibility (CSR), as outlined in Article 74 of Law No. 40 of 2007 on Limited Liability Companies. Additionally, there is a requirement for companies to establish Environmental Guarantee Funds, mandated by the Environmental Law. Hence, the social demand in Indonesia for banks to disclose their non-financial report can also be used as a proof why the report disclosure surrounding SDGs material can help to improve the materiality of Banks' Sustainability Report. According the Bank Indonesia regulation number 14/26/PBI/2012 about the business activities and office network based on core capital of the bank, the grouping of banks based on Commercial Bank Group of Business Activities (BUKU), the categories are such:

1. BUKU 4: Banks with core capital exceeding IDR 30 trillion. BUKU 4 banks can invest up to 35% in domestic or offshore equity.
2. BUKU 3: Banks with core capital ranging from IDR 5 trillion to IDR 30 trillion. BUKU 3 banks have unrestricted banking activities but are limited to 25% ownership in domestic or Asian financial institutions.
3. BUKU 2: Banks with core capital between IDR 1 trillion and IDR 5 trillion. BUKU 2 banks have broader business activities than BUKU 1, including derivatives trading. They can hold up to 15% ownership in domestic financial institutions.
4. BUKU 1: Banks with core capital below IDR 1 trillion.

These banks primarily offer Rupiah-based services, with foreign exchange trans-actions limited to money changer activities. Materiality analysis is a relatively recent concept in sustainability reporting literature [15]. It involves identifying and prioritizing the most pertinent sustainability issues, considering their impact on an organization and its stakeholders. According to Accountability [1], a topic is deemed material when it significantly influences and affects the assessments, decisions, actions, and performance of an organization and/or its stakeholders in the short, medium, and/or long term. To contribute to this underexplored field, we have developed an SDG Materiality Analysis Framework (SDGs_MAF) aimed at elucidating the content of sustainability reports in the banking sector. This framework employs scoring–rating techniques suitable for ana-lyzing sustainability report content [10]. As the nuance of this research, the objective of the study is to investigate which aspect of sustainability is emphasized by the material issues, considering the specific characteristics and nature of the examined sector. In the context of this academic research, the SDGs_MAF was implemented in the analysis of a sample comprising Indonesian commercial banks listed in BUKU 4 and BUKU 3. As this study uses Sardianou et al., [10] as the reference, this study will also develop similar research question with some modification to suit the object of this study. (R1) How do Indonesian commercial banks listed under BUKU 4 and 3 incorporate SDGs into their sustainability reports, and how does this incorporation interrelate with the materiality assessment? (R2)Which SDGs emerge as focal points and are deemed crucial through the materiality analysis framework? (R3) What is the quality of the sustainability reports concerning the SDGs? (R4) Which dimension of sustainability is most relevant to the material issues?

2   Theoretical Background

2.1   Context and Framework of Sustainable Development Goals

The Sustainable Development Goals (SDGs) stand as a transformative global agenda set forth by the United Nations in 2015. Emerging from the shortcomings of the Millennium Development Goals, the SDGs represent a comprehensive framework addressing interconnected challenges spanning poverty, inequality, climate change, and environmental degradation. As articulated by the United Nations [12], these 17 goals serve as a universal call to action to end poverty, protect the planet, and ensure prosperity for all. Their historical context reveals a culmination of international efforts to create a roadmap for a more equitable and sustainable future [6].

The source of these goals lies in the inclusive and collaborative process involving governments, non-governmental organizations, and other stakeholders. The "Transforming our world: the 2030 Agenda for Sustainable Development" document, adopted by the UN General Assembly in September 2015, serves as the foundational source for these goals. This international consensus underscores the imperative of collective action, emphasizing the need for collaboration between governments, businesses, and civil society to achieve these ambitious targets.

2.2   The Current Experience of Commercial Bank Business Operation and SDGs

Commercial banks play a central role in the global economy as intermediaries facilitating the flow of capital. Beyond financial transactions, their operations have far-reaching implications for both the economic and social landscapes. Examining the current landscape of commercial banks involves an exploration of their diverse business models and core activities [7]. Real-world examples and case studies, such as Kumar & Prakash [4] highlight instances where leading banks in India have successfully integrated sustainable practices into their operations.

However, the integration of Sustainable Development Goals (SDGs) into commercial bank operations is not without its challenges. Regulatory constraints, as discussed in the "Sustainable Banking with the Poor" report by the Consultative Group to Assist the Poor (CGAP) [3], financial considerations, and evolving customer expectations present barriers that require careful consideration. The banking sector's experience with SDGs reflects an ongoing journey, one that necessitates a delicate balance between profitability and responsible, sustainable practices.

2.3   The Level of Alignment of Sustainability Reporting with SDGs

Sustainability reporting serves as a vital tool for organizations to disclose their economic, environmental, and social impacts. In the context of commercial banks, this section explores how these institutions align their sustainability reporting with the broader framework of SDGs [10]. Akhter & Dey [2] sheds light on the depth of integration of sustainability reporting practices among companies in Bangladesh, revealing a predominant focus on community development within the ESG factors. This highlights the challenges in adopting frameworks and the willingness of companies to disclose their sustainability activities, particularly concerning SDGs.

Examining best practices and frameworks adopted by leading banks globally involves referencing sources like the Global Reporting Initiative (GRI) Standards and the Sustainability Accounting Standards Board (SASB). These frameworks offer guidelines for organizations, including banks, to effectively integrate SDGs into their reporting structures. The implications of transparent reporting on accountability, credibility, and stakeholder trust underscore the crucial role of communication in advancing sustainable development goals within the banking sector [5].

2.4 The Materiality Status of Sustainability Reporting

In the contemporary business landscape, organizations navigate a complex web of internal and external factors, necessitating a comprehensive approach to decision-making. According to Busco et al., [16], this approach integrates strategic planning, performance measurement, and reporting, aligning with stakeholders' expectations. Companies must articulate their "Corporate Purpose" to bolster social legitimacy and maintain their "license to operate," as highlighted by the board of directors. This articulation typically takes the form of ESG Reporting, blending environmental, social, and governance factors, as outlined by the UN Principles for Responsible Investment [17]. Frameworks, such as those proposed by Bose [17], structure and regulate non-financial information disclosure to ensure its materiality. Materiality, a critical aspect of sustainability reporting, involves assessing the relevance and significance of specific issues to organizations and stakeholders. In commercial banks, this section explores how institutions identify, prioritize, and report material SDGs, in line with insights from Proxy Season Preview 2022 [8], which emphasizes the growing importance of ESG factors in materiality assessments.

3 Research Framework

The primary data source for this research will be the annual sustainability reports of Indonesian commercial banks listed under BUKU 4 and 3. These reports provide comprehensive information about the banks' sustainability initiatives, including their approach to integrating SDGs into their operations.

The study will focus on a purposive sample of banks listed in BUKU 4 and 3, ensuring representation from both categories. Banks in these categories are significant players in the Indonesian financial sector, making them crucial subjects for analysis. The sample selection will be based on the latest available sustainability reports, ensuring the most current data is considered (Table 1).

3.1 The Structure of SGDs_MAF

The research will employ a customized Materiality Analysis Framework (SDGs_MAF) designed specifically for evaluating the integration of SDGs in the banking sector's sustainability reports [10]. The SDGs_MAF will encompass key criteria derived from the Sustainable Development Goals, focusing on both qualitative and quantitative aspects of SDGs implementation. This framework will be structured to assess the relevance, specificity, measurability, and alignment of reported initiatives with each SDG.

Table 1. Sample Characteristic

Bank	BUKU	Description	Total assets
BI_1	4	Bank Central Asia (BCA)	IDR 179.16 trillion
BI_2	4	Bank Mandiri	IDR 179.16 trillion
BI_3	4	Bank Rakyat Indonesia (BRI)	IDR 159 trillion
BI_4	4	Bank Negara Indonesia (BNI)	IDR 139.35 trillion
BI_5	3	Bank Tabungan Indonesia (BTN)	IDR 21 trillion
BI_6	3	Bank Permata	IDR 21 trillion
BI_7	3	Bank Mega	IDR 14 trillion
BI_8	3	Bank DBS Indonesia (BTN)	IDR 7.6 trillion

3.2 Key Materiality Item

Key materiality items will be identified based on a thorough analysis of SDGs_MAF and the context brought up by the SDGs. These items will serve as the focal points for evaluating the banks' sustainability reports. Criteria such as social impact, environmental responsibility, and economic sustainability will be considered to ensure a comprehensive evaluation.

3.3 Scoring and Rating System

A scoring mechanism will be established to assess the extent of integration of key materiality items within sustainability reports. Criteria for scoring will focus on the clarity, depth, and efficacy of SDGs integration strategies outlined in the reports. The scoring system will assign numerical values to different aspects of SDGs implementation, facilitating quantitative analysis.

Based on the scores attained, a rating system will be developed to categorize banks' performance regarding SDGs integration. Ratings like 'Highly Integrated,' 'Moderately Integrated,' and 'Low Integration' will be assigned, offering a comparative assessment of how effectively banks in BUKU 4 and 3 align their efforts with the SDGs. This system will provide clarity on their performance levels (Table 2).

In accordance to Sardianou et al. [10], a weighting criterion was introduced based on recommendations from sustainability and CSR experts, guided by report evaluations. This criterion reflects the sample's emphasis on their disclosures. For instance, if a materiality item receives a high score due to its strong link with a Sustainable Development Goal (SDG), it will receive half that score if the bank deems it less significant. The weighting calculation involves a ratio between two factors, A and B, where A equals 2 times B (or B equals A divided by 2). A represents the higher score for more significant materiality items, while B denotes the score for less significant ones.

To gauge the materiality score from corporate social responsibility or sustainability reports regarding SDGs, a composite index was devised. Equation (1) outlines the total

Table 2. Description of Scoring System

Scale	Description
0	No relationship exists between materiality items and SDGs
1	A minimum relationship exists between a materiality item and SDGs
2	A weak relationship exists between a materiality item and SDGs
3	A moderate relationship exists between a materiality item and SDGs
4	A strong relationship exists between a materiality item and SDGs
5	A very strong relationship exists between a materiality item and SDGs

score assigned to each SDG for an organization, calculated by summing individual scores of materiality items associated with the respective SDG.

$$SDGi = \sum_{i=1}^{z} SDGi * A + \sum_{p=z+1}^{n} SDGp * B \tag{1}$$

where:

- $(i = 1,\ldots, 16\backslash)$ denotes the number assigned to the examined Sustainable Development Goals (SDGs).
- $(k = 1,\ldots, z\backslash)$ represents the range for materiality items deemed of higher significance.
- $(p = z + 1,\ldots, n\backslash)$ signifies the range for materiality items considered of lesser significance.

The maximum achievable score for the SDGs is contingent upon the number of materiality items encompassed in corporate social responsibility or sustainability reports. It's important to highlight that the 17th sustainable development goal, focused on fostering worldwide partnerships among governments and organizations to advance sustainable development goals [11], has been omitted from the proposed framework. The exclusion is attributed to the challenge, if not impossibility, of quantifying this goal through corporate and sustainability reports.

4 Result and Discussion

In this section, we implement the proposed methodology to address the research questions. The first question (RQ1) focuses on the materiality analysis process, which identifies various sustainable items relevant to an organization's sustainability. Each item contributes to the organization's materiality, resulting in distinct maximum scores for each financial institution. To ensure comparability, SDGi scores are normalized by dividing them by the SDGi max scores, scaling them between 0 and 1. The total score of each SDG (TSDG) across all firms ranges from 0 to 37, with higher scores indicating a stronger relationship between materiality items and SDGs. The exploration of the second (RQ2) and third research questions (RQ3) is depicted in Figs. 1 and 2. Figure 1 illustrates the cumulative scores of surveyed banks across different Sustainable Development

474   S.-H. Lin et al.

Goals (SDGs), showing a focus on 'Peace and Justice Strong Institutions' and 'Decent Work and Economic Growth' (SDGs 16 and 8). Notably, top scores were achieved by all surveyed banks and specific institutions. While several banks emphasize 'Affordable, reliable, sustainable, and modern energy' (SDG_7), many register lower scores for SDG 15 and SDG 14, indicating a lack of focus on land and sea conservation in their strategic planning and core business considerations. These findings suggest a moderate relationship with the corresponding SDGs.

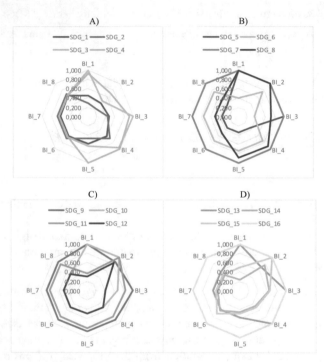

Fig. 1. A) The emphasis of sampled banking institutions per SDG 1, 2, 3, 4. B) The emphasis of sampled banking institutions per SDG 5, 6, 7, 8. C) The emphasis of sampled banking institutions per SDG 9, 10, 11, 12. D) The emphasis of sampled banking institutions per SDG 13, 14, 15, 16.

The findings corroborate previous research, highlighting a perceived inadequacy in disclosing sustainable development goals within published reports. This study introduces an approach addressing the intricate relationship between sustainable development goals and organizational strategy, aiming for improved financial returns and asset quality via enhanced reputation. While banks increasingly incorporate sustainability into non-financial reports to mitigate reputation risks and strengthen stakeholder relationships, their contribution to sustainable development, as outlined by the SDGs, remains somewhat limited.

Transitioning to RQ4, Fig. 5 delves into this research question, rooted in the triple bottom line approach encompassing economic, environmental, and social dimensions. Building on previous classifications, the economic dimension may encompass SDGs 8,

9, and 12, while the environmental dimension could include SDGs 6, 7, 11, 13, 14, and 15. Finally, the social dimension may involve the remaining SDGs 1, 2, 3, 4, 5, 10, and 16.

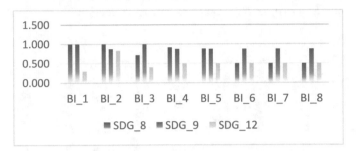

Fig. 2. SDG analysis with emphasis on the economic dimension.

Figure 2 depicts the economic priorities of the surveyed banks, highlighting their focus on sustainable economic growth and improving working conditions. This shift is driven by the need for enhanced productivity and skilled employees, moving away from traditional production line setups. Sustainability efforts aim to generate new knowledge and resources, emphasizing the crucial role of employees in achieving both sustainability and institutional objectives, particularly SDGs 8 and 9. However, areas like sustainable consumption and production (SDG 12) receive less attention, reflecting the banking sector's focus on financial services rather than direct involvement in production or consumption sustainability (Fig. 3).

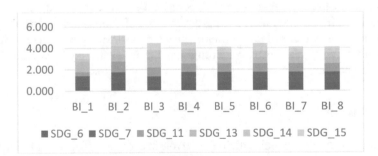

Fig. 3. SDGs sustainability analysis with emphasis on the environmental dimension.

In the environmental category, surveyed banks concentrate on urban sustainability and climate change, reflecting their societal impact and indirect contributions to climate change. Many banks address climate change directly through impact reduction practices or indirectly by integrating environmental criteria into lending procedures. However, providing affordable energy for all citizens and protecting land and sea (SDGs 14 and 15) receive lower priority due to the banking sector's focus on financial services, rather than environmental protection.

In the social sphere, banks prioritize promoting peaceful, just, and strong institutions (SDG 16), alongside efforts in health, education, and inequality reduction (SDGs 3, 4, and 10). This aligns with central banks globally offering financial products to support educational programs. However, issues like poverty (SDG 1), hunger (SDG 2), and gender equality (SDG 10) receive comparatively less attention.

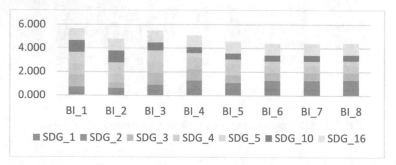

Fig. 4. SDGs sustainability analysis with emphasis on the social dimension.

Figure 4 illustrates a hierarchy of emphasis among economic, social, and environmental dimensions, with the economic dimension receiving the most focus, followed by the social dimension, and the environmental dimension receiving the least attention. This hierarchy is expected given the banking sector's inherent prioritization of economic concerns, followed by social issues related to human resources. The findings suggest that banking institutions place relatively less emphasis on environmental issues, primarily through indirect impact reduction efforts. This may be due to the sector's limited direct impact on the natural environment compared to other industries. Despite some environmental efforts, such as reducing environmental footprint and financing green investments, there remains room for improvement in disclosing environmental information, consistent with previous research highlighting its exclusion from banks' annual reports and websites.

5   Conclusion

Companies worldwide recognize their pivotal role in economic, social, and environmental responsibilities to achieve sustainable development goals. Collaborating with governments, NGOs, and local authorities, the banking sector plays a vital role in addressing sustainability issues. Beyond profit, banks act as crucial mediators in promoting sustainable business practices. Focusing on Indonesian conventional banks listed in BUKU 4 and BUKU 3, this study evaluates the alignment between sustainability disclosures and Sustainable Development Goals (SDGs). It introduces a tailored framework for assessing the materiality of sustainability reporting. Findings reveal a strong emphasis on specific SDGs aligning with banks' core missions. The study underscores the importance of fair working conditions and ethical practices to maintain productivity and customer trust. Despite some SDGs receiving high scores, there is a weak link between environmental concerns and SDGs in banking, emphasizing the need for further research across diverse sectors for validation.

References

1. Accountability (2013). https://www.accountability.org/insights/?gclid=CjwKCAiA0syqBhB xEiwAeNx9N5NBZrmYQSuxeiazC4lL490jheGluqGfcrKkIBDkf1Mvw1Z4eYmkTBoCu OgQAvD_BwE
2. Akhter, S., Dey, P.K.: Sustainability reporting practices: evidence from Bangladesh. Int. J. Account. Financ. Reporting **7**(2), 61 (2017). https://doi.org/10.5296/IJAFR.V7I2.11659
3. Bennett, L., Cuevas, C.E.: Sustainable banking with the poor. J. Int. Dev. **8**(2), 145–152 (1996). https://doi.org/10.1002/(SICI)1099-1328(199603)8:2
4. Kumar, K., Prakash, A.: Developing a framework for assessing sustainable banking performance of the Indian banking sector. Soc. Responsib. J. **15**(5), 689–709 (2019). https://doi. org/10.1108/SRJ-07-2018-0162
5. Making the Business Case for Sustainability (n.d.). https://online.hbs.edu/blog/post/business-case-for-sustainability. Accessed 15 Nov 2023
6. Midgley, J., Midgley, J.: Social development in historical context. In: Future Directions in Social Development, pp. 21–40 (2017). https://doi.org/10.1057/978-1-137-44598-8_2
7. Park, H., Kim, J.D.: Transition towards green banking: role of financial regulators and financial institutions. Asian J. Sustain. Soc. Responsib. **5**, 1–25 (2020). https://doi.org/10.1186/s41180-020-00034-3
8. Proxy Season Preview 2022: ESG – Growing Importance, Developments and Perspectives from Stakeholders. Timely disclosure. (n.d.). https://www.timelydisclosure.com/2022/02/14/proxy-season-preview-2022-esg-growing-importance-developments-and-perspectives-from-stakeholders/. Accessed 15 Nov 2023
9. Raith, M.G., Siebold, N.: View of building business models around sustainable development goals. J. Bus. Models **6**(2), 71–77 (2018). https://journals.aau.dk/index.php/JOBM/article/view/2467/1940
10. Sardianou, E., Stauropoulou, A., Evangelinos, K., Nikolaou, I.: A materiality analysis framework to assess sustainable development goals of banking sector through sustainability reports. Sustain. Prod. Consumption **27**, 1775–1793 (2021). https://doi.org/10.1016/J.SPC. 2021.04.020
11. Stafford-Smith, M., et al.: Sustainability science and implementing the sustainable development goals integration: the key to implementing the sustainable development goals. Sustain. Sci. **12**, 911–919 (2017). https://doi.org/10.1007/s11625-016-0383-3
12. THE 17 GOALS: Sustainable Development (n.d.). https://sdgs.un.org/goals. Accessed 15 Nov 2023
13. Tsalis, T.A., Malamateniou, K.E., Koulouriotis, D., Nikolaou, I.E.: New challenges for corporate sustainability reporting: United Nations' 2030 Agenda for sustainable development and the sustainable development goals. Corp. Soc. Responsib. Environ. Manag. **27**(4), 1617–1629 (2020). https://doi.org/10.1002/CSR.1910
14. UN Global Compact (2015). https://unglobalcompact.org/sdgs/about
15. Whitehead, J.: Prioritizing sustainability indicators: using materiality analysis to guide sustainability assessment and strategy. Bus. Strateg. Environ. **26**(3), 399–412 (2017). https://doi. org/10.1002/BSE.1928
16. Busco, C., Consolandi, C., Eccles, R.G., Sofra, E.: A preliminary analysis of SASB reporting: disclosure topics, financial relevance, and the financial intensity of ESG materiality. J. Appl. Corp. Financ. **32**(2), 117–125 (2020)
17. Bose, S.: Evolution of ESG reporting frameworks. In: Values at Work: Sustainable Investing and ESG Reporting, pp. 13–33 (2020)

Author Index

L. Barolli (Ed.): IMIS 2024, LNDECT 214, pp. 479–480, 2024.
https://doi.org/10.1007/978-3-031-64766-6